# The Prefrontal Cortex as an Executive, Emotional, and Social Brain

Masataka Watanabe
Editor

# The Prefrontal Cortex as an Executive, Emotional, and Social Brain

Springer

*Editor*
Masataka Watanabe
Department of Physiological Psychology
Tokyo Metropolitan Institute of Medical Science
Tokyo, Japan

ISBN 978-4-431-56506-2 ISBN 978-4-431-56508-6 (eBook)
DOI 10.1007/978-4-431-56508-6

Library of Congress Control Number: 2017931562



Printed on acid-free paper

This Springer imprint is published by Springer Nature
The registered company is Springer Japan KK
The registered company address is: Chiyoda First Bldg. East, 3-8-1 Nishi-Kanda, Chiyoda-ku, Tokyo 101-0065, Japan

# Preface

The prefrontal cortex (PFC) is phylogenetically most developed in humans, occupying about one-third of the cerebral cortex. It takes more than 20 years to fully develop and is the first region of the brain to decline. Until recently, this part of the brain was called the "silent areas" or "uncommitted cortex", because neither stimulation nor injury to this area induces noticeable sensorimotor signs. About 50 years ago, H-L. Teuber wrote a chapter in the still very influential book on the PFC *The Frontal Granular Cortex and Behavior* (JM Warren and K Akert, eds., New York, McGraw Hill, 1964) titled "The Riddle of Frontal Lobe Function in Man" (pp. 410–444). In his words: "There certainly is no other cerebral structure in which lesions can produce such a wide range of symptoms, from cruel alternations in character to mild changes in mood that seem to become undetectable in a year or two after the lesion". Twenty years later, PS Goldman-Rakic called PFC regions "uncharted provinces of the brain" (*Trends in Neuroscience*, 1984, 7:426–429), indicating how this brain area had not been fully explored.

At least until the end of the twentieth century, although clinical, animal lesions, and neurophysiological studies on the PFC had accumulated, the PFC was still not well charted and was full of riddles. Thanks to the development of neuroimaging techniques, such as positron emission tomography (PET), functional magnetic resonance imaging (fMRI), and diffusion tensor imaging (DTI), and also of noninvasive brain stimulation techniques, such as repetitive transcranial magnetic stimulation (rTMS) and transcranial direct current stimulation (tDCS), many attempts have been made to chart PFC regions, and several riddles appear to have been solved.

However, compared with other brain areas, the PFC is the area that is still the least explored and is still full of enigmas. For example, there have long been, and still are, hot discussions regarding whether the PFC has a unitary function or is functionally segregated, and when it is functionally segregated, how it is segregated.

It has been well documented that the PFC plays the most important role for executive control. Also, it is well known that the PFC is important for emotional and motivational behavior. Recent studies also indicate that the PFC is actively involved in social cognition and social behavior. However, it is still not well known how these functions are (if they are) segregated within the PFC. I have long been concerned with investigating cognitive and motivational operations in the PFC by recording neuronal activities from task-performing monkeys. I have noticed that each single PFC neuron is related to several aspects of task events, such as stimulus, delay, response, and reward/no reward, as well as to several different kinds of functions such as perception, memory, and action. Also, such multi-aspect-related neurons are observed across almost all PFC regions, indicating no clear functional differentiation among different PFC regions. Human neuroimaging studies also have shown that several regions of the PFC are activated in relation to a certain function, and a certain PFC region is activated in relation to several functions. For example, in both neuroimaging and neurophysiological studies, motivation/emotion-related neural activities are observed in all the lateral, medial, and orbital PFC regions. And, for example, in the lateral PFC, neural activity is observed in relation to all the cognition, emotion, and social behavior. Furthermore, I have observed in our neurophysiological studies that many neurons are concerned not only with, for example, cognition and motivation, but also with the integration of cognition and motivation.

Although there is a consensus that there is some functional segregation within the PFC with the lateral region more concerned with executive control, the orbital region more concerned with motivation/emotion, and the medial region more concerned with self and social cognition/behavior, the segregation is hardly absolute, and only a kind of "gradient" in functional differentiation is observed across PFC regions. Indeed, recent studies indicate an anterior-posterior gradient in information processing where the more abstract the representation is, the more anterior the region that is active in the PFC. Taking these results into consideration, it appears to be more appropriate to suppose that the PFC has a unitary function as the integrator of executive, emotional, and social functions where different regions are concerned with the integration of different kinds of functions in an overlapping manner.

In this book, thus, I have focused on this aspect of "integration" of different kinds of function within the PFC. In Part I, "Functional Organization of the Prefrontal Cortex in Human and Nonhuman Primates", how the PFC is functionally organized rather than segregated is described. In Chap. 1, Seo et al. describe how primate PFC regions including the frontopolar and dorsomedial regions are functionally differentiated/organized in relation to strategic decision making, and indicate that each of the PFC regions makes multiple contributions to improving the strategies of decision makers through experience. In Chap. 2, Tanaka et al. describe how different regions of the primate PFC differentially contribute to the performance of an analog of the Wisconsin Card Sorting task, introducing the unique role of the frontal pole in exploratory behavior, and arguing that the overall performance of the PFC goes beyond a mere sum of each subarea's elementary function. In Chap. 3, Postle challenges the commonly accepted idea regarding the working memory function of the PFC, indicating that the PFC is not so much concerned with maintaining the information in working memory but is more concerned with attention and behavioral inhibition, by reviewing monkey neurophysiological studies as well as introducing the author's fMRI and TMS studies. In this chapter, the author proposes a new idea regarding the temporal organization of the PFC in relation to working memory task performance.

In Part II, "The Prefrontal Cortex as an Integrator of Executive and Emotional Function", how the PFC is involved in the integration of executive and emotional function is described. In Chap. 4, Barbas et al. show that PFC regions associated with emotions and cognition are strongly linked and influence each other according to principles based on the structural organization of the cortex, stressing the importance of the connection of the posterior orbitofrontal cortex/anterior cingulate cortex with the lateral PFC, whose connection is essential for the integration of emotion and cognition. In Chap. 5, Kodama and Watanabe introduce their primate microdialysis studies and show how the neurotransmitters dopamine and glutamate interact in the primate PFC for the integration of cognition (working memory) and motivation (reward). In Chap. 6, Schultz introduces human neuroimaging and monkey neurophysiological studies regarding how risk (which both cognition and motivation are required to process) is represented in the PFC, and indicates the existence of the neural activity that signals reward risk distinct from reward value. In Chap. 7, Dixon et al. describe how goals at different temporal scales are represented in a hierarchical anterior-posterior gradient in the human PFC, referring also to the functional significance of the default mode of brain activity in goal-directed behavior.

In Part III, "The Prefrontal Cortex as an Integrator of Executive and Social Function", how the PFC is related to the integration of executive and social function is introduced. In Chap. 8, Isoda describes neuronal activity of the primate medial PFC observed while two monkeys were monitoring each other's actions. The author indicates that the medial PFC plays pivotal roles in differentiating between one's own actions and others' actions and in monitoring the correctness of others' actions for adaptive social decision. In Chap. 9, Hosokawa describes neuronal activity of the primate lateral PFC during a face-to-face computer video game, introducing PFC neurons' differential coding of the reward/no-reward depending on the presence/absence of the competitor and on the animacy of the opponent. In Chap. 10, Yaoi et al. describe their fMRI studies indicating how self-recognition is realized in the medial PFC. They propose that by accessing internal representation of self and others, the medial PFC would support both our mental self and a wide variety of social activities by managing internal representation and make us social beings. In Chap. 11, Sadato introduces fMRI brain-hyperscanning methods in studying real-time interaction between two persons and indicates that the shared attention induces inter-individual neural synchronization in the right inferior PFC.

Part IV "Default Mode of Brain Activity and the Prefrontal Cortex" is concerned with the "default mode of brain activity" which is observed while the subject is in a resting state. The PFC is considered to be concerned with active processing of cognitive, motivational, and social information. However, recent studies indicate that there are PFC regions that are active during the resting state. Furthermore, there are reports indicating the co-activation of the so-called executive PFC region and default PFC region in relation to certain mental operations. In Chap. 12, Watanabe introduces the default mode of brain activity observed in the monkey and discusses the functional significance of the co-activation of the executive and default regions in relation to the integration of cognitive and motivational operations. In Chap. 13, Koshino describes the co-activation of human default mode and executive network regions in relation to cognitive task performance, stressing that the relationship between the two networks changes dynamically during different phases within a task.

The PFC is investigated by using many kinds of methodology. In this book, a variety of methods in investigating the integrative function of the PFC are described: neuroanatomy (Chap. 4), human neuroimaging (fMRI, EEG) (Chaps. 3, 6, 7, 10, 11, and 13), human stimulation (Chap. 3), monkey neuropsychology (Chap. 2), monkey neurophysiology (Chaps. 1, 2, 3, 6, 8, and 9), monkey neuroimaging (Chap. 12), and monkey neurochemistry (Chap. 5).

I hope readers of this book will obtain useful ideas about how the PFC is currently studied with a variety of methods, and how the PFC is concerned with the integration of cognitive, emotional/motivational, and social information for a better way of life.

I express my great thanks to all authors for contributing exciting chapters. I also wish to thank Dr. Yasutaka Okazaki and Ms. Momoko Asawa of Springer Japan for their patient help and proficient editing.

Tokyo, Japan Masataka Watanabe

# Contents

**Part I Functional Organization of the Prefrontal Cortex in Human and Nonhuman Primates**

1 **Neural Correlates of Strategic Decision-Making in the Primate Prefrontal Cortex** . . . 3
Hyojung Seo, Soyoun Kim, Xinying Cai, Hiroshi Abe, Christopher H. Donahue, and Daeyeol Lee

2 **Functional Division Among Prefrontal Cortical Areas in an Analog of Wisconsin Card Sorting Test** . . . 17
Keiji Tanaka, Mark J. Buckley, and Farshad A. Mansouri

3 **Working Memory Functions of the Prefrontal Cortex** . . . 39
Bradley R. Postle

**Part II The Prefrontal Cortex as an Integrator of Executive and Emotional Function**

4 **Prefrontal Cortex Integration of Emotion and Cognition** . . . 51
Helen Barbas and Miguel Ángel García-Cabezas

5 **Interaction of Dopamine and Glutamate Release in the Primate Prefrontal Cortex in Relation to Working Memory and Reward** . . . 77
Tohru Kodama and Masataka Watanabe

6 **Neuronal Risk Processing in Human and Monkey Prefrontal Cortex** . . . 103
Wolfram Schultz

7 **Hierarchical Organization of Frontoparietal Control Networks Underlying Goal-Directed Behavior** . . . 133
Mathew L. Dixon, Manesh Girn, and Kalina Christoff

**Part III The Prefrontal Cortex as an Integrator of Executive and Social Function**

**8 Self–Other Differentiation and Monitoring Others' Actions in the Medial Prefrontal Cortex of the Monkey** . . . . . . . . . . . . . . . 151
Masaki Isoda

**9 Neural Correlates of Competition in the Primate Prefrontal Cortex** . . . . . . . . . . . . . . . . . . . . . . . . . . . . . . . . . . . . . . . . . 169
Takayuki Hosokawa

**10 Self-Recognition Process in the Human Prefrontal Cortex** . . . . . . . 187
Ken Yaoi, Mariko Osaka, and Naoyuki Osaka

**11 Shared Attention and Interindividual Neural Synchronization in the Human Right Inferior Frontal Cortex** . . . . . . . . . . . . . . . . . 207
Norihiro Sadato

**Part IV Default Mode of Brain Activity and the Prefrontal Cortex**

**12 Default Mode of Brain Activity Observed in the Lateral, Medial, and Orbital Prefrontal Cortex in the Monkey** . . . . . . . . . . . . . . . . 229
Masataka Watanabe

**13 Coactivation of Default Mode Network and Executive Network Regions in the Human Brain** . . . . . . . . . . . . . . . . . . . . . . . . . . . . 247
Hideya Koshino

# Part I
# Functional Organization of the Prefrontal Cortex in Human and Nonhuman Primates

# Chapter 1
# Neural Correlates of Strategic Decision-Making in the Primate Prefrontal Cortex

**Hyojung Seo, Soyoun Kim, Xinying Cai, Hiroshi Abe, Christopher H. Donahue, and Daeyeol Lee**

**Abstract** The prefrontal cortex of primates is well poised for carrying out multiple types of functions related to strategic decision-making. For example, outcomes of many strategic decisions can be observed only after substantial delays. The prefrontal cortex might play a key role in incorporating such delays into decision-making by representing the subjective value of delayed outcomes. In addition, the prefrontal cortex is likely to make multiple contributions to improving the strategies of decision-makers through experience. For trial-and-error learning, signals related to the decision-maker's previous choices and their outcomes must be combined properly, and this might be implemented flexibly in different regions of the prefrontal cortex according to the demands of specific tasks. How the brain predicts the outcomes of hypothetical actions based on its internal model of the environment is less well understood, but the arbitration and switching between different learning

---

H. Seo
Department of Neuroscience, Yale University School of Medicine, New Haven, CT 06510, USA

S. Kim
Center for Functional Connectomics, Korea Institute of Science and Technology, Seoul 136-791, Republic of Korea

X. Cai
NYU-ECNU Institute of Brain and Cognitive Science, New York University Shanghai, Shanghai, China

H. Abe
RIKEN Brain Science Institute, Wako, Saitama, Japan

C.H. Donahue
The Gladstone Institutes, San Francisco, CA 94158, USA

D. Lee (✉)
Department of Neuroscience, Yale University School of Medicine, New Haven, CT 06510, USA

Kavli Institute for Neuroscience, Yale University School of Medicine, New Haven, CT 06510, USA

Department of Psychology, Yale University, New Haven, CT 06520, USA
e-mail: daeyeol.lee@yale.edu



M. Watanabe (ed.), *The Prefrontal Cortex as an Executive, Emotional, and Social Brain*, DOI 10.1007/978-4-431-56508-6_1

algorithms might also rely on specific regions of the prefrontal cortex, including the frontopolar cortex and dorsomedial prefrontal cortex.

**Keywords** Game theory • Intertemporal choice • Mental simulation • Reinforcement learning • Reward

## 1.1 Introduction

An important function of the brain is to allow the animal to choose a motor response that is most likely to produce an outcome most beneficial to it. This process of decision-making covers a wide range of behaviors, especially in primates, which have a complex motor system with a large degree of freedom. In addition, decision-making is not a unitary process but instead consists of several different distinct types of computations. First, the likelihoods and desirabilities of outcomes expected from each available action must be evaluated and compared to the expected effort or cost necessary to complete the action. In most cases, the desired outcomes do not immediately follow the action chosen for them, and often there is a substantial delay between the time of an action and when its outcomes are observed. The preference for such a delayed outcome is often diminished according to its delay, which is referred to delay or temporal discounting (Kable and Glimcher 2007; Kim et al. 2008; Cai et al. 2011). Second, when the actual outcome from the chosen action is observed, this might be different from what was expected at the time when the action was taken. This discrepancy is often known as prediction error and should be incorporated into the decision-maker's future strategies (Sutton and Barto 1998). Depending on the type of information about the outcomes utilized to adjust the strategies of the decision-maker, the future strategies might be determined by different computational algorithms. For example, in many animal species, including primates, the outcome of a choice is often influenced by the actions of other animals, and predicting the outcomes of an action during such social decision-making is especially challenging (Byrne and Whitten 1988).

The primate prefrontal cortex consists of a large portion of the frontal cortex that lies anterior to the primary motor and premotor cortex (Passingham and Wise 2012). Many different but closely related functions, such as working memory and attention, have been attributed to this region of the cortex, but its anatomical connectivity with high-order sensory cortical areas as well as the basal ganglia suggests that various aspects of decision-making might be localized in the prefrontal cortex. In addition, such processes as working memory and attention are important for decision-making. In fact, a large number of neurophysiological studies in the primate prefrontal cortex during the last decade have demonstrated that various formal frameworks of decision sciences are quite useful in revealing the functional properties of neural circuits in the prefrontal cortex. Many of these studies converge on several key findings. In particular, there is clear regional specialization in that particular types of signals related to values or outcomes might be observed more frequently in one cortical area than in another, although many different types of decision-related signals are observed in multiple subdivisions of the prefrontal cortex (Lee et al. 2007; Wallis and Kennerley 2010). For example, signals related to choices and values in the lateral prefrontal cortex tend to be spatially selective, whereas neuronal activity recorded in the orbitofrontal cortex tend to be less spatially selective (Tremblay and Schultz 1999; Wallis and Miller 2003; Padoa-Schioppa and Assad 2006; Kim et al. 2008; Abe and Lee 2011; Donahue and Lee 2015). By contrast, signals related to decision outcomes are more prevalent in the medial prefrontal cortex, including the anterior cingulate cortex (Seo and Lee 2007, 2008; Hayden et al. 2009, 2011; Donahue et al. 2013). This chapter will focus on three aspects of prefrontal cortex related to decision-making, including intertemporal choice, feedback-driven learning algorithms, and strategic decision-making in a social setting. We found that the prefrontal cortex harbors signals related to the subjective values of delayed rewards (Kim et al. 2008, 2012). In addition, during decision-making in a virtual social environment, neural signals related to hypothetical outcomes and strategic selection of learning algorithms tend to be localized in specific regions of the prefrontal cortex (Donahue et al. 2013; Seo et al. 2014).

## 1.2 Prefrontal Cortex and Intertemporal Choice

The ability to anticipate and evaluate the outcomes of previous actions has an important role in decision-making, especially when actions produce their desired consequences after substantial delay. Previous neuroimaging studies have revealed that multiple cortical and subcortical areas, including the prefrontal cortex and striatum, change their level of activation according to the subjective value of a delayed reward (McClure et al. 2004; Kable and Glimcher 2007). To investigate further the nature of signals transmitted by individual neurons in these brain areas, we have trained monkeys to indicate their choice between a small but more immediate reward and a larger but more delayed reward in an oculomotor intertemporal choice task (Kim et al. 2008; Hwang et al. 2009; Cai et al. 2011). During this task, one red and one green target were presented, one on each side of the central target fixated by the animal at the beginning of each trial (Fig. 1.1a). After a brief delay, the central target was extinguished, and the animal indicated its choice by shifting its gaze toward one of the two peripheral targets. The size of the juice reward delivered to the animal was determined by the color of the target chosen by the animal (0.4 ml for the large reward, 0.2 or 0.26 ml for the small reward), and its delay was indicated by the number of small yellow disks presented around each target (Fig. 1.1a). The positions of the targets for small and large rewards as well as their corresponding delays were randomized across trials, making it impossible for the animals to predict their values in advance. In addition, if the animal chose the target for a small reward, the onset of the subsequent trial was delayed according to the difference in the delays for the small and large reward,

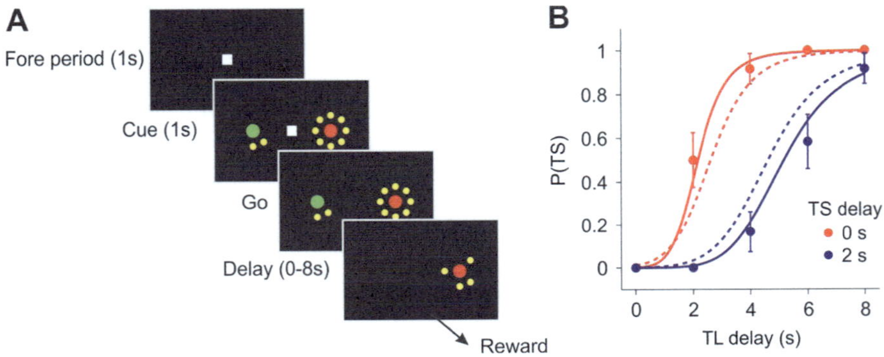


**Fig. 1.1** Spatiotemporal sequence of the intertemporal choice task (**a**) and behavior (**b**). (**a**) Target *colors* indicate the reward magnitude, whereas the reward delay is indicated by the number of small *yellow* disks surrounding each target. (**b**) The probability of choosing the small reward target, P(TS), is *plotted* as a function of the delays for the large (TL) and small (TS) rewards. *Dashed* and *solid lines* show the predictions from the best fitting exponential and hyperbolic discount functions, respectively

so that the animal does not choose the small reward target simply in order to maximize the rate of rewards (c.f. Blanchard et al. 2013).

As expected, we found that the animal's choice was systematically influenced by the reward delays for both targets (Fig. 1.1b). Intertemporal choices are typically analyzed by assuming that the subjective value of a delayed reward is given by a product of the reward magnitude and a temporal discount function (Frederick et al. 2002). The shape of the temporal discount function determines how steeply the subjective value of reward decreases with its delay and whether the rate of temporal discounting remains fixed or changes as the reward delay increases. For example, the rate of temporal discounting is constant for an exponential discount function, whereas it decreases with delay for a hyperbolic discount function. Consistent with the findings from previous studies in both humans and animals, we found that the animal's choice was better accounted for by a hyperbolic discount function than by an exponential discount function (Hwang et al. 2009; Fig. 1.1b).

We also analyzed the activity of individual neurons recorded from the dorsolateral prefrontal cortex (DLPFC) and the striatum using a series of regression models to identify the factors that significantly modulated the activity of neurons in these areas (Kim et al. 2008, 2012; Cai et al. 2011). We found that neurons in both DLPFC and striatum often change their activity according to the magnitude and/or delay of the reward expected from a particular location. Moreover, activity of some neurons in both areas was also significantly influenced by the delay specifically associated with a particular target color (hence the magnitude of the corresponding reward; Fig. 1.2). This suggests that neurons in these two brain areas might utilize spatial or object frame of reference to encode the subjective (temporally discounted) values of delay rewards. Indeed, this was confirmed using a regression model that included the separate terms related to the temporally discounted values of the rewards expected from different locations and different target colors (Kim

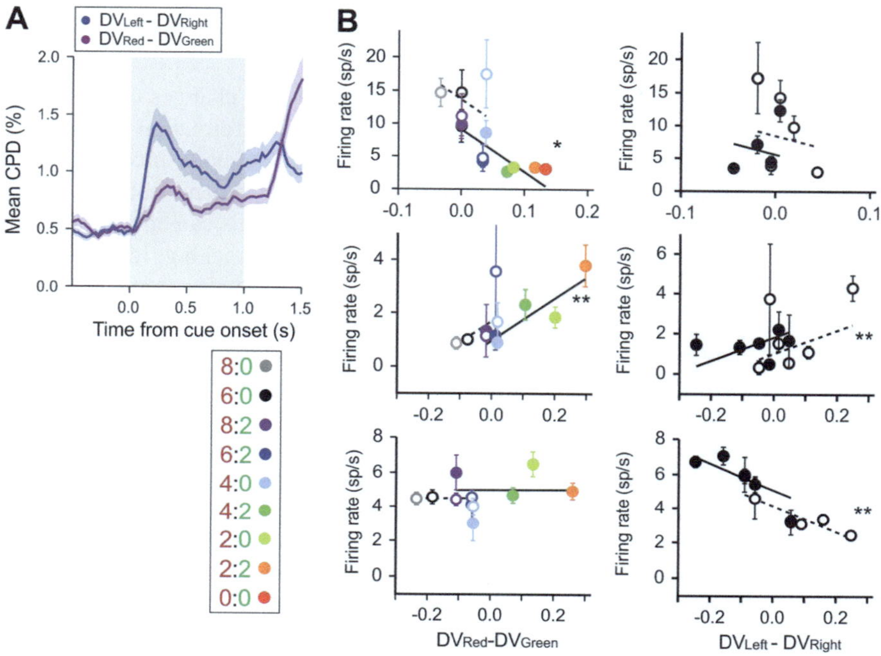


**Fig. 1.2** Neural signals related to temporally discounted values (DV) in the dorsolateral prefrontal cortex (DLPFC). (**a**) The time course of the coefficient of partial determination (CPD) related to the difference in the DV for the targets in different locations (*left* vs. *right*) and for the targets associated with different reward magnitudes (i.e., target colors; *red* vs. *green*). (**b**) Three example neurons in the DLPFC that modulated their activity significantly according to the difference in the DV for colors (*top*), locations (*bottom*), or both (*middle*). Symbols correspond to the average spike rate during the 1-s cue period (*gray* background in panel **a**), and they are *plotted* as a function of the difference in the value functions for the *red* vs. *green targets* (*left*) for different combinations of rewards for *red* and *green* targets as shown in the inset or as a function of the difference in the value functions for the left vs. right target (*right*). Filled and empty symbols correspond to the trials in which the animal chose the *red* and *green* targets (*left*) or the leftward and rightward targets (*right*), respectively. *Dashed* and *solid lines* correspond to the best fitting regression lines for two alternative choices

et al. 2012). These results suggest that the prefrontal cortex and basal ganglia are likely to play an important role in mediating the decisions involving rewards expected after unequal delays. Given that delayed rewards are also common during social interactions, such as cooperation (Stephens et al. 2002), this also suggests that the neural processes in these brain areas related to temporal discounting might also contribute to strategic decision-making in a social setting.

## 1.3 Prefrontal Cortex and Reinforcement Learning

The animal's environment is seldom fully known and also changes often. Therefore, new information about the environment, especially information about unexpected outcomes from previous actions, needs to be continually incorporated to update the animal's decision-making strategies. Reinforcement learning theory provides a rich set of computational algorithms as to how this can be done (Sutton and Barto 1998). Similar to economic theories of decision-making, reinforcement learning theory assigns a numerical quantity referred to as value function to each action. In model-free reinforcement learning algorithms, the value function for each action is updated exclusively based on the discrepancy between the actual outcome resulting from that action and the outcome expected from the current value functions. This discrepancy is known as reward prediction error and is encoded by neurons throughout many cortical and subcortical areas, including the dopamine neurons in the brainstem (Schultz 1998; Cohen et al. 2012; Lee et al. 2012; Fiorillo 2013). By contrast, model-based reinforcement learning algorithms update the action values, not based on reward prediction errors, but according to the results of simulating the outcomes of hypothetical actions. Therefore, when the animal acquires new information about its environment or when the subjective value of a particular outcome changes, for example, as a result of satiation, the value functions can be updated more flexibly in model-based reinforcement learning algorithms than in model-free learning algorithms. Previous studies have found that choice behaviors of human decision-makers are most consistent with a mixture of model-free and model-based reinforcement learning algorithms (Daw et al. 2011; Eppinger et al. 2013; Lee et al. 2014).

Although neuronal signals related to reward prediction errors have been relatively well characterized, how these signals are utilized to update the value functions in the brain remains poorly understood. Several lines of evidence suggest that this process might be distributed broadly in multiple brain areas and changed flexibly depending on the nature of the task performed by the animal (Lee and Seo 2016). First, signals related to the actions chosen by the animal and their outcomes are encoded persistently by the neurons in multiple cortical areas (Lee et al. 2012, Donahue et al. 2013; Fig. 1.3). These signals might have an important role for updating the value functions by providing appropriate memory signals about previous actions, often referred to as an eligibility trace (Sutton and Barto 1998). Such persistent choice signals might contribute to resolving the so-called temporal credit assignment problem, by linking a particular action and its sequent outcome even after a significant temporal delay. Second, it was found that the time scale of memory signals related to previous choices and outcomes is heterogeneous and varies broadly across different neurons according to a distribution given by a power function (Bernacchia et al. 2011). Computationally, updating the value functions can be understood as appropriate integration of signals related to outcomes resulting from previous actions with a time constant chosen appropriately for the stability or volatility of a specific environment. Therefore, memory signals

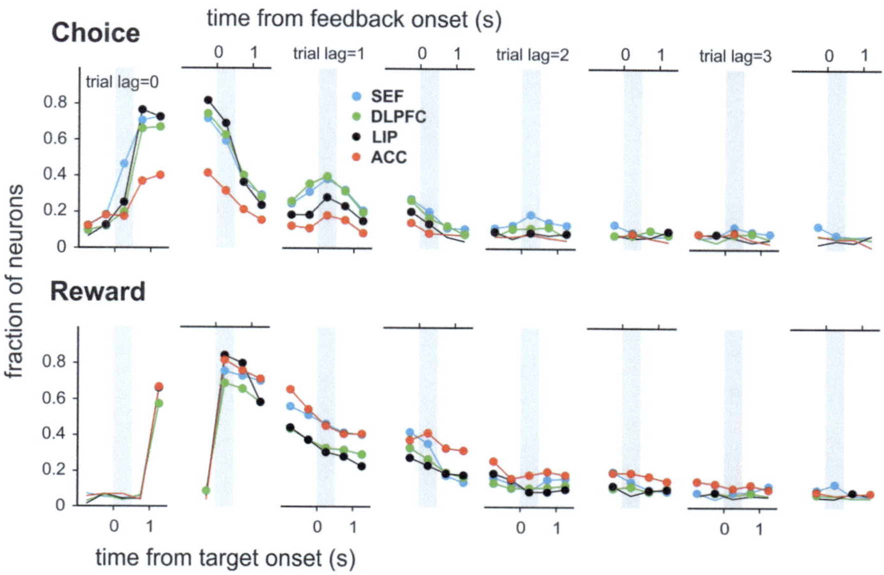


**Fig. 1.3** Time course of neural signals related to previous choices and outcomes. Each panel shows the fraction of neurons recorded in the supplementary eye field (SEF), DLPFC, lateral intraparietal cortex (LIP), and anterior cingulate cortex (ACC) that modulated their activity significantly according to the animal's previous choices in the current and previous trials (*top*) or their outcomes (*bottom*). Large symbols indicate that the proportion is significantly higher than the chance level (0.05)

related to previous choices and outcomes with multiple time scales might provide the necessary neural substrates for updating value functions flexibly in multiple environments. Finally, neuroimaging studies in humans have also revealed that signals related to reinforcement and punishment are present practically in the entire cerebral cortex (Vickery et al. 2011, 2015), also suggesting that the process of updating value functions might be broadly distributed.

Compared to model-free reinforcement learning model, the range of learning algorithms that can be considered as model based is more open-ended, and therefore the investigation of its neural substrates is also at a relatively early stage. Nevertheless, a hallmark of model-based reinforcement learning is the ability to estimate the outcomes from hypothetical actions without directly observing the actual outcomes from previously chosen actions (Lee and Seo 2016). This process is closely related to episodic future thinking and scene construction (Atance and O'Neill 2001; Hassabis and Maguire 2007; Corballis 2013) and might therefore rely on the hippocampus (Johnson and Redish 2007; Simon and Daw 2011; Pezzulo et al. 2014). Analogous to reward prediction errors, hypothetical outcomes estimated by such mental simulations can be compared to the actual outcomes, and the resulting discrepancy, often referred to as fictive or hypothetical reward prediction errors, can be used to update the value functions for hypothetical actions. Neuroimaging studies and, more recently, voltammetric measurements of dopamine

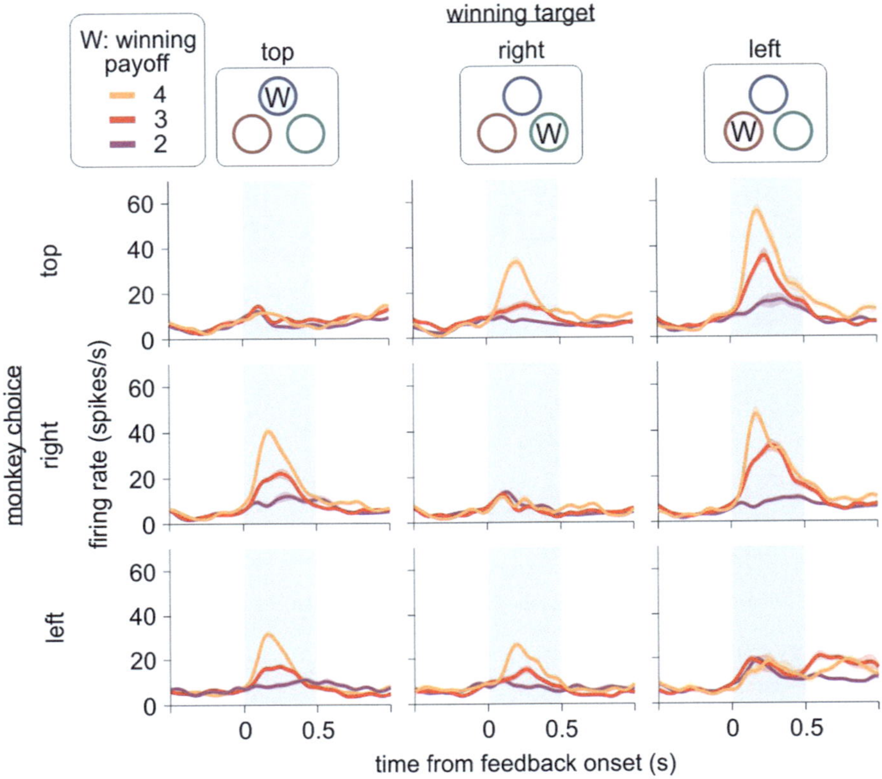


**Fig. 1.4** Example neuron in the orbitofrontal cortex with hypothetical outcome signals. Each panel shows the spike density functions aligned to the time of feedback onset for three different winning outcomes during a computer-simulated rock-paper-scissors task, separately according to the animal's choice (*rows*) and the position of the winning target (*columns*; Abe and Lee 2011)

concentrations in humans have shown that signals related to fictive reward prediction errors might be processed in the same brain regions innervated by dopamine neurons, such as the striatum (Lohrenz et al. 2007; Daw et al. 2011; Kishida et al. 2016). Neurophysiological recordings in nonhuman primates have also identified signals related to hypothetical rewards in multiple areas of the prefrontal cortex, including the anterior cingulate cortex (Hayden et al. 2009) as well as the orbital and dorsolateral prefrontal cortex (Abe and Lee 2011; Fig. 1.4). Therefore, although how the brain actually estimates the hypothetical outcomes from unchosen actions remains unknown, the signals related to hypothetical outcomes are represented in multiple brain areas and might contribute to updating the value functions according to hypothetical reward prediction errors.

## 1.4 Prefrontal Cortex and Strategic Decision-Making

Many primate species, including humans, live in social groups and, therefore, often make decisions in social settings where the outcome of an animal's decision is also influenced by the actions taken by other animals. Such social decision-making is formally analyzed by game theory. In game theory, the so-called payoff matrix displays the amount of reward or utility given to each decision-maker or player for each combination of actions chosen by all the players in the group. For a given player, their optimal strategy can be easily determined once the probabilities of choosing various actions for all the other players are known and is referred to as a best response. However, since the strategies of other players are generally unknown in advance, trying to arrive at an optimal strategy by applying best strategies for all players iteratively often leads to an infinite regress, as occurs for the rock-paper-scissors game. Instead, the so-called Nash equilibrium is defined as a set of strategies from which no individual players can deviate to increase their payoffs. For example, the Nash equilibrium for the rock-paper-scissors is for everyone to choose each option equally often. Any deviation from this so-called mixed strategy can be exploited by other players. It was John Nash who proved that there is at least one such equilibrium for every game (Nash 1950), so such equilibrium is named after him.

When choice behaviors of humans and other animals are carefully analyzed, it is often shown that the predictions of Nash equilibrium are frequently violated, suggesting that they lack the cognitive capabilities required for calculating the Nash equilibrium strategies (Camerer 2003). Instead, previous studies have consistently shown that strategies of human and animal decision-makers during social interaction tend to be adjusted dynamically according to the outcomes of their previous choices (Mookherjee and Sopher 1994; Erev and Roth 1998; Lee and Seo 2016), consistent with the predictions of reinforcement learning theory (Sutton and Barto 1998). Similar to nonsocial decision-making such as a multistage multiarmed bandit task (Daw et al. 2011; Lee et al. 2014), humans and animals tend to rely on a mixture of model-free and model-based reinforcement learning algorithms during social interactions (Camerer and Ho 1999; Lee et al. 2005; Abe and Lee 2011; Zhu et al. 2012).

In order to investigate the nature of learning algorithms and underlying neural mechanisms used by the animal during social decision-making, we have trained rhesus monkeys to play a virtual biased matching pennies game against a computerized opponent (Seo and Lee 2009; Seo et al. 2014). Throughout this experiment, a set of six red circles were displayed at the center of a computer screen as placeholders for tokens collected by the animal (Fig. 1.5a). Tokens were red disks, and their number was adjusted after each trial according to the payoff matrix of a biased matching pennies (Fig. 1.5b), namely, the number of tokens increased, decreased, or remained unchanged, when the animal won, lost, or tied with the computer opponent, respectively. The behavior of monkeys during this biased matching pennies were qualitatively similar to the results from previous studies (Lee et al.

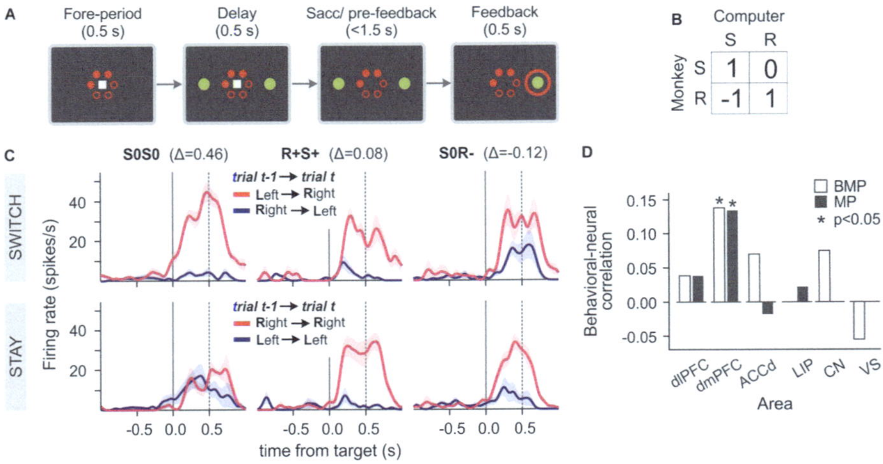


**Fig. 1.5** Signals related to switching away from a model-free reinforcement learning. (**a**) Spatio-temporal sequence of the token-based biased matching pennies task. (**b**) Payoff matrix for the biased matching pennies game. S and R refer to the safe and risky targets, respectively. (**c**) Switching-related activity shown for three example sequences of choices and outcomes in the two previous trials. S0 and R− indicate the trials resulting in neural outcome and loss, whereas S+ and R+ indicate those resulting in gains after choosing safe and risky targets. Spike density functions show the activity averaged according to the animal's choices in the current and previous trial. The value of Δ shows the difference in the accuracy of decoding the animal's choice depending on whether the animal's current choice is different (switch) from the previous trial or not (stay). (**d**) Correlation coefficient between the decoding accuracy related to switching (Δ in panel **c**) and the tendency to deviate from model-free reinforcement learning across various choice-outcome sequences was significant only for the dorsomedial prefrontal cortex. This was true during the biased matching pennies task (BMP) as well as during the symmetric matching pennies task (MP; Donahue et al. 2013)

2005; Abe and Lee 2011). First, the overall probability of choosing each target was not substantially different from the Nash equilibrium. Second, the trial-by-trial dynamics in the animal's choice behavior were largely consistent with the patterns expected for a model-free reinforcement learning algorithm, including a significant tendency to use the win-stay-lose-switch strategy. This suggests that similar to humans (Mookherjee and Sopher 1994; Erev and Roth 1998), monkeys might approximate the Nash equilibrium using a learning algorithm. We also found that monkeys could successfully counterexploit the computer's strategy in some trials by deviating systematically from the strategies predicted by a model-free reinforcement learning algorithm (Seo et al. 2014). In contrast to the signals necessary for implementing model-free reinforcement learning algorithms, such as the information about the animal's previous choices and their outcomes, that are distributed broadly in association cortical areas and basal ganglia (Lee et al. 2012; Vickery et al. 2011, 2015), neural signals related to switching away from such simple heuristic learning algorithms were found specifically in the dorsomedial prefrontal cortex, suggesting that this area might play a critical role in arbitrating between different learning algorithms (Seo et al. 2014; Lee and Seo 2016; Fig. 1.5c, d).

## 1.5 Conclusions

The primate prefrontal cortex consists of multiple subdivisions that can be defined by specific patterns of anatomical connectivity with sensory and motor cortical areas and various subcortical areas, such as the amygdala, hippocampus, and basal ganglia. Nevertheless, precisely whether and how these different regions of the prefrontal cortex contribute to different cognitive functions remains incompletely understood. Recently, a number of neuroimaging studies in human subjects and neurophysiological recordings in nonhuman primates have begun to elucidate the contribution of the prefrontal cortex related to specific aspects of reinforcement learning and decision-making. The results from these studies suggest that the prefrontal cortex is well positioned for strategic decision-making. For example, the subjective values of delayed outcomes are robustly represented in the prefrontal cortex. In addition, the process related to a model-free reinforcement learning might be implemented broadly in many areas of the brain, including the prefrontal cortex. The signals necessary for model-free reinforcement learning, such as previous choices and outcomes, must be integrated flexibly according to the demands of specific tasks, and the prefrontal cortex is likely to have an important role in this regard (Donahue and Lee 2015). Although the contribution of the prefrontal cortex for model-based reinforcement learning is less well understood, the results from neuroimaging studies suggest that the frontopolar cortex might play an important role in evaluating the accuracies of different learning algorithms (Lee et al. 2014). In addition, the dorsomedial prefrontal cortex might also contribute to switching away from the model-free reinforcement learning algorithm that is likely to be used as a default strategy (Seo et al. 2014).

## References

Abe H, Lee D (2011) Distributed coding of actual and hypothetical outcomes in the orbital and dorsolateral prefrontal cortex. Neuron 70:731–741

Atance CM, O'Neill DK (2001) Episodic future thinking. Trends Cogn Sci 5:533–539

Bernacchia A, Seo H, Lee D, Wang XJ (2011) A reservoir of time constants for memory traces in cortical neurons. Nat Neurosci 14:366–372

Blanchard TC, Pearson JM, Hayden BY (2013) Postreward delays and systematic biases in measures of animal temporal discounting. Proc Natl Acad Sci U S A 110:15491–15496

Byrne R, Whitten A (1988) Machiavellian intelligence. Oxford University Press, Oxford

Cai X, Kim S, Lee D (2011) Heterogeneous coding of temporally discounted values in the dorsal and ventral striatum during intertemporal choice. Neuron 69:170–182

Camerer CF (2003) Behavioral game theory. Princeton University Press, Princeton

Camerer C, Ho TH (1999) Experience-weighted attraction learning in normal form games. Econometrica 67:827–874

Cohen JY, Haesler S, Vong L, Lowell BB, Uchida N (2012) Neuron-type-specific signals for reward and punishment in the ventral tegmental area. Nature 482:85–88

Corballis MC (2013) Mental time travel: a case for evolutionary continuity. Trends Cogn Sci 17:5–6

Daw ND, Gershman SJ, Seymour B, Dayan P, Dolan RJ (2011) Model-based influences on humans' choices and striatal prediction errors. Neuron 69:1204–1215
Donahue CH, Lee D (2015) Dynamic routing of task-relevant signals for decision making in dorsolateral prefrontal cortex. Nat Neurosci 18:295–301
Donahue CH, Seo H, Lee D (2013) Cortical signals for rewarded actions and strategic exploration. Neuron 80:223–234
Eppinger B, Walter M, Heekeren HR, Li SC (2013) Of goals and habits: age-related and individual differences in goal-directed decision-making. Front Neurosci 7:253
Erev I, Roth AE (1998) Predicting how people play games: reinforcement learning in experimental games with unique, mixed strategy equilibria. Am Econ Rev 88:848–881
Fiorillo CD (2013) Two dimensions of value: dopamine neurons represent reward but not aversiveness. Science 341:546–549
Frederick S, Loewenstein G, O'Donoghue T (2002) Time discounting and time preference: a critical review. J Econ Lit 40:351–401
Hassabis D, Maguire EA (2007) Deconstructing episodic memory with construction. Trends Cogn Sci 11:299–306
Hayden BY, Pearson JM, Platt ML (2009) Fictive reward signals in the anterior cingulate cortex. Science 324:948–950
Hayden BY, Heilbronner SR, Pearson JM, Platt ML (2011) Surprise signals in anterior cingulate cortex: neuronal encoding of unsigned reward prediction errors driving adjustment in behavior. J Neurosci 31:4178–4187
Hwang J, Kim S, Lee D (2009) Temporal discounting and inter-temporal choice in rhesus monkeys. Front Behav Neurosci 3:9
Johnson A, Redish AD (2007) Neural ensembles in CA3 transiently encode paths forward of the animal at a decision point. J Neurosci 27:12176–12189
Kable JW, Glimcher PW (2007) The neural correlates of subjective value during intertemporal choice. Nat Neurosci 10:1625–1633
Kim S, Hwang J, Lee D (2008) Prefrontal coding of temporally discounted values during intertemporal choice. Neuron 59:161–172
Kim S, Cai X, Hwang J, Lee D (2012) Prefrontal and striatal activity related to values of objects and locations. Front Neurosci 6:108
Kishida KT et al (2016) Subsecond dopamine fluctuations in human striatum encode superposed error signals about actual and counterfactual reward. Proc Natl Acad Sci U S A 113:200–205
Lee D, Seo H (2016) Neural basis of strategic decision making. Trends Neurosci 39:40–48
Lee D, McGreevy BP, Barraclough (2005) Learning and decision making in monkeys during a rock-paper-scissors game. Cogn Brain Res 25:416–430
Lee D, Rushworth MF, Walton ME, Watanabe M, Sakagami M (2007) Functional specialization of the primate frontal cortex during decision making. J Neurosci 27:8170–8173
Lee D, Seo H, Jung MW (2012) Neural basis of reinforcement learning and decision making. Annu Rev Neurosci 35:287–308
Lee SW, Shimojo S, O'Doherty JP (2014) Neural computations underlying arbitration between model-based and model-free learning. Neuron 81:687–699
Lohrenz T, McCabe K, Camerer CF, Montague PR (2007) Neural signature of fictive learning signals in a sequential investment task. Proc Natl Acad Sci U S A 104:9493–9498
McClure SM, Laibson DI, Loewenstein G, Cohen JG (2004) Separate neural systems value immediate and delayed monetary rewards. Science 306:503–507
Mookherjee D, Sopher B (1994) Learning behavior in an experimental matching pennies game. Game Econ Behav 7:62–91
Nash JF (1950) Equilibrium points in n-person games. Proc Natl Acad Sci U S A 36:48–49
Padoa-Schioppa C, Assad JA (2006) Neurons in the orbitofrontal cortex encode economic values. Nature 441:223–226
Passingham RE, Wise SP (2012) The neurobiology of the prefrontal cortex. Oxford University Press, Oxford

Pezzulo G, MA v M, Lansink CS, Pennartz CM (2014) Internally generated sequences in learning and executing goal-directed behavior. Trends Cogn Sci 18:647–657
Schultz W (1998) Predictive reward signals of dopamine neurons. J Neurophysiol 80:1–27
Seo H, Lee D (2007) Temporal filtering of reward signals in the dorsal anterior cingulate cortex during a mixed-strategy game. J Neurosci 27:8366–8377
Seo H, Lee D (2008) Cortical mechanisms for reinforcement learning in competitive games. Philos Trans R Soc Lond Ser B Biol Sci 363:3845–3857
Seo H, Lee D (2009) Behavioral and neural changes after gains and losses of conditioned reinforcers. J Neurosci 29:3627–3641
Seo H, Cai X, Donahue CH, Lee D (2014) Neural correlates of strategic reasoning during competitive games. Science 346:340–343
Simon DA, Daw ND (2011) Neural correlates of forward planning in a spatial decision task in humans. J Neurosci 31:5528–5539
Stephens DW, McLinn CM, Stevens JR (2002) Discounting and reciprocity in an iterated prisoner's dilemma. Science 298:2216–2218
Sutton RS, Barto AG (1998) Reinforcement learning. MIT Press/Cambridge University Press, Oxford
Tremblay L, Schultz (1999) Relative reward preference in primate orbitofrontal cortex. Nature 398:704–708
Vickery TJ, Chun MM, Lee D (2011) Ubiquity and specificity of reinforcement signals throughout the human brain. Neuron 72:166–177
Vickery TJ, Kleinman MR, Chun MM, Lee D (2015) Opponent identity influences value learning in simple games. J Neurosci 35:11133–11143
Wallis JD, Kennerley SW (2010) Heterogeneous reward signals in prefrontal cortex. Curr Opin Neurobiol 20:191–198
Wallis JD, Miller EK (2003) Neuronal activity in primate dorsolateral and orbital prefrontal cortex during performance of a reward preference task. Eur J Neurosci 18:2069–2081
Zhu L, Mathewson KE, Hsu M (2012) Dissociable neural representations of reinforcement and belief prediction errors underlie strategic learning. Proc Natl Acad Sci U S A 109:1419–1424

# Chapter 2
# Functional Division Among Prefrontal Cortical Areas in an Analog of Wisconsin Card Sorting Test

**Keiji Tanaka, Mark J. Buckley, and Farshad A. Mansouri**

**Abstract** The primate prefrontal cortex (PFC) is composed of several different areas, which have different anatomical connections with other brain structures. We have studied the functional divisions among the prefrontal areas by combining lesion-behavioral experiments with single-cell recordings in intact monkeys using an analog of the Wisconsin Card Sorting Test as a behavioral paradigm. Our results suggest that the PFC is composed of multiple functional units, each of which plays different elementary roles in the performance of cognitively demanding tasks. These elementary functions are mutually dependent on one another, and then the overall performance of the PFC goes beyond a mere sum of the elementary functions. We have also obtained results suggesting that the control of cognitively demanding tasks depends on the posterior parts of the PFC when they are well learned. The most anterior part of PFC, the frontopolar cortex, starts to play another role, i.e., exploration of other possibilities than those pursued by the current task. By having this function of the frontopolar cortex, primates may have increased the flexibility and adaptability of their behaviors in changing environments.

**Keywords** Anterior cingulate sulcus area • Principal sulcus area • Orbitofrontal cortex • Frontopolar cortex • Double dissociation • Rule-based behavior • Working memory of rule • Response-selection mode • Rapid learning • Exploration

The primate prefrontal cortex (PFC) is composed of several different areas, which have different anatomical connections with other brain structures, such as sensory

K. Tanaka (✉)
RIKEN Brain Science Institute, Hirosawa, Wako, Japan
e-mail: keiji@riken.jp

M.J. Buckley
University of Oxford, Oxford, UK

F.A. Mansouri
Monash University, Clayton, VIC, Australia



M. Watanabe (ed.), *The Prefrontal Cortex as an Executive, Emotional, and Social Brain*, DOI 10.1007/978-4-431-56508-6_2

and motor association cortical areas, the basal ganglia, limbic structures, and hypothalamus (Barbas 2000; Fuster 2008; Goldman-Rakic 1987; Ongür and Price 2000). The functional divisions among the prefrontal areas, however, remain largely uncovered. While the elucidation of functional divisions will provide us a functional mapping in the prefrontal cortex, it also helps us decompose apparently complicated cognitive functions into more elementary processes.

The selection of task paradigm is important in the study of functional divisions in animals' brains, as the number of tasks on which we can train the same individual animals is quite limited, especially when the tasks are demanding. To show double or triple dissociation among areas, the task has to include the functions mediated by two or more areas. The Wisconsin Card Sorting Test (WCST) (Berg 1948; Heaton 1981) satisfies this requirement. WCST is a neuropsychological assessment measure frequently used in the human clinical field. It is sensitive to dysfunction/damages in various parts of PFC (Milner 1963, 1995; Stuss et al. 2000). The subject takes cards, one at a time, from a card deck and sorts them according to the number, shape, or color of symbols drawn on each card (Fig. 2.1). The correct sorting rule is not instructed to the subject, and only the feedback of whether the response was correct or wrong is provided. Therefore, the subject has to infer the current rule based on a combination of the rule that the subject has just applied in the last response and the feedback given to the response. The rule is consistent across trials, but it changes, without a notice given to the subject, when the performance of the

**Fig. 2.1** Wisconsin Card Sorting Test (WCST). The subject (*left*) takes cards from a deck and sorts them according to the match in color, shape, or number of symbols. The experimenter (*right*) provides only correct or wrong feedback to individual responses of the subject. The sorting rule is consistent across trials, but it suddenly changes when the subject reaches a certain performance criterion

subject on one rule reaches a certain level so that the subject has to adapt to the rule changes as well. The performance of WCST, thus, requires multiple cognitive functions, including the determination of the currently relevant rule based on the own response and feedback, the maintenance of the currently relevant rule in the working memory, and the suppression of perseverative responses driven by the previously relevant rule. This requirement of multiple functions may make the task sensitive to damages in various parts of PFC.

Because monkeys are relatively poor at dealing with numbers, we dropped the dimension of number in our WCST analog. We implemented the WCST analog test on a computer-controlled display with a touch screen (Fig. 2.2) (Mansouri and Tanaka 2002). A sample stimulus first appeared at the center of the display. When the monkey touched it, three test stimuli were added around the sample. One of the test stimuli matched the sample in color, another matched the sample in shape, and the third one did not match the sample in either color or shape. When the color-matching rule was relevant, the monkey had to touch the test stimulus that matched the sample in color. When the shape-matching rule was relevant, the monkey had to touch the test stimulus that matched the sample in shape. Correct responses were rewarded with a small primate food reward pellet, and erroneous responses were not rewarded and were followed instead by the presentation of a visual signal alerting the animal that an error response has been made. The stimuli in each trial were selected from a large set of 36 stimuli composed of all combinations of 6 colors and 6 simple shapes (Fig. 2.2). The matching rule was consistent within a block of trials, but when the monkey reached 85 % correct responses in 20 consecutive trials, the

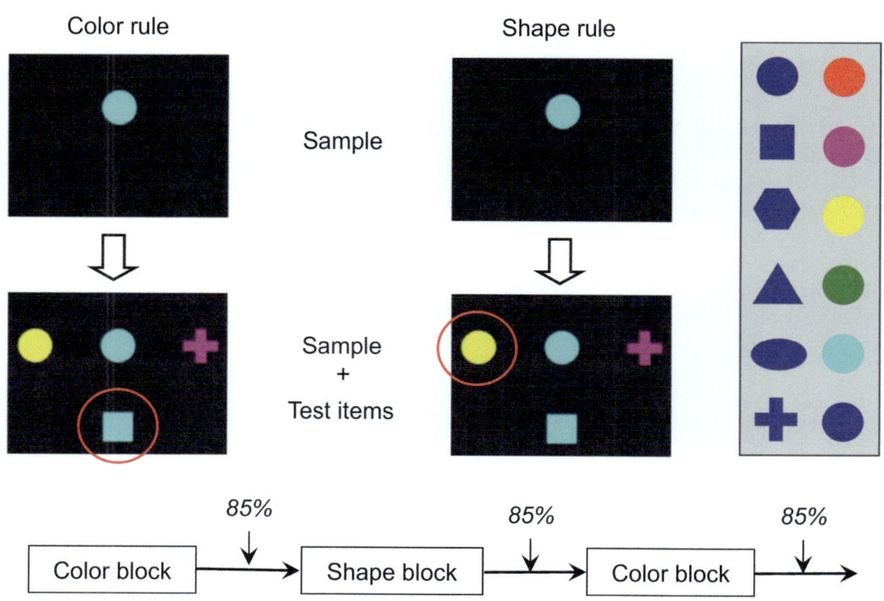


**Fig. 2.2** Monkey version of WCST (WCST analog). The stimuli are presented and responses made on a display with a touch screen

rule changed. There was no cue for the currently relevant rule or for the switch of the relevant rule. The centrally positioned sample in each trial was randomly selected from the 36 stimuli, and the three surrounding test stimuli were also randomly selected from the 36 stimuli with the above-described constraints (Fig. 2.2). The arrangement of the three test stimuli at the three possible test stimulus positions was also randomized. Because of the large number of stimuli used in the task and the randomization of test-item locations, the monkeys had to use the abstract rule of color matching or shape matching to follow the rule changes within several trials.

In the first series of experiments (Buckley et al. 2009), 14 macaque monkeys learned the task to high proficiency so that they achieved 11 rule switches, on average, per daily session of 300 trials. Figure 2.3 shows the average percentage of erroneous responses in the trials after the rule changes. Because the rule changes were not cued, the error percentage was naturally as high as 90% at the first trial after the rule change. Because there were only two rules, subjects should be able to find the correct rule just after one error. Indeed, intact adult humans can completely adapt to the rule change after one mistake. However, macaque monkeys were poorer than intact adult humans on average. The percentage of errors gradually decreased after the rule change and reached a semi-stationary low level only after several trials on average. Figure 2.3 also shows a clear bias in the type of errors. There were two types of erroneous responses: a response that would have been correct according to the other matching rule and a response that was wrong according to either rule. Most of the erroneous responses were the former type, which indicates that the monkeys had learned the color- and shape-matching rules well in their long-term memory and they switched between the two rules with the currently correct rule held in short-term memory.

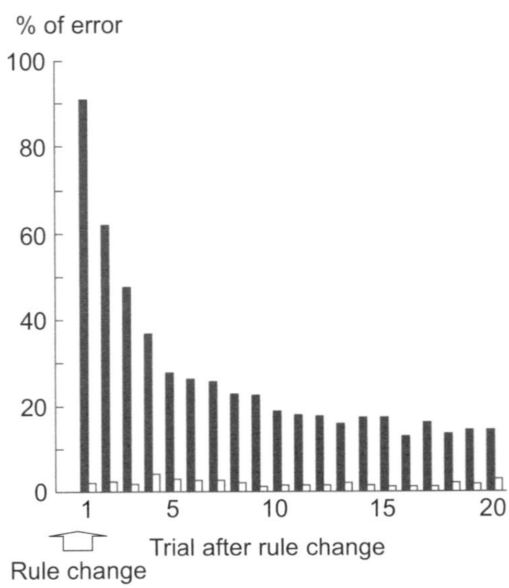


**Fig. 2.3** Decrease of error proportion after rule switches. The *black bars* represent perseverative errors that would have been correct according to the other rule. The *white bars* represent completely wrong errors

After the monkeys had learned the task well, we made bilateral lesions by the aspiration of the gray matter in four different parts of PFC: the superior dorsolateral area (sdlPFC), the principal sulcus area (PS) located more inferiorly within dorsolateral PFC, the orbital area (OFC), and the anterior cingulate sulcus area (ACS) (Fig. 2.4). The PS and ACS lesions were made in the first stage, and the monkeys that were used as an intact control group in the first stage later received the sdlPFC or OFC lesion. Each group had three or four monkeys.

The PS, ACS, and OFC lesions degraded the performance of the monkeys in the WCST analog (Fig. 2.5). The number of rule changes per daily session decreased by about 40–50%. The sdlPFC-lesion group maintained the original performance. Note that the number of rule switches per daily session is a good parameter of the overall performance. As the total number of trials given per daily session was fixed to 300, there were more rule changes when the monkey followed the rule changes quickly, and fewer rule changes occurred when more trials were required for the monkey to recover a high performance level after a rule change.

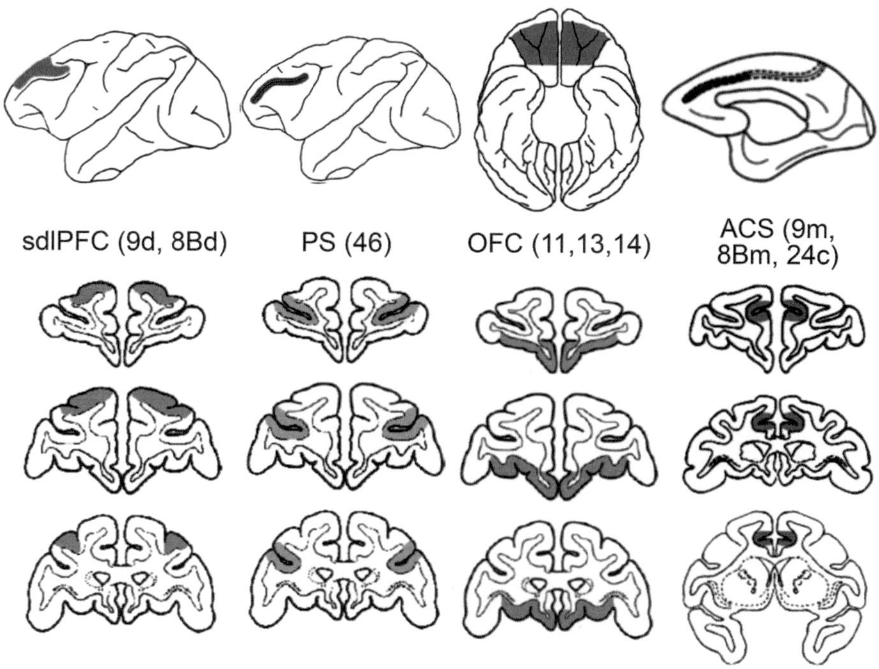


**Fig. 2.4** Intended extent of lesions in the superior dorsolateral (*sdlPFC*), principal sulcus (*PS*), orbitofrontal (*OFC*), and anterior cingulate sulcus (*ACS*) areas of the PFC. Diagrams at the *top line* indicate the lateral (*left two*), *bottom* (*second right*), and medial (*the rightmost*) views of the brain. The *bottom three lines* indicate frontal sections, in the order of from anterior to posterior. The lesion extent is indicated by *gray* shadings. The parentheses show the corresponding Brodmann's areas. The ACS lesion included the most ventral parts of the presupplementary motor area (*preSMA*) and supplementary motor area (SMA) as well as areas 9m, 8Bm and 24c (Cited from Buckley et al. (2009) with a modification)

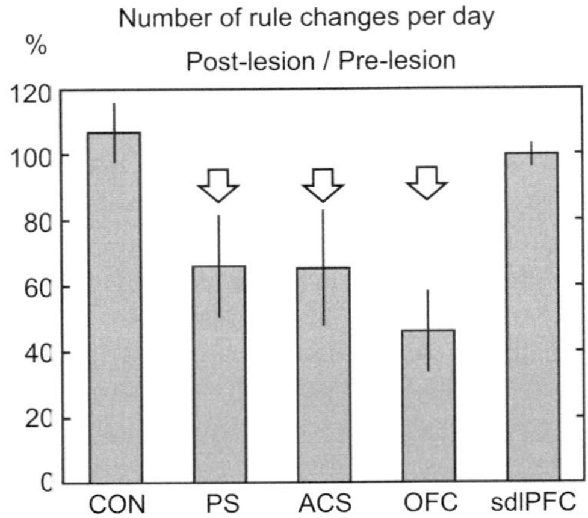


**Fig. 2.5** Changes in the number of rule changes per day after the lesions. As the number of trials given per day was fixed to 300, the number of rule changes per day was a good measure of the overall performance of monkeys. The number of rule changes per day in the post-lesion test was divided by that in the pre-lesion test in each monkey and then averaged over the monkeys in each group. CON represents the intact control group of monkeys, which only took a rest between the two tests. The error bars indicate the SEM across monkeys in this and all the following figures (Cited from Buckley et al. (2009) with a modification)

Although the overall performance in the WCST analog degraded by the PS, ACS, and OFC lesions, the results did not necessarily indicate that the three areas played similar functional roles. Rather, the following observations of various aspects of the monkeys' behavior in the WCST analog and other probe tests as well as single-cell activities recorded from the regions in other intact monkeys performing the WCST analog indicated that the reason of the degradation was different in different areas.

## 2.1 Contribution of the Principal Sulcus Area (PS)

We first examined the robustness of maintaining the currently relevant rule in working memory. Both color- and shape-matching rules were repeatedly used each day, and therefore long-term memory does not help specify the currently relevant rule. The currently relevant rule had to be maintained in working memory. The vulnerability of the maintenance of the currently relevant rule was examined by occasionally increasing the intertrial interval (ITI) by about 100% (originally 6 s) after the monkey reached 85% correct performance. The rule was not changed across this increased ITI. The average performance in the first trial after the long ITI dropped to 73% in the intact control monkeys (Fig. 2.6). As the monkeys had been accustomed to the fixed 6-s delay, the long delay may have irritated them. They moved around within the test cage and often hit the wall of the cage. The irritation

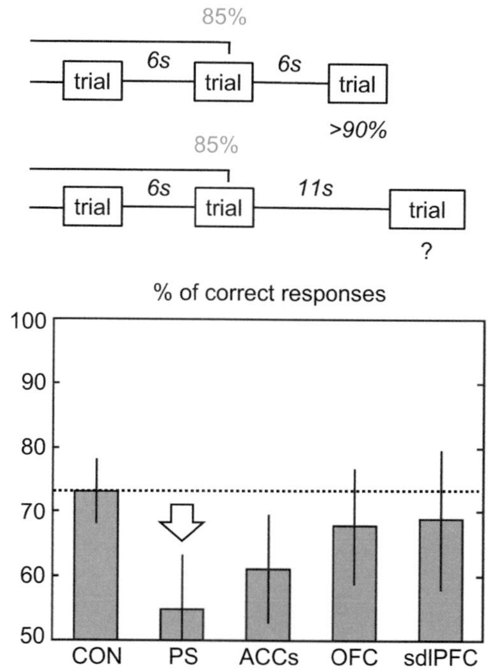


**Fig. 2.6** Vulnerability of the working memory of the currently relevant rule to an unusual interruption. The intertrial interval (ITI) between trials had been fixed to 6 s during the training. The monkeys were thus disturbed by rarely introduced longer ITIs (11 s). The percentage of correct responses in the trials immediately after the longer ITI was measured (Cited from Buckley et al. (2009) with a modification)

might cause some forgetting of working memory of the currently relevant rule. More importantly, the average performance of the PS-lesioned monkeys further dropped to 55%, which was indistinguishable from the practical chance level (50%) and significantly lower than that in the intact control monkeys. The averaged percentages of correct responses in the other lesion groups were not significantly lower than that in the intact control group. These results suggest that working memory of the currently relevant rule was more sensitive to interruption in the PS-lesioned monkeys and that the PS area is essential for the maintenance of the currently relevant rule.

The version of the WCST analog task used for single-cell recordings (Mansouri et al. 2006; Kuwabara et al. 2014) was similar to that used for lesion and behavioral experiments except a few details. Firstly, the monkeys for single-cell recordings sat in a monkey chair with their head restrained in a fixed position. Secondly, the monkeys had to fixate their eye gaze at the center of the display for 700 ms before the sample appeared. They had to maintain gaze fixation during the sample presentation until the test items appeared. Thirdly, the monkeys were not required to touch the sample. Fourthly, we used only four test-item arrangements, and each test-item arrangement was associated with a fixed set of samples. The monkeys, thus, could know the direction of the correct response when the sample appeared. Finally, the reward was a drop of water.

Single-cell recordings from the PS area of intact monkeys performing the WCST analog (Mansouri et al. 2006) showed that about 30% of the task-relevant cells in

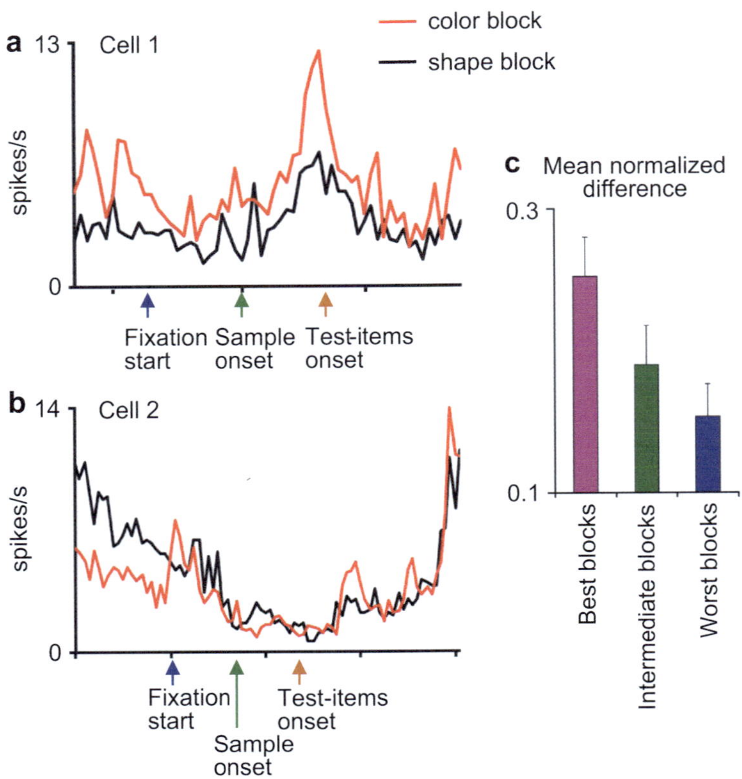


**Fig. 2.7** Differential activities of neurons in the principal sulcus area (PS). (**a**, **b**) Activities in two examples of PS cells. The *red* and *black lines*, respectively, show mean firing rates in *color* and *shape blocks*. The *arrows* along the x-axis indicate the onsets of fixation and sample and test-item presentation. The interval between tacs on the x-axis is 1 s. (**c**) Deviation of mean activities in individual trials in the three groups of blocks in which the monkeys reached the high correct percentage within the smallest numbers of trials (*best blocks*), intermediate trial numbers (*intermediate blocks*), and the largest numbers of trials (*worst blocks*). The deviation was measured by the difference from the moving average of activities in the previous, current, and next blocks (Cited from Mansouri et al. (2006) with a modification)

the PS area showed significantly differential activities between the color and shape blocks. The number of cells that showed higher activities in color blocks was comparable to that of cells that showed higher activities in shape blocks. Some of them showed differential activities throughout the trial (as Cell 1 in Fig. 2.7), but the majority showed differential activities only in part of a trial. The period in a trial where the activity was different between color and shape blocks varied from cell to cell and across cells covered the whole extent of a trial; for example, Cell 2 in Fig. 2.7 showed differential activities in the ITI. Moreover, the magnitude of differential activities (deviation from the average) in each block correlated with the monkey's averaged performance in the block (measured by the length of the block) (Fig. 2.7, right). These differential activities may underlie the maintenance of the currently relevant rule in the PS area.

The PS area has traditionally been associated with spatial-working memory (Goldman-Rakic 1996). Recent single-unit recording studies show that the information carried by neuronal activities in the PS area is not limited to the working memory of spatial locations but can cover the working memory of objects (Miller et al. 1996; Rao et al. 1997). The neural activities in the PS area even reflect rules (Assad et al. 2000; Wallis et al. 2001; White and Wise 1999). However, there have been no prior reports of lesions restricted to the PS area impairing tasks other than spatial memory. Our findings expand the role of the PS area to include working memory for abstract rules (Mansouri et al. 2015b).

## 2.2 Contribution of the Orbitofrontal Cortex (OFC)

Learning from errors is important in WCST (and WCST analog) performance. We examined the averaged performance in those trials that followed erroneous selections. Before lesions were introduced, the averaged performance on trials following an error trial was roughly at chance level (50%, Fig. 2.8). Because there are only two rules in the WCST analog, in theory, after an erroneous selection, the subject can always behave according to the correct rule by suppressing selection according to the rule used in the previous selection. That performance was at chance levels on trials after errors suggests that the monkeys were not necessarily retaining information of the rule they applied in the previous selection through to the time when the feedback was provided. They might instead retain only more general information about rule, e.g., information pertaining to which rule they applied in most, or a majority, of recent trials. Then, because there were two types of errors, ones caused

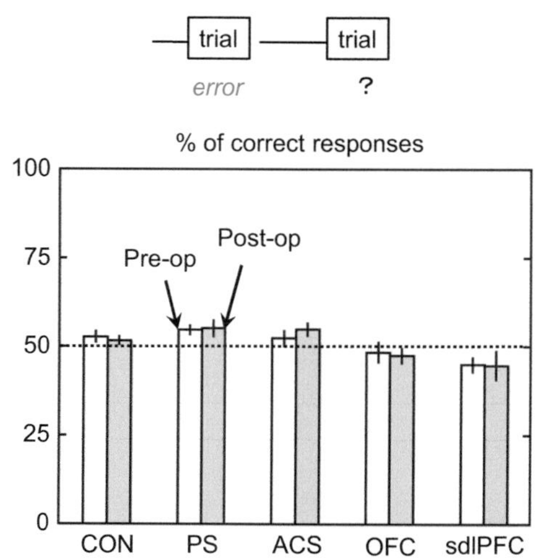


**Fig. 2.8** Performance after an error trial. The averaged percentage of correct responses in the trials after an error trial following a correct trial. The left open and right gray bars, respectively, show the values in the pre- and post-lesion tests (Cited from Buckley et al. (2009) with a modification)

by rule changes and others caused by the monkey's mistakes, it was not advantageous for the monkeys to take the other rule than the one that they used in a majority of recent trials. The monkeys might have learned instead to adopt a trial-and-error strategy, i.e., start off with a random selection between the two rules in the first trial after a mistake.

Learning from successful and rewarded choice outcomes is also important in the WCST performance. Therefore, we also examined the influence of a single correct (and rewarded) response following an erroneous response on performance of the subsequent trial. Preoperatively, the averaged performance in the trials that followed a sequence of erroneous and correct trials was around 75% (Fig. 2.9). The performance of the OFC-lesioned group dropped to chance level, whereas the other lesion groups did not show significant degradation (Fig. 2.9). However, it was not the case that the OFC-lesioned monkeys were totally insensitive to success experiences. When the next response after an error-correct sequence happened to be correct, the percentage of correct responses in the following trial became significantly higher than chance level (Fig. 2.10). These results suggest that there were two types of learning from success experiences in the WCST analog. One is a slow learning following consecutive success experiences and the other is a quick learning from a single success. The OFC area plays an essential role in the latter, i.e., rapid increase of the value of a rule after a single success.

Previous studies suggested that the OFC is important for flexibly following changes in stimulus-reward association: monkeys with bilateral OFC lesions show problems in object discrimination reversal (Iversen and Mishkin 1970; Meunier et al. 1997). Although these effects may be explained by possible damage to passing fibers (Rudebeck et al. 2013), a recent fMRI monkey study showed reversal-related activities in the lateral part of OFC (Chau et al. 2015). A probabi-

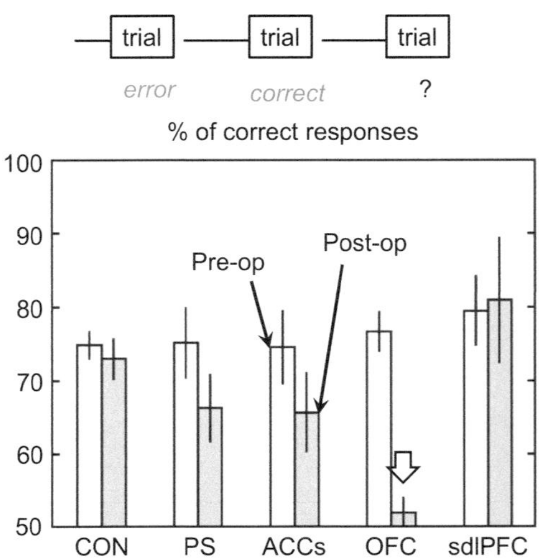


**Fig. 2.9** Performance after a single correct trial. The averaged percentage of correct responses in the trials after a correct trial that was preceded by an error trial (Cited from Buckley et al. (2009) with a modification)

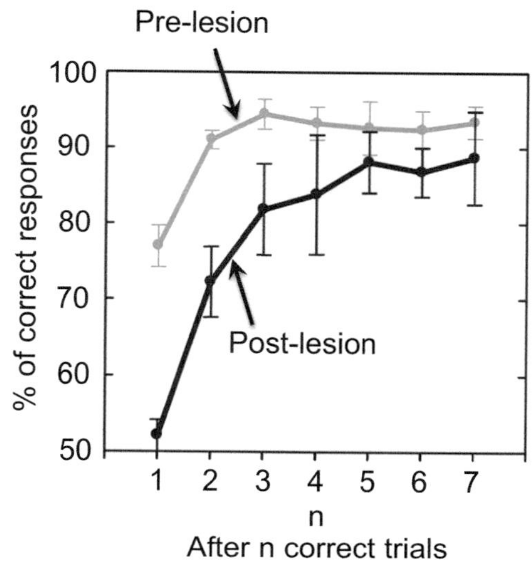


**Fig. 2.10** Recovery of performance after an error trial in the OFC-lesioned monkeys. The averaged percentage of correct responses in the trials after n times of consecutive correct trials following an error trial. The *gray* and *black lines*, respectively, show the values in the pre- and post-lesion tests (Cited from Buckley et al. (2009) with a modification)

listic reward assignment to alternative stimuli also reveals problems in OFC-lesioned monkeys (Rudebeck et al. 2008). A more recent study, using ever-changing reward probability, shows that the problem in OFC-lesioned monkeys exists in assigning the reward to the immediately preceding stimulus selection (Walton et al. 2010). The failure in the rapid learning from a single success in the present study may be also explained by a problem in assigning the reward to the rule used in the last response selection that provided the reward. However, as we discussed above, even intact monkeys unlikely carried the information of the specific rule used in the last response selection to the time of feedback. We rather think that the OFC area registers the trial-and-error strategy used in the trials after an erroneous trial as a cognitive state, maintains the rule applied as a trial under the trial-and-error strategy, and increases the value of the rule when the reward is provided.

## 2.3 Contribution of the Anterior Cingulate Sulcus Area (ACS)

The most prominent change after the ACS lesions appeared in response times. We measured the response time from the onset of the test items to the monkey's first touch on the screen. In intact control monkeys, the response time was significantly longer in the trials after an error trial than that in the trials after consecutive correct trials (Fig. 2.11). The difference was larger than 1 s. After the ACS lesions, this difference in response time disappeared: both responses after an error and those

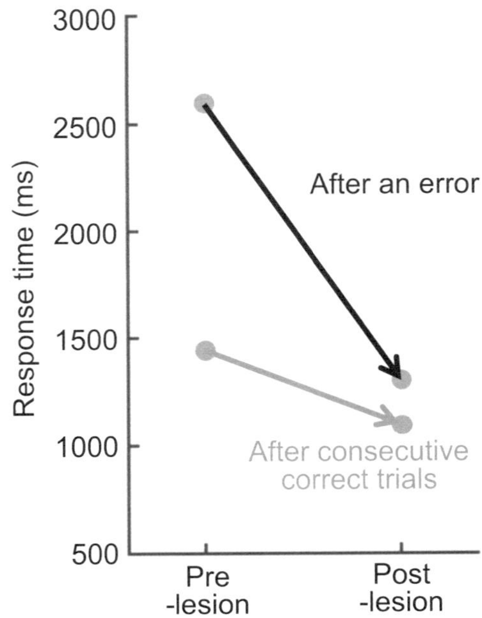


**Fig. 2.11** Mean response times in the trials following an error trial (*black line*) and in the trials following consecutive (more than two) correct trials (*gray line*) in the ACS-lesioned monkeys (Cited from Kuwabara et al. (2014) with a modification)

after consecutive correct trials became quick. As we discussed, the monkeys likely took a trial-and-error strategy in the trials after an error trial. The long response time in the trials after an error trial may be due to a demanding cognitive control required for the trial-and-error strategy. The shortening of response time in such trials suggests that ACS was essential for the implementation of the trial-and-error strategy.

Single-cell recordings from intact monkeys performing the WCST analog showed that about 45% of cells in the ACS area showed activities significantly dependent on the recent history of trials. Half of them showed higher activities in the trials after an error trial (Population 1 in Fig. 2.12), and the remaining cells showed higher activities in the trials after consecutive success trials (Population 2 in Fig. 2.12). The differential activities in some cells appeared immediately after the feedback in the previous trial and maintained through the current trial. Other cells started to show the differential activities sometime in the ITI between the previous and current trials.

We reason that intact monkeys may use two modes of response selections in the WCST analog: one based on working memory of the currently relevant rule and the other based on trial and error of selecting one of two possible responses indicated by the two rules. The working-memory-dependent mode was maintained in consecutive correct trials, whereas the trial-and-error mode was used in after-error trials. The monkeys were not simply responding without cognitive control in the trial-and-error mode, because they overcame the perseverative tendency (Passingham 1972; Roberts et al. 1988) and the response time was longer. The differential activities of the two populations of neurons in the ACS area might be involved in supporting one

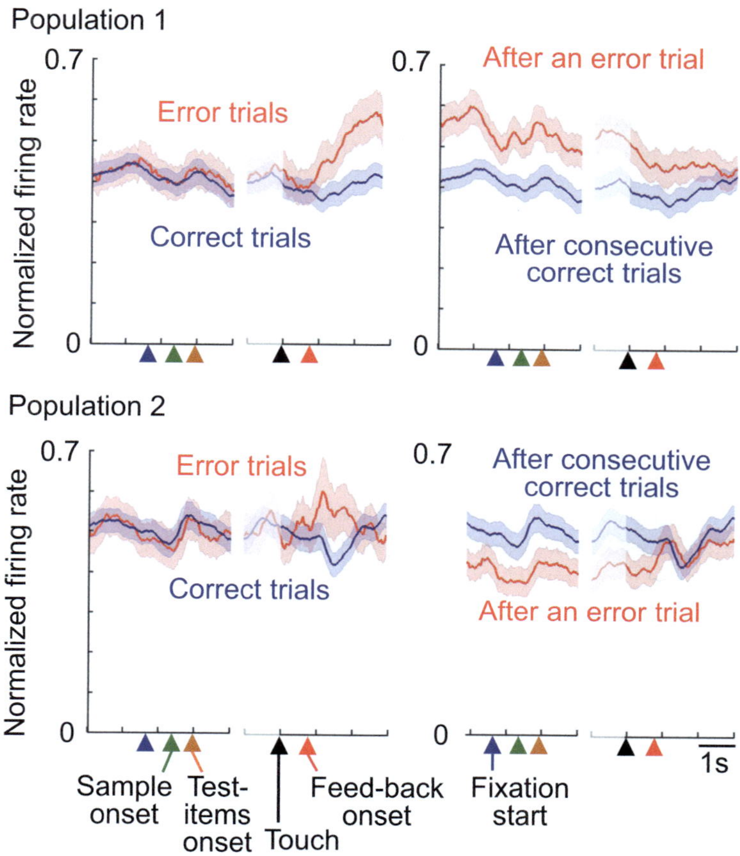


**Fig. 2.12** Activities of two populations of ACS cells. The graphs in the right show averaged firing rates in the trials following an error trial (*red*) and in the trials following consecutive (more than two) correct trials (*blue*). The graphs in the left show averaged firing rates in error trials (*red*) and correct trials following one or more correct trials (*blue*). The shadings show SEM across cells (*shading*). Populations 1 and 2 constituted 73 and 80 cells among the 343 cells recorded from the ACS area

of the two modes in a given situation. The ACS lesions, by removing these neuronal activities, resulted in two types of changes in the percentages of correct responses: the recovery from an error trial became slower and more occasional errors were committed even after several successive correct trials. The slower recovery could be caused by a deficit in reestablishing the working-memory-dependent mode after the trial-and-error mode. The occasional commission of errors after successive correct trials could represent easier slips from the working-memory-dependent mode into the trial-and-error mode. The control of response speed is assumed to be coupled with the two modes. In summary, the results suggest that the ACS area supports the context-dependent transition between the working-memory-dependent mode and trial-and-error mode of response selection, both of which are cognitively demanding. Previous studies have shown that the ACS supports the goal-directed response selection by generating a response plan based on the expectation of reward conditions associated with the responses (Matsumoto et al. 2003) and represents the negative and positive outcomes after internally driven selection of actions (Matsumoto et al. 2007). Hence the ACS area also contributes to controlling individual actions when they are internally driven.

## 2.4 Functional Division Among the PS, OFC, and ACS Areas

Our results thus show that the PS, OFC, and ACS areas play different roles in WCST performance (Fig. 2.13). They demonstrate the complementary strength of combining lesion and single-cell recording methods. In contrast to neuropsychological studies of human brain-damaged patients, in studies with experimental monkeys, the extent of the lesion can precisely follow an experimental plan. Lesion studies examine the causality between the assumed function in an area and the behavior, while they do not directly show the way in which the area performs the function. Single-cell recording studies reveal the representation of information by individual neurons and neuronal populations, while they cannot examine the causality.

## 2.5 Contribution of the Frontopolar Cortex (FPC)

We have expanded the studies by making bilateral lesions in the FPC of macaque monkeys after they had well learned the WCST analog. The FPC corresponds to the cytoarchitectural area 10 and is located at the most rostral part of the PFC (Petrides

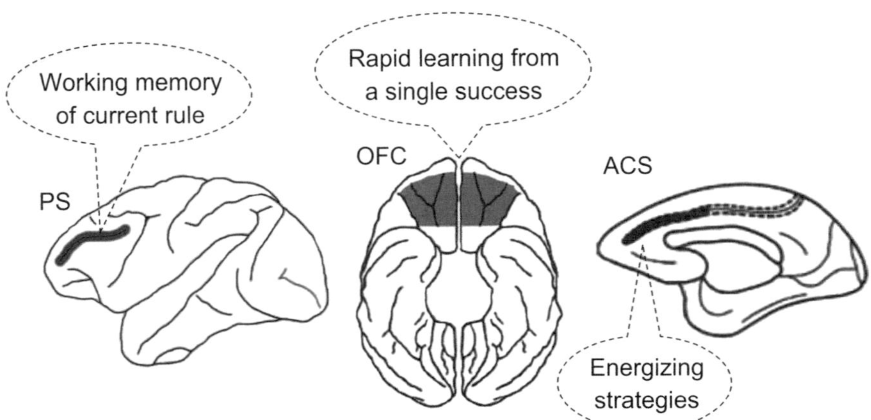


**Fig. 2.13** Elementary functions played by the PS, OFC, and ACS areas in the WCST analog

and Pandya 1994). Based on its anatomical connections, the FPC is thought to be located at the highest hierarchical level in the PFC (Barbas and Pandya 1989; Petrides and Pandya 2007; Burman et al. 2011). Because the WCST analog is highly demanding, it was expected that the FPC plays an essential role in the performance of the task. Seven macaque monkeys were trained with the WCST analog, and then four of them received bilateral lesions of the FPC (Fig. 2.14). The remaining three served as an intact control group.

Against our expectation, the FPC lesion did not degrade the overall performance in the WCST analog: the number of rule changes per day in the postoperative test was comparable to that in the preoperative test (Fig. 2.15). This absence of any degradation in performance after the FPC lesions made a prominent contrast with the significant degradations after the PS, OFC, and ACS lesions described above.

Rather, the monkeys' behavior in the WCST analog became better after the FPC lesions in a few aspects. Firstly we found an increase of conflict adaptation. In addition to the basic version of the WCST analog, we had trained the monkeys preoperatively with another version of WCST analog (WCST conflict), in which trials without conflict were intermingled with trials with conflict (Fig. 2.16). In trials without conflict, one of the test items was identical to the sample. Thus, both the color- and shape-matching rules indicated the same target, and so there was no conflict between the two well-learned rules. These trials without conflict were named "low-conflict trials" following the tradition in the field, while the trials in which the color- and shape-matching rules indicated different targets were named "high-conflict trials." The monkeys' response times were longer in high-conflict trials than that in low-conflict trials. The magnitude of this difference in response times for high versus low conflict, which has been named "conflict cost," was not altered by the FPC lesions. The experience of conflict also has an influence on the response time in the next trial. The response time in high-conflict trials following a high-conflict trial is shorter than that in high-conflict trials following a low-conflict trial. This difference, which has been named "conflict adaptation," is thought to be a result of a cognitive control, in which a focus on the relevant rule, or attention to the relevant stimulus dimension, is enhanced by the experience of the conflict in the previous trial. It has been found that the conflict adaptation in the WCST conflict was reduced by the PS lesions (Mansouri et al. 2007, 2009) or by the OFC lesions (Mansouri et al. 2014). However, in contrast, the conflict adaptation was augmented after the FPC lesions (Fig. 2.16).

The conflict adaptation effect is explained as follows: experience of high conflict promotes the system to increase the level of cognitive control exerted. With greater concentration on the currently relevant matching rule, or greater attention to the relevant stimulus dimension, the conflict in the following trial is resolved more efficiently as the cognitive control carries over to this trial, thereby decreasing the response time to the second high-conflict trial in the pair. Accordingly the FPC lesion results suggest the counter-intuitive hypothesis that the intact FPC actually acts to degrade the cognitive adaptation process; hence we assumed that the

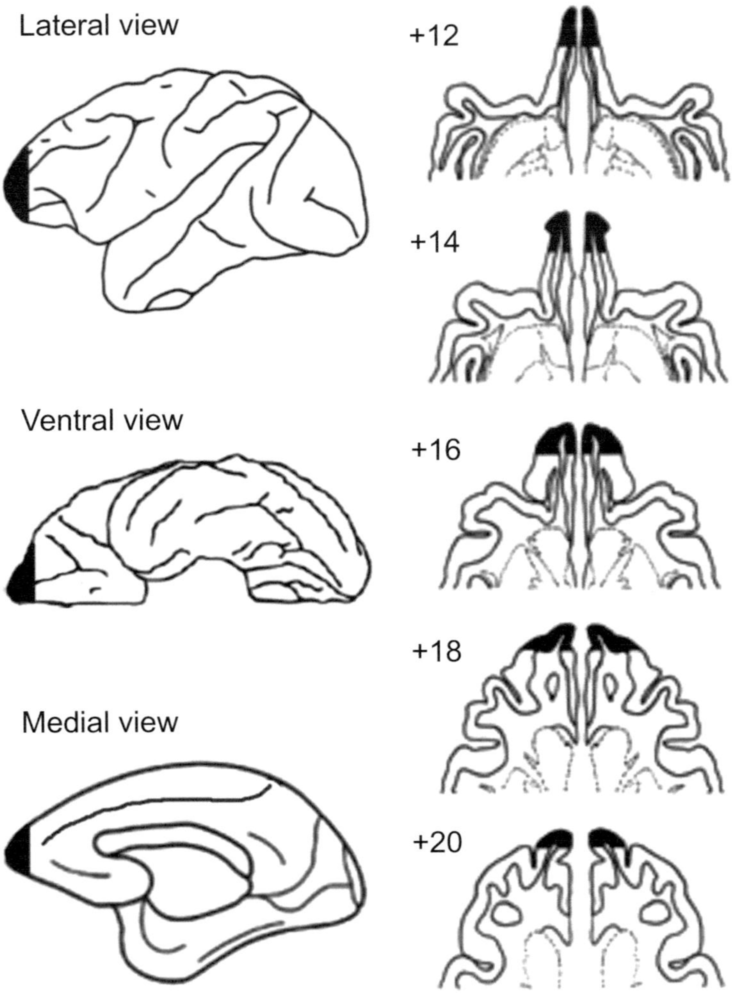


**Fig. 2.14** Intended extent of lesions in the frontopolar cortex (FPC). The numbers attached to drawings of the horizontal sections indicate the height of the section above the AC-PC line

augmenting effect of the FPC lesion generally reflects increased concentration on the current task in FPC-lesioned monkeys.

To examine this working hypothesis, we postoperatively introduced two additional tests. Both tests examined effects of interruption on the performance of the WCST analog. After the monkey reached 85% correct level, we introduced an interruption and then returned to the WCST analog without a rule change. The interruption was a face detection task in the first test. The image of a face and that of a non-face object were presented side by side, and the monkey had to touch on the face. The monkeys learned this task easily within 3 days when the task was given by itself. We had never given the face detection task intermingled with trials of the

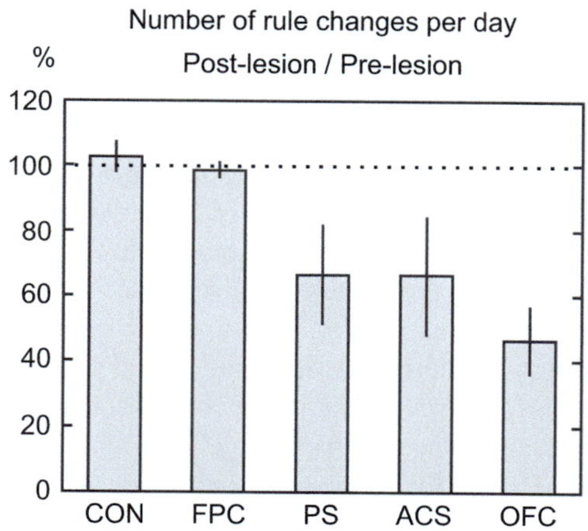


**Fig. 2.15** Changes in the number of rule changes per day after the lesions. The data shown for the PS, ACS, and OFC lesions are the same as those shown in Fig. 2.5 (Cited from Mansouri et al. (2015a) with a modification)

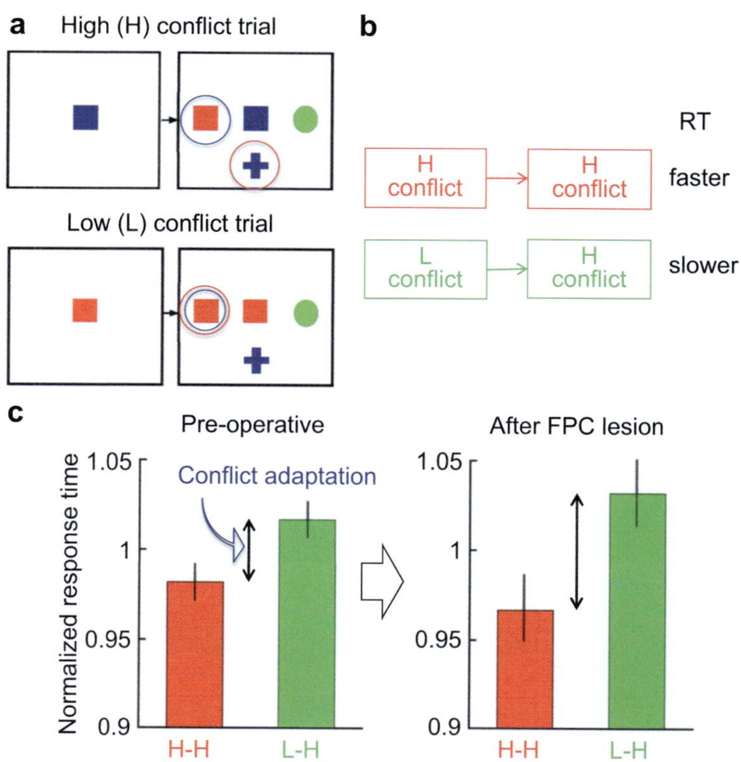


**Fig. 2.16** Increase of conflict adaptation after the FPC lesions. (**a**) Examples of high- and low-conflict trials. (**b**) Comparisons to evoke conflict adaptation. (**c**) Averaged normalized response times in high-conflict trials following a high-conflict trial (H-H) and in high-conflict trials following a low-conflict trial (L-H). The difference between the response times in two types of trials is the conflict adaptation

WCST analog. When we introduced a trial of the face detection task as an interruption to the WCST analog, the monkeys had no problem in conducting the face detection task. However, when they returned to the WCST analog, the performance of the intact control monkeys dropped to 57%, which is indistinguishable from chance level (Fig. 2.17). By performing the rare and unexpected task, the monkeys might lose the working memory of the relevant rule. However, the averaged performance of the FPC-lesioned monkeys in the first WCST trials after the interruption was 78%, significantly higher chance level (Fig. 2.17). In the second interruption test, the unexpected interruption was even simpler, i.e., a free reward (two food reward pellets) was delivered on rare occasions in the interval between trials. The performance of the intact control monkeys in the first WCST trials after the interruption dropped to 61%, which was again indistinguishable from chance level (Fig. 2.17). The FPC-lesioned monkeys again performed much better (81% on average). Thus, the FPC-lesioned monkeys were less disturbed by the interruption, regardless of the nature of the interruption.

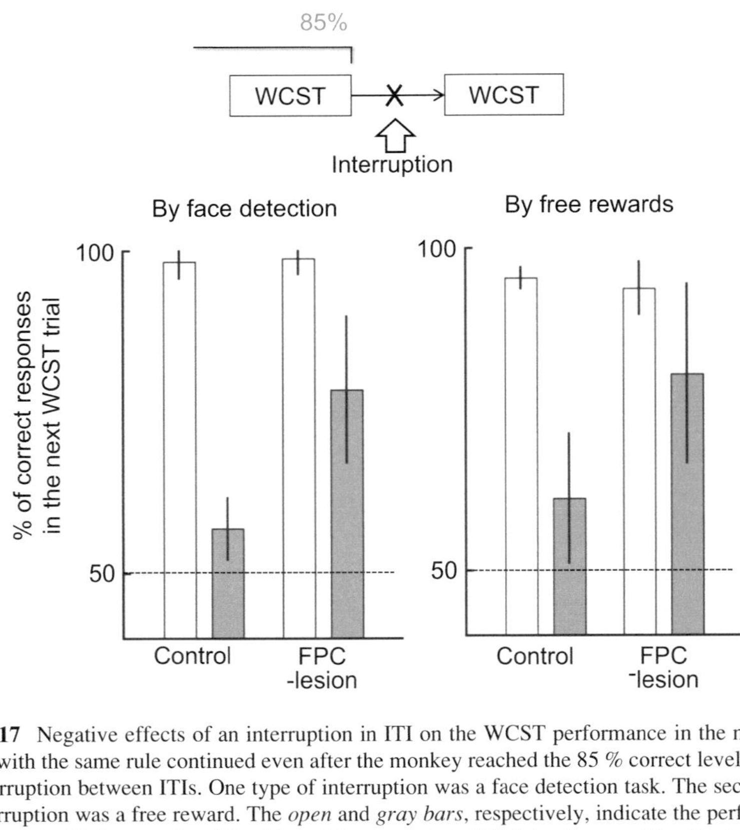


**Fig. 2.17** Negative effects of an interruption in ITI on the WCST performance in the next trial. Trials with the same rule continued even after the monkey reached the 85 % correct level but with an interruption between ITIs. One type of interruption was a face detection task. The second type of interruption was a free reward. The *open* and *gray bars*, respectively, indicate the performance without and with interruption (Cited from Mansouri et al. (2015a) with a modification)

## 2.6 Functional Division Between Anterior and Posterior PFC Areas

The above-described findings show that the FPC-lesioned monkeys were unimpaired in performing the WCST analog. This absence of impairment made a striking contrast with the clear deficits observed after the lesions in PS, OFC, or ACS. Although there was no impairment in the FPC-lesioned monkeys, there were significant differences in the behavior of the FPC-lesioned and intact monkeys. The conflict adaptation effect was significantly augmented after the FPC lesions, and the FPC-lesioned monkeys were less disturbed by the unusual interruptions, i.e., the insertion of face detection task or free reward during the ITI between WCST trials. A better than intact performance after FPC damage has also been reported in humans when the subjects repeated the same task (Rowe et al. 2007). Although FPC-damaged humans performed better when the same task was repeated, they were impaired when they were required to shift between different tasks. FPC-lesioned monkeys were also impaired in learning a new task rule (Boschin et al. 2015).

We have proposed a hypothesis of functional division between the FPC and more posterior PFC areas (Fig. 2.18). In this hypothesis, the performance of a well-learned task, even if it is complicated, is supported by posterior parts of the PFC, including the PS, OFC, and ACS areas, and probably premotor areas as well, whereas the FPC plays a crucial role in redistribution of cognitive resources from the current task to other potential tasks that have existed or appeared in the environment. In complex changing environments, it is not necessarily advantageous to fully focus on the current task. It is better to occasionally explore other possibilities to catch new chances. Thus, the hypothesis proposes a balance between

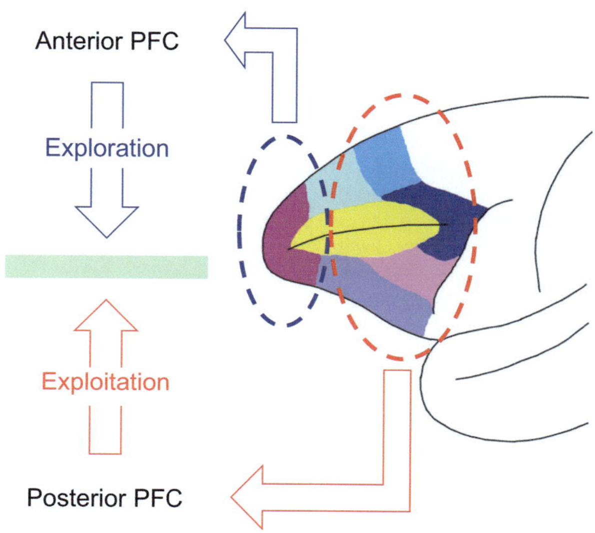


**Fig. 2.18** Functional differentiation between posterior and anterior parts of the PFC (Cited from Mansouri et al. (2015a) with a modification)

"exploitatory" drive from the posterior part of the PFC and "exploratory" drive from the FPC.

## 2.7 Conclusion

The results introduced above suggest that the PFC is composed of multiple functional units, each of which plays different elementary roles in the performance of cognitively demanding tasks. In the WCST analog, the PS area maintains the working memory of the currently relevant rule, the OFC area supports the rapid updating of the value of the relevant rule after an error trial, and the ACS area supports the working-memory-dependent and trial-and-error modes of response selection depending on the context. These elementary functions are different but they are mutually dependent on one another. For example, the information of the response-selection mode represented in the ACS area has to be conveyed to the PS area to maintain or turn down the working memory of the rule. The same information has also to be conveyed to the OFC area to support the rapid learning after a success in the trial-and-error mode. Thus, the overall performance of the PFC goes beyond a mere sum of the elementary functions.

The results also suggest that the control of cognitively demanding tasks depends on the posterior parts of the PFC when they are well learned. The anterior part of PFC, especially the FPC, starts to play another role, i.e., exploration of other possibilities than those pursued by the current task. By having this function of the FPC, primates may have increased the flexibility and adaptability of their behaviors in changing environments.

## References

Assad WF, Rainer G, Miller EK (2000) Task-specific neural activity in the primate prefrontal cortex. J Neurophysiol 84:451–459

Barbas H (2000) Connections underlying the synthesis of cognition, memory, and emotion in primate prefrontal cortices. Brain Res Bull 52:319–330

Barbas H, Pandya DN (1989) Architecture and intrinsic connections of the prefrontal cortex in the rhesus monkey. J Comp Neurol 286:353–375

Berg EA (1948) A simple objective technique for measuring flexibility in thinking. J Gen Psychol 39:15–22

Boschin EA, Piekema C, Buckley MJ (2015) Essential functions of primate frontopolar cortex in cognition. Proc Natl Acad Sci U S A 112:E1020–E1027

Buckley MJ, Mansouri FA, Hoda H, Mahboubi M, Browning PG, Kwok SC, Phillips A, Tanaka K (2009) Dissociable components of rule-guided behavior depend on distinct medial and prefrontal regions. Science 325:52–58

Burman KJ, Reser DH, Yu HH, Rosa MG (2011) Cortical input to the frontal pole of the marmoset monkey. Cereb Cortex 21:1712–1737

Chau BK, Sallet GK, Papageorgiou GK, Noonan MP, Bell AH, Walton ME, Rushworth MF (2015) Contrasting roles of orbitofrontal cortex and amygdala in credit assignment and learning in macaques. Neuron 87:1106–1118

Fuster JM (2008) The prefrontal cortex. Academic Press, London

Goldman-Rakic PS (1987) Circuitry of primate prefrontal cortex and regulation of behavior by representation memory. In: Plum F (ed) Handbook of physiology, the nervous system, higher functions of the brain, section I, vol V. American Physiological Society, Bethesda, pp 373–417

Goldman-Rakic PS (1996) The prefrontal landscape: implications of functional architecture for understanding human mentation and the central executive. Phil Trans R Soc London B351:1445–1453

Heaton RK (1981) A manual for the Wisconsin Card Sorting Test. Psychological Assessment Resources, Odessa

Iversen SD, Mishkin M (1970) Perseverative interference in monkeys following selective lesions of the inferior prefrontal convexity. Exp Brain Res 11:376–386

Kuwabara M, Mansouri FA, Buckley MJ, Tanaka K (2014) Cognitive control functions of anterior cingulate cortex in macaque monkeys performing a Wisconsin Card Sorting Test analog. J Neurosci 34:7531–7547

Mansouri FA, Tanaka K (2002) Behavioral evidence for working memory of sensory dimension in macaque monkeys. Behav Brain Res 136:415–426

Mansouri FA, Matsumoto K, Tanaka K (2006) Prefrontal cell activities related to monkeys' success and failure in adapting to rule changes in a Wisconsin Card Sorting Test analog. J Neurosci 26:2745–2756

Mansouri FA, Buckley MJ, Tanaka K (2007) Mnemonic function of lateral prefrontal cortex in conflict-induced behavioral adjustment. Science 318:987–990

Mansouri FA, Tanaka K, Buckley MJ (2009) Conflict-induced behavioural adjustment: a clue to the executive functions of prefrontal cortex. Nat Rev Neurosci 10:141–152

Mansouri FA, Buckley MJ, Tanaka K (2014) The essential role of primate orbitofrontal cortex in conflict-induced executive control adjustment. J Neurosci 34:11016–11031

Mansouri FA, Buckley MJ, Mahboubi M, Tanaka K (2015a) Behavioral consequences of selective damage to frontal pole and posterior cingulate cortices. Proc Natl Acad Sci U S A 112:E3940–E3949

Mansouri FA, Rosa M, Atapour N (2015b) Short-term memory in the service of executive control functions. Front Syst Neurosci 9:166

Matsumoto K, Suzuki W, Tanaka K (2003) Neuronal correlates of goal-based motor selection in the prefrontal cortex. Science 301:229–232

Matsumoto M, Matsumoto K, Abe H, Tanaka K (2007) Medial prefrontal cell activity signaling prediction errors of action values. Nat Neurosci 10:647–656

Meunier M, Bachevalier J, Mishkin M (1997) Effects of orbital frontal and anterior cingulate lesions on object and spatial memory in rhesus monkeys. Neuropsychologia 35:999–1015

Miller EK, Erickson CA, Desimone R (1996) Neural mechanisms of visual working memory in prefrontal cortex. J Neurosci 16:5154–5167

Milner B (1963) Effects of different brain lesions on card sorting: the role of frontal lobes. Arch Neurol 9:90–100

Milner B (1995) Aspects of human frontal lobe function. Adv Neurol 66:67–81

Ongür D, Price JL (2000) The organization of networks within the orbital and medial prefrontal cortex of rats, monkeys and humans. Cereb Cortex 10:206–219

Passingham R (1972) Non-reversal shifts after selective prefrontal ablations in monkeys (*Macaca mulatta*). Brain Res 92:89–102

Petrides M, Pandya DN (1994) Comparative architectonic analysis of the human and the macaque frontal cortex. In: Boeller F, Grafman J (eds) Handbook of Neuropsychology. Elsevier, Amsterdam, pp 17–58

Petrides M, Pandya DN (2007) Efferent association pathways from the rostral prefrontal cortex in the macaque monkey. J Neurosci 24:11573–11586

Rao SC, Rainer G, Miller EK (1997) Integration of what and where in the primate prefrontal cortex. Science 276:821–824
Roberts AC, Robbins TW, Everitt BJ (1988) The effects of intradimensional and extradimensional shifts on visual discrimination learning in humans and non-human primates. Q J Exp Psychol 40:321–341
Rowe JB, Sakai K, Lund TE, Ramsøy T, Christensen MS, Baare WF, Paulson OB, Passingham RE (2007) Is the prefrontal cortex necessary for establishing cognitive sets? J Neurosci 28:13303–13310
Rudebeck PH, Behrens TE, Kennerley SW, Baxter MG, Buckley MJ, Walton ME, Rushworth MF (2008) Frontal cortex subregions play distinct roles in choices between actions and stimuli. J Neurosci 28:13775–13785
Rudebeck PH, Saunders RC, Prescott AT, Chau LS, Murray EA (2013) Prefrontal mechanisms of behavioral flexibility, emotion regulation and value updating. Nat Neurosci 16:1140–1145
Stuss DT, Levine B, Alexander MP, Hong J, Palumbo C, Hamer L, Murphy KJ, Izukawa D (2000) Wisconsin Card Sorting Test performance in patients with focal frontal and posterior brain damage: effects of lesion location and test structure on separable cognitive processes. Neuropsychologia 38:388–402
Wallis JD, Anderson KC, Miller EK (2001) Single neurons in prefrontal cortex encode abstract rules. Nature 411:953–956
Walton ME, Behrens TE, Buckley MJ, Rudebeck PH, Rushworth MF (2010) Separable learning systems in the macaque brain and the role of orbitofrontal cortex in contingent learning. Neuron 65:927–939
White IM, Wise SP (1999) Rule-dependent neuronal activity in the prefrontal cortex. Exp Brain Res 126:315–335

# Chapter 3
# Working Memory Functions of the Prefrontal Cortex

**Bradley R. Postle**

**Abstract** The prefrontal cortex (PFC) plays an important role in many behaviors, including in situations in which actions must be guided by information that is not currently accessible in the environment. Although the construct of "working memory" is often invoked in association with the PFC, imprecise or erroneous specification of which computations relate to which aspect of anatomy or physiology has been the basis of many erroneous ideas about the functional organization of the PFC. Indeed, the manner in which working memory has been related to the PFC over the past 75 years offers several cautionary tales about the difficulty of relating brain function to behavior. This proposition is supported by consideration of data from lesions and physiological measurements from human and nonhuman primates.

**Keywords** Working memory • Prefrontal cortex • Frontal eye fields • Short-term memory • Central executive • Lesion • Interference • Functional magnetic resonance imaging • Transcranial magnetic stimulation • Intraparietal sulcus

## 3.1 Introduction

Science progresses through the articulation of models, using these models to generate predictions, and testing predictions with experiments. Most often the outcomes of experiments, whether confirming or disconfirming a prediction, are used to refine the theoretical framework within which the experiments were carried out. Periodically, however, the model itself can be superceded by a different model that provides a better account of the phenomenon under study. One example comes from physics, in which the Newtonian model that held sway during the nineteenth and early twentieth centuries has given way to a theory of general relativity. It is noteworthy in this example that, even though physicists no longer use the Newtonian framework to guide their thinking about the physical universe, Newton's laws still capture the naïve intuition of most nonphysicists about "how the world works."

B.R. Postle (✉)
Departments of Psychology and Psychiatry, University of Wisconsin–Madison, Madison, WI, USA
e-mail: postle@wisc.edu



M. Watanabe (ed.), *The Prefrontal Cortex as an Executive, Emotional, and Social Brain*, DOI 10.1007/978-4-431-56508-6_3

If one can overlook concerns about grandiosity on the part of this author, a useful analogy might be drawn to the study of working memory: Although scientific thinking has been dominated by a model of working memory as a multicomponent cognitive system, one that may correspond nicely with intuitions about "how the mind works," it is being superceded by a new framework, one that understands working memory as an "emergent property" (Postle 2006) arising from the attentional selection of information that is relevant for the current behavioral context (e.g., Anderson 1983; Cowan 1995, 1988; Sreenivasan et al. 2014; Desrochers et al. 2015; Lara and Wallis 2015). From this perspective, the label "working memory" applies to a category of behaviors, and to the tasks that are used to measure performance on these behaviors, but not to a unitary cognitive system whose engagement can be inferred from the "first-order" inspection of levels of activity in one or more regions in the brain.

The prefrontal cortex (PFC) plays an important role in many behaviors, including in situations in which actions must be guided by information that is not currently accessible in the environment. Although the construct of "working memory" is often invoked in association with the PFC, imprecise or erroneous specification of which computations relate to which aspect of anatomy or physiology has also been the basis of many erroneous ideas about the functional organization of the PFC. Indeed, the manner in which working memory has been related to the PFC over the past 75 years offers several cautionary tales about the difficulty of relating brain function to behavior. This chapter will be organized in three sections. The first will address the construct of working memory, and how one's conceptualization of the architecture of high-level cognition can constrain how one goes about studying the brain. The second will address the phenomenon of sustained activity and how a priori notions of "what a memory signal must look like" can lead to flawed inference about brain-behavior relations. Finally, the third will review some experiments that provide a framework within which we might make further progress in studying the working memory functions of the PFC.

## 3.2 "Working Memory" as a Cognitive Construct

### *3.2.1 Working Memory as RAM*

The idea that the cognitive system requires a working memory derives directly from the metaphor of the brain as a computing machine, with working memory carrying out the function of maintaining multiple action plans in a rapidly accessible state (Miller et al. 1960), as does random access memory (RAM) in many von Neumann computing architectures. Pribram et al. (1964), in building on this idea, were the first associate working memory with the PFC. They found that poor performance by frontally lesioned monkeys, on a variety of tasks imposing a delay between cue and response, was better explained as impaired control of behavior, rather than forgetting, per se. For example, one task required first searching through a set of "junk" objects to learn which covered a reward, then returning to the rewarded object until a criterion level of five consecutive correct choices was achieved, at which time a different object in the set would be selected by the experimenter (baiting of food wells on all trials was concealed from the animal, thereby adding the "working memory" element). On the first set of trials, the frontal animals made more errors, a pattern that could have been due either to trial-to-trial forgetting of which object had been rewarded or by an inability to shift from an "explore" strategy to an "exploit" strategy. Once they achieved criterion, however, this ambiguity was resolved, because the frontal animals then also perseverated on the "exploit" strategy longer than did temporal lobe-lesioned and control animals. That is, their impairment wasn't in the ability to retain a small amount of trial-specific information over a short period of time but, rather, in the ability to use the discrepancy between the previous trial's stimulus-reward contingency vs. that of the present trial to change behavioral strategy (to change "set," in the parlance of mid-century neuropsychology). In their discussion of this and several other experimental findings, Pribram et al. (1964) drew from contemporaneous computer models of problem-solving to propose that, rather than reflecting "memory trace formation and decay," the deficits resulting from frontal lobe damage may have reflected a "mechanism of temporary, flexible stimulus compounding" (p. 51),[1] a hypothesized process that is reminiscent of contemporary ideas of establishing trial-unique "bindings" between stimulus features and behavioral repertoires (Oberauer 2013). Thus, in this first instance in the literature of an association between the construct of working memory and the PFC, the emphasis was on "working with memory" rather than on the storage, per se, of the remembered information

In the analogy to RAM, another factor is relevant, which is that RAM is not inherently time dependent. When, for example, while composing this chapter, I leave my word processing application to open a Web browser and access the precise wording of that quote from Pribram, the manuscript file running in the word processing application will remain immediately accessible whether I return to it as soon as I access the quote or, instead, if I set the computer down, make breakfast, walk to the beach, and then return to my computer several hours later to resume this work. In biological systems, it's also the case that trial-specific memories need not be temporally constrained. If a rat explores three arms of an eight-arm radial maze and is then returned to its home cage and only returned to the maze several hours later, it can "pick up where it left off," knowing which five arms remain baited. And this "working memory" of which three arms had been visited will be of no use, of course, once the remaining five arms are visited and the experimenter rebaits all eight arms of the maze (Olton et al. 1979). Similarly, a memory for where in the lot

---

[1]The clarity and prescience with which Pribram et al. (1964) relate this line of reasoning and, more generally, with which they advocate an approach of "simulation . . . with the use of computers" is remarkable. Although the edited volume in which their chapter appeared is no longer in print, at the time of this writing, a digitized copy was downloadable from http://www.karlpribram.com/wp-content/uploads/pdf/D-049.pdf

she parked the car on Monday is of little use to the office worker leaving at the end of the day Tuesday, assuming that Tuesday morning's choice of parking space was not influenced by previous choices. Furthermore, successful performance on such tasks is known to depend on the hippocampus, not on the PFC (as reviewed by Becker and Morris 1999).

These examples highlight several important points. The first is that there is not a principled *computational* reason for working memory to be time delimited. (Whether there may be biological factors, relating to, for example, decay or interference, will be taken up further along in this chapter.) A second is that there is no a priori reason why working memory functions need to be carried out by a specialized system that is distinct from other categories of cognition – in the "Honig-Olton" scheme (Becker and Morris 1999), for example, working memory and reference memory can both depend on medial temporal lobe neural systems. A third, as exemplified by the example from Pribram et al. (1964), is that many computationally distinct operations must be carried out in order to successfully execute even the simplest working memory task, and the retention of stimulus information is only one of these.

### 3.2.2 The Multiple Component Model and Its Relation to the Dorsolateral PFC

In the 1970s and 1980s, Baddeley and colleagues formulated an explicit cognitive model of working memory. It posited a multicomponent architecture whereby the storage function of domain-specific short-term memory buffers was controlled by a domain-general central executive (e.g., Baddeley and Hitch 1974; Baddeley 1986). Importantly, the central executive was construed as a general purpose controller, akin to Norman and Shallice's (1980) Supervisory Attentional System, and, as such, wouldn't only be engaged by tasks with an overt memory component. (Indeed, consistent with this idea, an early neuroimaging study designed to isolate brain activity attributable to the central executive (and identifying it in the PFC) employed a dual-task procedure in which neither of the individual tasks was a memory task (D'Esposito et al. 1995)). Of further importance is that one would also expect an attentional controller to be active even during the simplest tasks that, on the surface, would seem to only require the engagement of a short-term store. This is because, among other things, one can never know when an unexpected change in the environment might render the short-term retention of information, and/or the need to guide behavior with that information, more difficult (e.g., Malmo 1942; Chao and Knight 1995, 1998; Postle 2005). From this perspective, the inferential flaw in studies purporting to localize visual working memory storage-related activity to PFC was to assume that sustained, spatially tuned activity in this region corresponded to the operation of the "visuospatial sketchpad" buffer from the multiple component model, instead of to its central executive (for more developed argumentation on this point, see (D'Esposito and Postle 2015; Postle 2015b, c).

## 3.3 Sustained Activity in the PFC (and Elsewhere)

The idea that short-term and working memory might depend on sustained, elevated activity dates back at least to Hebb (1949) and is seen in many of Goldman-Rakic's influential writings (Goldman-Rakic 1987, 1990). However, the once-popular assumption that such activity in the PFC makes a necessary contribution to the short-term retention, per se, of sensory information is no longer tenable. Empirically, it is well established that sustained activity in the PFC is neither specific for (e.g., Curtis and Lee 2010; Riggall and Postle 2012; Emrich et al. 2013) nor necessary for (e.g., Zaksas and Pasternak 2006; Lara and Wallis 2014; Fuster 2016; Wimmer et al. 2016) the short-term retention of this information (also reviewed in Postle 2015a). To consider just one type of information in more detail, recent studies have been unsuccessful with multivariate decoding of the direction of motion from the dorsolateral (dl)PFC in humans (Riggall and Postle 2012; Emrich et al. 2013) and successful in the monkey (Mendoza et al. 2014), but, most tellingly, evidence from a lesion study suggests that functions other than sensory storage are supported by the dlPFC: "[B]ecause th[e] deficit [in dlPFC lesioned animals] was independent of stimulus features giving rise to the remembered direction and was most pronounced during rapid shifts of attention, [the] role [of dlPFC] is more likely to be attending and accessing the preserved motion signals rather than their storage (p. 7095)" (Pasternak et al. 2015). With regard to sustained activity, a critical role for the dlPFC may emerge when a task requires the transformation of trial-initiating sensory information into a format that is needed for subsequent guidance of behavior, as well as in the retention of that transformed information (Meyers et al. 2012; Lee et al. 2013; Stokes et al. 2013; Lee and Baker 2015).

Together with the findings that we have just reviewed, it is important to note that the very relation of sustained activity to working memory is undergoing reconsideration: Just as the intuition that the behavioral constructs of *short-term memory* and *working memory* are inherently time delimited turns out to be flawed (see, e.g., Postle 2015b, 2016), so, too, might be the assumed relation between sustained activity working memory. On theoretical grounds, it has been argued that the short-term retention of information might be accomplished via short-term synaptic reorganization (e.g., Mongillo et al. 2008; Barak and Tsodyks 2014; Stokes 2015), with elevated activity corresponding, instead, to the focus of attention (Lewis-Peacock et al. 2012, 2015; LaRocque et al. 2014). Empirical evidence that a transient, synaptic weight-based mechanism is the basis for working memory storage is difficult to assemble, but findings that are consistent with this idea are beginning to emerge (Sugase-Miyamoto et al. 2008; Hayden and Gallant 2013; Wolfe and Stokes 2015). Another mechanism for the short-term retention of information that differs from "elevated activity" as it is traditionally construed would be fluctuations in intracellular voltages that can be sustained over tens of seconds (Strowbridge 2012).

## 3.4 Working Memory Functions of the dlPFC

An expedient rhetorical device for launching a discussion of the working memory functions of the dlPFC is to consider the idea that a punctate lesion of this region will produce a "mnemonic scotoma," whereby memory-guided saccades to a restricted area of the visual field are impaired, despite the sparing of visually guided saccades into that same region (Funahashi et al. 1993). When an independent group of researchers (Wajima and Sawaguchi) sought to replicate this finding several years later, however, they obtained results that are intriguingly reminiscent of those from Pribram et al. (1964) that were reviewed earlier in this chapter – when testing and scoring procedures were refined, the impairment was revealed to be attributable to factors other than memory. Specifically, the procedure of Funahashi et al. (1993) was to score each trial in which the initial saccade did not land within the cued location as an error (S. Funahashi, personal communication). Wajima and Sawaguchi (reported in Tsujimoto and Postle 2012), in contrast, allowed their animals to make multiple saccades on each trial and rewarded them if they eventually landed in the target location. Although they replicated the earlier finding – that a disproportionate number of misguided initial saccades were made on trials targeting the critical region of the visual field – they also observed that erroneous initial saccades were almost invariably followed by a second, corrective saccade that acquired to-be-remembered target location. Furthermore, the erroneous saccades were noted to have often been made to a region of space that had been relevant on the previous trial, either as that trial's cued location or as the target of that trial's saccade (or both). Thus, the animal's errors were better classified as perseverative, or as influenced by proactive interference, than as mnemonic, per se. In a conceptually related finding, Mackey et al. (2016) have recently shown that deficits on the oculomotor delayed-response task are only seen in human patients when their dlPFC lesions invade the territory of the frontal eye fields. Errors on tests of working memory that result from damage to the dlPFC may be qualitatively similar to those that we know, from decades of behavioral neurology, are characteristic of these patients in situations that make no overt demands on working memory (Tsujimoto and Postle 2012).

Over the past decade, my group has used (f)MRI-guided repetitive transcranial magnetic stimulation (rTMS) to dissociate mnemonic from nonmnemonic factors in working and short-term memory performance. Several findings are consistent with the assertion that ended the previous paragraph. For example, rTMS of the dlPFC during a delay period does not affect delayed recognition for locations – nor does delay-period rTMS of the postcentral gyrus – whereas rTMS of the intraparietal sulcus and of the frontal eye fields[2] does affect performance (Hamidi et al. 2008). A key role for the dlPFC emerges on this task, however, when rTMS is instead delivered concurrent with the onset of the stimulus that initiates the memory-guided response (whether it be recall or recognition, Hamidi et al. 2009). Delay-period rTMS of the dlPFC also does not disrupt the simple short-term retention of verbal information, unless subjects are required to mentally reorder it during the delay period (Feredoes et al. 2006, 2007; Postle et al. 2006). When it is applied during the response period, in contrast, rTMS reveals important roles for subregions of the PFC in such functions as controlling the effects of proactive interference (Feredoes et al. 2006), perhaps by adjudicating the influence of various memory signals (e.g., familiarity vs. recollection) on decision processes (Feredoes and Postle 2010). These findings are consistent with more recent work in the monkey, which also emphasize the role of PFC dynamics in memory-guided decision-making and action planning (Wimmer et al. 2016).

## 3.5 Conclusion

And so, is the PFC important for working memory function? Without a doubt. But a clear understanding of the ways in which PFC circuits do, versus do not, contribute to these behaviors will be important if we are to make meaningful progress in addressing "the riddle of frontal lobe function in man" (Teuber 1964).

## References

Anderson JR (1983) The architecture of cognition. Harvard University Press, Cambridge, MA

Baddeley AD (1986) Working memory. Oxford University Press, London

Baddeley AD, Hitch GJ (1974) Working memory. In: Bower GH (ed) The psychology of learning and motivation, vol 8. Academic, New York, pp 47–89

Barak O, Tsodyks M (2014) Working models of working memory. Curr Opin Neurobiol 25:20–24

Becker JT, Morris RG (1999) Working memory(s). Brain Cogn 41:1–8

Chao LL, Knight RT (1995) Human prefrontal lesions increase distractibility to irrelevant sensory inputs. Neuro Rep 6:1605–1610

Chao L, Knight R (1998) Contribution of human prefrontal cortex to delay performance. J Cogn Neurosci 10:167–177

Cowan N (1995) Attention and memory: an integrated framework. Oxford University Press, New York

---

[2]Although there are compelling reasons to classify the frontal eye fields as "prefrontal" from the standpoint of the evolution of neural systems (Passingham and Wise 2012), this author nonetheless finds it obfuscating when physiological studies that are limited to the frontal eye fields are labeled as studies of "prefrontal cortex." The fact is that the properties described in these reports would almost surely not be observed in circuits in the vicinity of the principal sulcus.

Cowan N (1988) Evolving conceptions of memory storage, selective attention, and their mutual constraints within the human information processing system. Psychol Bull 104:163–171

Curtis CE, Lee D (2010) Beyond working memory: the role of persistent activity in decision making. Trends Cogn Sci 14:216–222

D'Esposito M, Detre JA, Alsop DC, Shin RK, Atlas S, Grossman M (1995) The neural basis of the central executive system of working memory. Nature 378:279–281

D'Esposito M, Postle BR (2015) The cognitive neuroscience of working memory. Annu Rev Psychol 66:115–142

Desrochers TM, Burk DC, Badre D, Sheinberg DL (2015) The monitoring and control of task sequences in human and non-human primates. Front Syst Neurosci 9:185

Emrich SM, Riggall AC, Larocque JJ, Postle BR (2013) Distributed patterns of activity in sensory cortex reflect the precision of multiple items maintained in visual short-term memory. J Neurosci 33(15):6516–6523

Feredoes E, Postle BR (2010) Prefrontal control of familiarity and recollection in working memory. J Cogn Neurosci 22:323–330

Feredoes E, Tononi G, Postle BR (2006) Direct evidence for a prefrontal contribution to the control of proactive interference in verbal working memory. Proc Natl Acad Sci U S A 103:19530–19534

Feredoes E, Tononi G, Postle BR (2007) The neural bases of the short-term storage of verbal information are anatomically variable across individuals. J Neurosci 27:11003–11008

Funahashi S, Bruce C, Goldman-Rakic P (1993) Dorsolateral prefrontal lesions and oculomotor delayed-response performance: evidence for mnemonic "scotomas". J Neurosci 13:1479–1497

Fuster JM (2016) Fragmented LFPs and NIRS. J Cogn Neurosci 13:1479–1497

Goldman-Rakic PS (1987) Circuitry of the prefrontal cortex and the regulation of behavior by representational memory. In: Mountcastle VB, Plum F, Geiger SR (eds) Handbook of neurobiology. American Physiological Society, Bethesda, pp 373–417

Goldman-Rakic PS (1990) Cellular and circuit basis of working memory in prefrontal cortex of nonhuman primates. In: Uylings HBM, Eden CGV, DeBruin JPC, Corner MA, Feenstra MGP (eds) Progress in brain research. Elsevier Science Publishers, Amsterdam 85, pp 325–336

Hamidi M, Tononi G, Postle BR (2008) Evaluating frontal and parietal contributions to spatial working memory with repetitive transcranial magnetic stimulation. Brain Res 1230:202–210

Hamidi M, Tononi G, Postle BR (2009) Evaluating the role of prefrontal and parietal cortices in memory-guided response with repetitive transcranial magnetic stimulation. Neuropsychologia 47:295–302

Hayden BY, Gallant JL (2013) Working memory and decision processes in visual area V4. Front Neurosci 7:18. doi:10.3389/fnins.2013.00018

Hebb DO (1949) The organization of behavior: a neuropsychological theory. Wiley, New York

Hyde RA, Strowbridge BW (2012) Mnemonic representations of transient stimuli and temporal sequendes in teh rodent hippocampus in vitro. Nat Neurosci 15:1430–1438

Lara AH, Wallis JD (2014) Executive control processes underlying multi-item working memory. Nat Neurosci 17:876–883

Lara AH, Wallis JD (2015) The role of prefrontal cortex in working memory: a mini review. Front Syst Neurosci 9:173

LaRocque JJ, Lewis-Peacock JA, Postle BR (2014) Multiple neural states of representation in shortterm memory? It's a matter of attention. Front Hum Neurosci 8. doi:10.3389/fnhum.2014.00005

Lee SH, Baker CI (2015) Multi-voxel decoding and the topography of maintained information during visual working memory. Front Syst Neurosci. doi:10.3389/fnsys.2016.00002

Lee SH, Kravitz DJ, Baker CI (2013) Goal-dependent dissociation of visual and prefrontal cortices during working memory. Nat Neurosci 16(8):997–999

Lewis-Peacock JA, Drysdale AT, Oberauer K, Postle BR (2012) Neural evidence for a distinction between short-term memory and the focus of attention. J Cogn Neurosci 24:61–79

Lewis-Peacock JA, Drysdale A, Postle BR (2015) Neural evidence for the flexible control of mental representations. Cereb Cortex 25:3303–3313
Mackey W, Devinsky O, Doyle W, Meager M, Curtis CE (2016) Human dorsolateral prefrontal cortex is not necessary for spatial working memory. J Neurosci 36:2847–2856
Malmo RB (1942) Interference factors in delayed response in monkey after removal of the frontal lobes. J Neurophysiol 5:295–308
Mendoza D, Torres S, Martinez-Trujillo J (2014) Sharp emergence of feature-selective sustained activity along the dorsal visual pathway. Nat Neurosci 17:1255–1262
Meyers EM, Qi XL, Constantinidis C (2012) Incorporation of new information into prefrontal cortical activity after learning working memory tasks. Proc Natl Acad Sci U S A 109:4651–4656
Miller GA, Galanter E, Pribram KH (1960) Plans and the structure of behavior. Henry Holt and Company, New York
Mongillo G, Barak O, Tsodyks M (2008) Synaptic theory of working memory. Science 319:1543–1546
Norman DA, Shallice T (1980) Attention to action: willed and automatic control of behavior. University of California, San Diego
Oberauer K (2013) The focus of attention in working memory – from metaphors to mechanisms. Front Hum Neurosci 7:673. doi:10.3389/fnhum.2013.00673
Olton DS, Becker JT, Handelmann GE (1979) Hippocampus, space, and memory. Behav Brain Sci 2:313–365
Passingham RE, Wise SP (2012) The neurobiology of the prefrontal cortex. Oxford University Press, Oxford, UK
Pasternak T, Lui LL, Spinelli PM (2015) Unilateral prefrontal lesions impair memory-guided comparison. J Neurosci 35:7095–7105
Postle BR (2005) Delay-period activity in prefrontal cortex: one function is sensory gating. J Cogn Neurosci 17:1679–1690
Postle BR (2006) Working memory as an emergent property of the mind and brain. Neuroscience 139:23–38
Postle BR (2015a) The cognitive neuroscience of visual short-term memory. Curr Opin Behav Sci 1:40–46. doi:10.1016/j.cobeha.2014.1008.1004
Postle BR (2015b) Essentials of cognitive neurocience. Wiley, Chichester
Postle BR (2015c) Neural bases of the short-term retention of visual information. In: Jolicoeur P, LeFebvre C, Martinez-Trujillo J (eds) Mechanisms of sensory working memory: attention & performance, XXV edn. Academic Press, London, pp 43–58
Postle BR (2016) The hippocampus, memory, and consciousness. In: Laureys S, Gosseries O, Tononi G (eds) Neurology of consciousness, 2nd edn. Elsevier, Amsterdam, pp 349–363
Postle BR, Ferrarelli F, Hamidi M, Feredoes E, Massimini M, Peterson MJ, Alexander A, Tononi G (2006) Repetitive transcranial magnetic stimulation dissociates working memory manipulation from retention functions in prefrontal, but not posterior parietal, cortex. J Cogn Neurosci 18:1712–1722
Pribram KH, Ahumada A, Hartog J, Roos L (1964) A progress report on the neurological processes disturbed by frontal lesions in primates. In: Warren JM, Akert K (eds) The frontal granular cortex and behavior. McGraw-Hill Book Company, New York, pp 28–55
Riggall AC, Postle BR (2012) The relationship between working memory storage and elevated activity as measured with funtional magnetic resonance imaging. J Neurosci 32:12990–12998
Sreenivasan KK, Curtis CE, D'Esposito M (2014) Revisiting the role of persistent neural activity in working memory. Trends Cogn Sci 18:82–89
Stokes MG (2015) 'Activity-silent' working memory in prefrontal cortex: a dynamic coding framework. Trends Cogn Sci 19:394–405
Stokes MG, Kusunoki M, Sigala N, Nili H, Gaffan D, Duncan J (2013) Dynamic coding for cognitive control in prefrontal cortex. Neuron 78(2):364–375
Strowbridge (2012). "get it." get it get it: get it.

Sugase-Miyamoto Y, Liu Z, Wiener MC, Optican LM, Richmond BJ (2008) Short-term memory trace in rapidly adapting synapses of inferior temporal cortex. PLoS Comput Biol 4(5): e1000073

Teuber H-L (1964) The riddle of frontal lobe function in man. In: Warren JM, Akert K (eds) The frontal granular cortex and behavior. McGraw Hill, New York, pp 410–444

Tsujimoto S, Postle BR (2012) The prefrontal cortex and delay tasks: a reconsideration of the "mnemonic scotoma". J Cogn Neurosci 24:627–635

Wimmer K, Ramon M, Pasternak T, Compte A (2016) Transitions between multiband oscillatory patterns characterize memory-guided perceptual decisions in prefrontal circuits. J Neurosci 36:489–505

Wolfe, J. M. and M. G. Stokes (2015). Reactivation. Frontiers.

Zaksas D, Pasternak T (2006) Directional signals in the prefrontal cortex and in area MT during a working memory for visual motion task. J Neurosci 26:11726–11742

# Part II
# The Prefrontal Cortex as an Integrator of Executive and Emotional Function

# Chapter 4
# Prefrontal Cortex Integration of Emotion and Cognition

**Helen Barbas and Miguel Ángel García-Cabezas**

**Abstract** Ideas from classical philosophy and psychology that emotion and cognition are distinct and separate have been challenged by evidence of their intricate interaction in the brain. Accumulated evidence shows that areas associated with emotions and cognition are strongly linked and influence each other according to principles based on the structural organization of the cortex. Subcortical structures that process information about needs and drives are old in phylogeny and innervate the prefrontal cortex, targeting strongly the posterior orbitofrontal cortex (pOFC) and the anterior cingulate cortex (ACC). These two prefrontal regions are part of the cortical component of the limbic system but are functionally distinct. Through widespread connections, the pOFC acts as an integrator of the internal and external environments for the perception of affective events. In contrast, the ACC specializes in the expression of emotions through robust pathways to central autonomic structures. Both prefrontal limbic regions are connected with lateral prefrontal areas associated with cognitive functions, effectively linking areas associated with emotion and cognition. Functional and circuit studies in animals and humans indicate that even simple tasks necessitate interaction between areas associated with cognition and emotions and suggest that their linkage is disrupted in several psychiatric and neurologic diseases.

**Keywords** Prefrontal cortex • Limbic cortex • Orbitofrontal cortex • Anterior cingulate cortex • Amygdala • Cortical systematic variation • Structural model • Prefrontal specialization • Thalamus • Psychiatric diseases

## 4.1 Overview

The idea of interaction of the processes of cognition and emotion in the brain is now widely accepted (e.g., Damasio 1994; Barbas 1995; Pessoa 2013). Decisions about appropriate action are embedded within an affective context, as people make

H. Barbas (✉) • M.Á. García-Cabezas
Neural Systems Laboratory, Boston University, 635 Commonwealth Ave., Room 431, Boston, MA 02215, USA
e-mail: barbas@bu.edu; http://www.bu.edu/neural



M. Watanabe (ed.), *The Prefrontal Cortex as an Executive, Emotional, and Social Brain*, DOI 10.1007/978-4-431-56508-6_4

choices from the mundane decision of choosing from a lunch menu to consequential decisions about family, work, and career. Individuals are often conflicted about what is desirable or reasonable, as they weigh the pros and cons of different scenarios before reaching a decision. Everyday decisions and choices thus exemplify the intimate influence of emotions on cognition and vice versa.

Discussions about the role of emotions on cognition preoccupied philosophers from antiquity to modern times. Plato relegated emotions to a lower status than cognition (Plato 1892/impression of 1931), and interpretations vary as to whether he ever put them on an equal footing. At the level of metaphysical abstraction, at least some psychological discussions have proceeded as though cognition and emotion are separate. At the level of the brain, the evidence shows that areas associated with cognition and emotion are inextricably linked (e.g., Nauta 1971, 1979; Yakovlev 1948).

What is the evidence for the interaction of areas associated with cognition and emotion in the brain? Evidence for their union ironically emerged from studies of the frontal lobe, the part of the brain thought to be the seat of human intellect. Cognitive operations were classically associated with the anterior half of the frontal lobe, the prefrontal cortex. The great anatomists, Nauta, Yakovlev, and MacLean, observed that the brain's chief executive, the prefrontal cortex, also receives information from areas associated with emotions (Nauta 1971, 1979; Yakovlev 1948; MacLean 1952).

Here we provide an overview of the evidence of the linkage of cognition and emotion in the brain from the perspective of circuits. It is not a comprehensive review of relevant connections on this topic, which have been described elsewhere (Barbas 1995, 2000a, b). Rather, the focus is on an overall scheme that makes the link between emotion and cognition intuitive and obligatory. Evidence about the interaction of neural pathways associated with cognition and emotion also suggests that their disconnection is at the root of the symptomatology in several psychiatric and neurologic diseases. Disconnection of cognition from emotion in the brain also has broad consequences for society at large.

We begin by considering some subcortical areas that underlie vital functions and are intricately associated with emotions and then identify their principal targets in the prefrontal cortex. We then consider the cortical connections of areas associated with emotions and cognition. In a later section, we contrast the connections of areas thought to have primary roles in cognitive processes and those associated with emotions and show that these regions are linked. Identifying the most prominent connections of functionally specialized but interacting prefrontal regions may help explain vulnerabilities when specific nodes are disrupted in psychiatric and neurologic diseases.

## 4.2 The Limbic System Is Broadly Associated with Emotions

Brain areas that are old in phylogeny have been classified under the rubric of the limbic system. The limbic system is perhaps the most misunderstood system in the brain, in spite of its centrality in vital functions. Some have thought that limbic areas are hard to define and not readily amenable to experimental study (e.g., LeDoux 2000). With regard to the functional role of emotions, investigators have focused on readily measureable behaviors, such as acquisition of fear and freezing responses in rodents, and the extinction of the behavior when stimuli no longer signal threat (Paré et al. 2004; Milad et al. 2007).

### *4.2.1 Subcortical Limbic Structures*

What are the commonalities and unique features of structures associated with emotions under the broad umbrella of the limbic system? With regard to subcortical structures, it may be safe to assume that if a structure is old in phylogeny and can be identified in a variety of species, it belongs to the limbic system. The comparative anatomy extends to a big range of extant species, but here we will restrict discussion to mammalian species. The hypothalamus and the amygdala are present in all mammalian species, ranging from lissencephalic rodents and early primates that have a smooth cortex to gyrencephalic primate species with a convoluted cerebrum (e.g., Johnston 1923; De Olmos 1990). Using the amygdala and the hypothalamus as examples, we review some of their major targets in the cortex and the implication of these connections for function and disruption in disease.

It is widely accepted that areas associated with emotions are old in phylogeny. They are the structures associated with basic needs and drives – seeking food and shelter, avoiding predators, and finding a mate. Subcortical areas, such as the hypothalamus and the amygdala, come readily to mind in the context of vital processes. The hypothalamus, for example, has groups of neurons engaged in feeding (e.g., Petrovich and Gallagher 2007). Damage to one part of the hypothalamus leads to inability to appreciate satiety, resulting in excessive eating. Damage to an adjacent hypothalamic region has the opposite effect: animals starve in the midst of plenty of food (Bereket et al. 2012; Brown et al. 2015).

The amygdala has a set of medially situated nuclei that have a role in endocrine and reproductive functions (De Olmos 1990). But the amygdala also has a set of basal nuclei, which are critical for forming associations between stimuli and their affective significance. Neurons in the basal nuclei of the amygdala are engaged as opportunities arise to forage for food. But neurons in the amygdala are also engaged when threats arise, when predators lurk in the brush, and when it is no longer safe to forage. Neurons that signal potential rewards and aversive outcomes or different task states are intermingled in the primate amygdala (e.g., Paton et al. 2006; Zhang et al. 2013; Nishijo et al. 2008; Mosher et al. 2010). It is the basal nuclei of the amygdala that have the greatest interactions with the cerebral cortex and especially the prefrontal cortex.

### 4.2.2 Expansion of the Limbic System to the Prefrontal Cortex

The limbic system may be old in phylogeny but it is neither primitive nor vestigial. In primates, the limbic system has expanded along with the evolved neocortex and maintains strong connections with it (Nauta 1979; Armstrong 1991). Darwin's (1872) account on emotions includes the readily understood emotions, such as anger and joy, but also shyness, guilt, and contempt (Darwin 1872). Some of these can be recognized in a large variety of species, including some complex emotions, such as jealousy and joy. But some other emotions may be uniquely human, including guilt, contempt, shame, and embarrassment (Table 4.1). One would be hard-pressed to identify embarrassment in nonhuman species, since it is intricately associated with societal norms, expectations, and context.

But even basic needs for survival, such as eating, have become elaborated in humans, as seen by the variety of foods offered and the extent to which humans will go in search for gastronomic excellence (Shepherd 2012). In this context, a special direct pathway from the primary olfactory cortex reaches the prefrontal cortex and especially the orbitofrontal cortex (Morecraft et al. 1992; Barbas 1993; Carmichael et al. 1994). The orbitofrontal cortex also receives projections from the gustatory cortex (reviewed in Barbas 2000a; Barbas et al. 2002). These findings show that the first and primary cortical destination of pathways from olfactory and gustatory areas is the prefrontal cortex. This pattern may help explain the emotions elicited when we remember fondly smells and tastes of favorite foods from childhood.

### 4.2.3 The Principal Cortical Targets of Some Subcortical Limbic Structures

Two main subcortical limbic structures exemplified here, the amygdala and hypothalamus, project widely to the cortex. But the major target of these structures is the prefrontal cortex. And among prefrontal cortices, the orbitofrontal cortex and the medial prefrontal cortex are the major recipients of pathways from the amygdala (Ghashghaei et al. 2007). Within these regions the amygdala innervates strongly the posterior parts of the orbitofrontal cortex (pOFC), found on the base of the frontal lobe. The amygdala also innervates another prefrontal region, situated at the posterior part of the medial prefrontal cortex. This region is often called the anterior cingulate cortex (ACC) and includes the cortex wrapped as a crescent above and

**Table 4.1** Darwin's emotion range

| | | | | | | | |
|---|---|---|---|---|---|---|---|
| Anger | Surprise | Determination | Anxiety | Joy | Hatred | Love | Patience |
| Horror | Astonishment | Reflection | Grief | High spirits | Distain | Tender feelings | Pride |
| Terror | | Meditation | Suffering | | Contempt | Devotion | Ill temper |
| Fear | | Shame | Low spirits | | Disgust | | Modesty |
| | | Shyness | Sulkiness | | | | |
| | | Guilt | Dejection | | | | |
| | | | Despair | | | | |
| | | | Helplessness | | | | |
| | | | Low spirits | | | | |

Compiled from discussion of emotions from Darwin's book (1872)

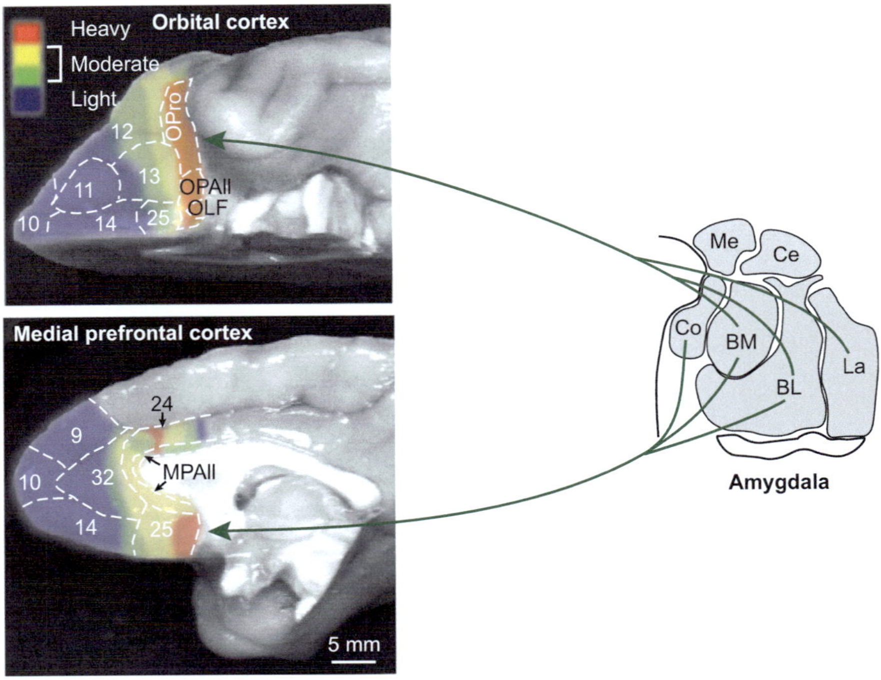


**Fig. 4.1** The principal targets of the amygdala in the prefrontal cortex. *Top*, the amygdala (*right*) innervates the entire prefrontal cortex but sends strongest pathways to the posterior orbitofrontal cortex (pOFC, which includes areas OPro, OPAll, and the caudal part of area 13); this pathway likely provides signals on the affective significance of events. *Bottom*, the amygdala (*right*) also innervates the anterior cingulate cortex (ACC, which includes areas MPAll, 25, the caudal part of area 32, and the rostral part of area 24). *Abbreviations*: *BL* basolateral nucleus, *BM* basomedial nucleus (also known as accessory basal), *Ce* central nucleus, *Co* cortical nucleus, *La* lateral nucleus, *Me* medial nucleus, *MPAll* medial periallocortex, *OLF* primary olfactory cortex, *OPAll* orbital periallocortex, *OPro* orbital proisocortex. Architectonic areas of the prefrontal cortex here and in other figures are according to the map of Barbas and Pandya (1989). Images on left were adapted from Ghashghaei et al. (2007)

below the front part of the corpus callosum (Fig. 4.1). Like the amygdala, the hypothalamus projects to all prefrontal areas but innervates preferentially the pOFC and the ACC (Rempel-Clower and Barbas 1998) (Fig. 4.2).

The pOFC and ACC, which are strongly connected with subcortical limbic structures, represent the prefrontal component of the cortical limbic system (Barbas and Pandya 1989). These connections suggest expansion of the limbic system at the level of the neocortex. This expansion has major influence on the rest of the neocortex by virtue of the connections of the pOFC and ACC with the rest of the cortex, in general, and the prefrontal cortex, in particular, as elaborated below.

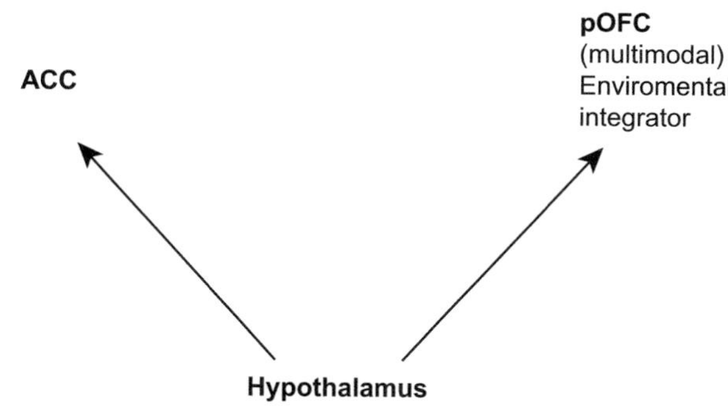


**Fig. 4.2** The principal targets of the hypothalamus in the prefrontal cortex. The hypothalamus innervates the entire prefrontal cortex but sends the strongest pathways to ACC and pOFC. *Abbreviations*: *ACC* anterior cingulate cortex, *pOFC* posterior orbitofrontal cortex

## 4.3 Systematic Structural Variation of the Cortex and Connections

Can cortical limbic areas be differentiated from other areas? Like subcortical limbic structures, limbic cortices are old in phylogeny. Limbic cortices can be defined objectively by their laminar structure. They are the areas that have fewer than six layers. Some limbic areas lack an inner granular layer IV and are called agranular; some have a rudimentary layer IV and are called dysgranular. The agranular and dysgranular areas are collectively called limbic cortices (reviewed in Barbas 2015). The term agranular does not refer to the motor or premotor areas, as often called in error in the literature (for discussion, see García-Cabezas and Barbas 2014; Barbas and García-Cabezas 2015).

If we examine the basic laminar structure of the cortex starting with the limbic cortices, we find that cortical lamination changes gradually from agranular and dysgranular cortices to areas that have six layers. We refer to all cortices with six layers as eulaminate. Eulaminate areas also show evidence of elaboration of their layers, with increase in the density of granular layer IV in sensory and high-order association cortices (reviewed in Pandya et al. 1988). In the prefrontal region, the limbic areas include the pOFC and ACC. Areas found at the posterior extent of the lateral prefrontal cortex have the most elaborate six-layer structure within the prefrontal region (caudal area 46 and caudal area 8). All other prefrontal areas have six layers, but an overall laminar structure that is between the dysgranular limbic areas, on one hand, and the best laminated lateral prefrontal areas, on the other hand (Fig. 4.3). The cortex thus is not homogeneous, as previously assumed (Rockel et al. 1980).

Why is the idea of systematic variation in the cortex useful? This analysis reveals that the connections of the cortex can be understood and summarized by their laminar structure. In the cortex limbic areas have other features in common. One of the most striking is myelination, which is sparse in limbic cortices and heavy in eulaminate areas, especially those with the most elaborate laminar structure within their cortical system (e.g., the primary visual or primary motor cortex).

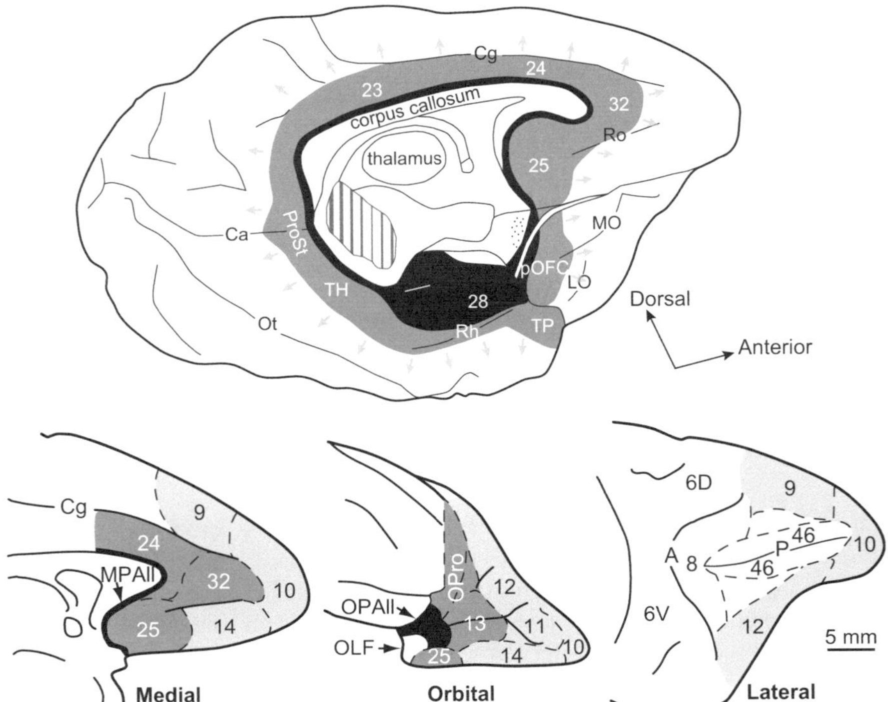


**Fig. 4.3** The entire cortical limbic system and its prefrontal components. *Top*, agranular (*black*) and dysgranular (*dark gray*) areas are considered to be limbic. The cortical limbic system is found at the base of the entire cortex. *Bottom*, the prefrontal components of the cortical limbic system in the anterior cingulate of the medial surface of the hemisphere (*left*, *black* and *dark-gray* areas) and on the orbital surface (*center*, *black* and *dark-gray* areas). *Light shades of gray* show eulaminate areas, defined as areas that have six distinct layers. A distinguishing feature of these areas is layer IV, which is absent in agranular areas, incipient in dysgranular areas, and progressively more elaborate in eulaminate areas. The *lightest gray shades* depict areas with the most elaborate laminar structure and thickest layer IV (*right*) (Adapted from Barbas 2015). *Abbreviations*: *A* arcuate sulcus, *Ca* calcarine fissure, *Cg* cingulate sulcus, *LO* lateral orbital sulcus, *MO* medial orbital sulcus, *MPAll* medial periallocortex, *OLF* primary olfactory cortex, *OPAll* orbital periallocortex, *OPro* orbital proisocortex, *Ot* occipitotemporal sulcus, *P* principal sulcus, *pOFC* posterior orbitofrontal cortex, *ProSt* area prostriata, *Rh* rhinal sulcus, *Ro* rostral sulcus, *TH* medial temporal area TH, *TP* temporal pole

The prefrontal limbic areas are not the only limbic areas of the cortex, but are part of a continuous ring at the base of the entire cortex, as shown by Broca, Yakovlev, and Sanides (e.g., (Broca 1878; Sanides 1970; Yakovlev 1959). The limbic cortex is situated at the foot of each and every cortical system (Fig. 4.3, top). On the basal surface, the pOFC is continuous caudally with the temporal pole, which leads to the rhinal region, and more posteriorly to the parahippocampal region. On the medial surface of each hemisphere, the ACC is continuous with the cingulate cortex above the corpus callosum, as it spans through the medial sectors of premotor, motor, somatosensory, and occipital cortices. At its most caudal extent, the limbic ring includes the prostriate cortex, and inferiorly it meets the parahippocampal region, completing the ring. In the prefrontal region, the inferior part of the ACC includes the ventromedial prefrontal cortex, which is continuous with the limbic areas of the pOFC. The entire limbic ring thus is composed of areas that are either agranular or dysgranular. The areas of the limbic ring occupy a thin stripe or are wider in extent, depending on the region (reviewed in Barbas 2015).

Analysis of the cortex by structure is important for several reasons. First, the gradations in the laminar elaboration of cortical areas beyond the limbic areas suggest gradual changes of the cortex in evolution (Sanides 1970; Pandya et al. 1988). Second, the structure of cortices underlies the organization of their connections (Barbas 1986; Barbas and Rempel-Clower 1997; Medalla and Barbas 2006; Barbas et al. 2005). Third, the connections of limbic cortices with eulaminate areas reveal the intricate relationship between areas associated with emotion and cognition (Barbas 1995, 2000b).

### 4.3.1 Specialization of Prefrontal Limbic Cortices: Their Principal Cortical Connections

The prefrontal component of the cortical limbic system shows specialization, like the rest of the cortical mantle. Here we restrict discussion to the prefrontal components of the limbic system, the pOFC and ACC (Barbas 1997). The pOFC is distinguished as the most multimodal region within the prefrontal cortex (Fig. 4.4). Sensory pathways to pOFC take origin in high-order visual, auditory, and somatosensory association cortices. The pathways from visual association areas originate in anterior inferior temporal cortices, where neurons have large receptive fields (Gross et al. 1969). The nature of the input suggests that the pOFC receives an overview of the visual environment rather than the detail of objects and scenes. In addition, the visual areas that project to pOFC have a role in visual memory (Gross 1992; Barbas 1993). The pathways from other sensory association cortices to pOFC have comparable organization. Projections from auditory association cortices to pOFC, in particular, originate from areas that have a role in species-specific vocalizations (Poremba and Mishkin 2007), known to have emotional significance within conspecifics. Pathways from gustatory areas also project to the orbitofrontal cortex, which is also the strongest recipient of pathways from the primary olfactory cortex (Morecraft et al. 1992; Barbas 1993; Carmichael et al. 1994). A variety of temporal multimodal areas project to pOFC as well (Barbas 1993; Carmichael and Price 1995; Cavada et al. 2000).

Pathways from sensory association cortices thus provide a broad overview of the external environment to the pOFC. The pOFC is also connected with other components of the cortical limbic system, including temporal polar areas and the

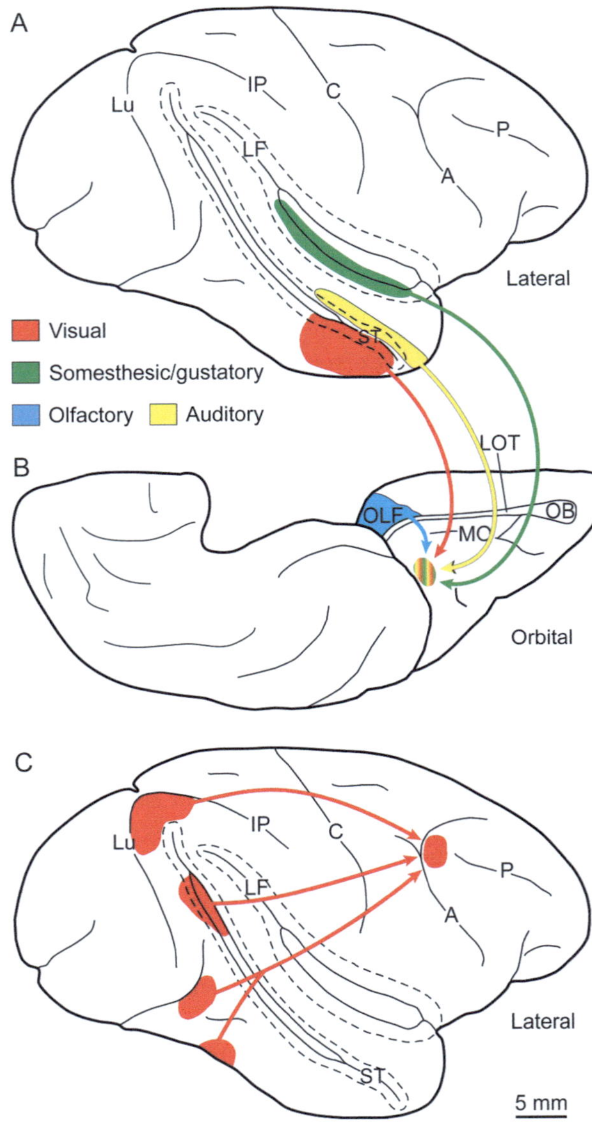


**Fig. 4.4** Differences in pathways from sensory association cortices to the pOFC and the frontal eye field (*FEF*). (**a**, **b**) The pOFC (**b**, orbital surface, *rainbow circle*) is the most multimodal region within the prefrontal region. Note the strong pathways from olfactory (**b**, *blue*), somatosensory and gustatory (**a**, *green*), auditory (**a**, *yellow*), and visual (**a**, *red*) to pOFC (**b**, *rainbow circle*). (**c**) Projections from sensory association cortices to the FEF in the confluence of the upper and lower limbs of the arcuate sulcus originate mostly from visual association cortices (*red* sites in occipital and temporal cortices). *Abbreviations*: *A* arcuate sulcus, *C* central sulcus, *IP* intraparietal sulcus, *LF* lateral fissure, *LOT* lateral olfactory tract, *Lu* lunate sulcus, *MO* medial orbital sulcus, *OB* olfactory bulb, *OLF* primary olfactory cortex, *P* principal sulcus, *ST* superior temporal sulcus

cingulate cortex. Projections from the cortical limbic ring provide information on the status of the internal, or emotional, environment to pOFC. The latter is enriched by input from the amygdala, which originates the densest pathways to pOFC in the prefrontal region (Ghashghaei et al. 2007). The pathways from the amygdala innervate all layers of pOFC, rivaling and surpassing the pathways from the thalamus to pOFC in terms of density and size of axon terminals, suggesting high synaptic efficacy (Timbie and Barbas 2014).

One interesting aspect of the connections described above is that the amygdala is as multimodal as the pOFC. The same high-order sensory and multimodal association areas that project to pOFC also project to the amygdala (Barbas 1995). These patterns suggest that sensory signals reach the pOFC directly through the cortex and indirectly through the amygdala (Barbas 1995). The indirect pathway through the amygdala may provide additional information about the affective significance of stimuli and events.

The ACC region in the frontal lobe shares the overall structural characteristics that identify limbic cortices, namely, an agranular or dysgranular laminar architecture. But the ACC differs from the pOFC by its predominant set of connections. The ACC does not have the wealth of direct projections from sensory association cortices. The only significant input from sensory cortices to the ACC originates from auditory association cortices (Barbas et al. 1999). Like the pOFC, the ACC receives projections from areas within the limbic ring, as well as dense projections from the amygdala. However, the ACC and the pOFC differ in the relative input from the amygdala in comparison with the output to the amygdala. While connections with the amygdala of the two limbic regions are dense in both directions, internal comparisons show marked differences in their relative input-output relationships (Ghashghaei et al. 2007). Thus, the pOFC is more of a receiver of pathways from the amygdala, while the ACC is more of a sender of pathways to the amygdala.

The pOFC and ACC also differ in their predominant innervation of nuclei in the amygdala. In nonhuman primates, the pOFC has strong bidirectional connections with the basolateral nucleus of the amygdala, but sends a uniquely strong pathway to the intercalated masses (IM) of the amygdala (Ghashghaei and Barbas 2002), which is not reciprocated. The IM is composed entirely of inhibitory neurons, which are squeezed in narrow corridors found between and above the basal nuclei of the amygdala. The IM neurons do not project to the cortex but have connections within the amygdala and especially with the central nucleus (Paré and Smith 1993). The latter is an output nucleus of the amygdala that innervates hypothalamic and brain stem autonomic structures (Jongen-Relo and Amaral 1998; Saha et al. 2000). The pOFC is thus positioned to influence the internal processing of the amygdala (Fig. 4.5).

The ACC projects most robustly to the amygdala and also targets different sites than the pOFC. The IM is not a significant target of ACC. However, the ACC has stronger projections to the central nucleus of the amygdala and thus can influence directly the output of the amygdala to autonomic structures. In addition, the ACC innervates other parts of the amygdala that project to central autonomic structures (Ghashghaei and Barbas 2002) (Fig. 4.5).

Both pOFC and ACC also have bidirectional connections with the hypothalamus but again differ in the relative innervation of the hypothalamus: pathways from the ACC to the hypothalamus are stronger than pathways from pOFC (Rempel-Clower and Barbas 1998), as are pathways to other brain stem and spinal autonomic centers. In aggregate, the cortical and subcortical connections suggest that the pOFC is poised to act as a cortical sensor for emotions. In contrast, the more robust

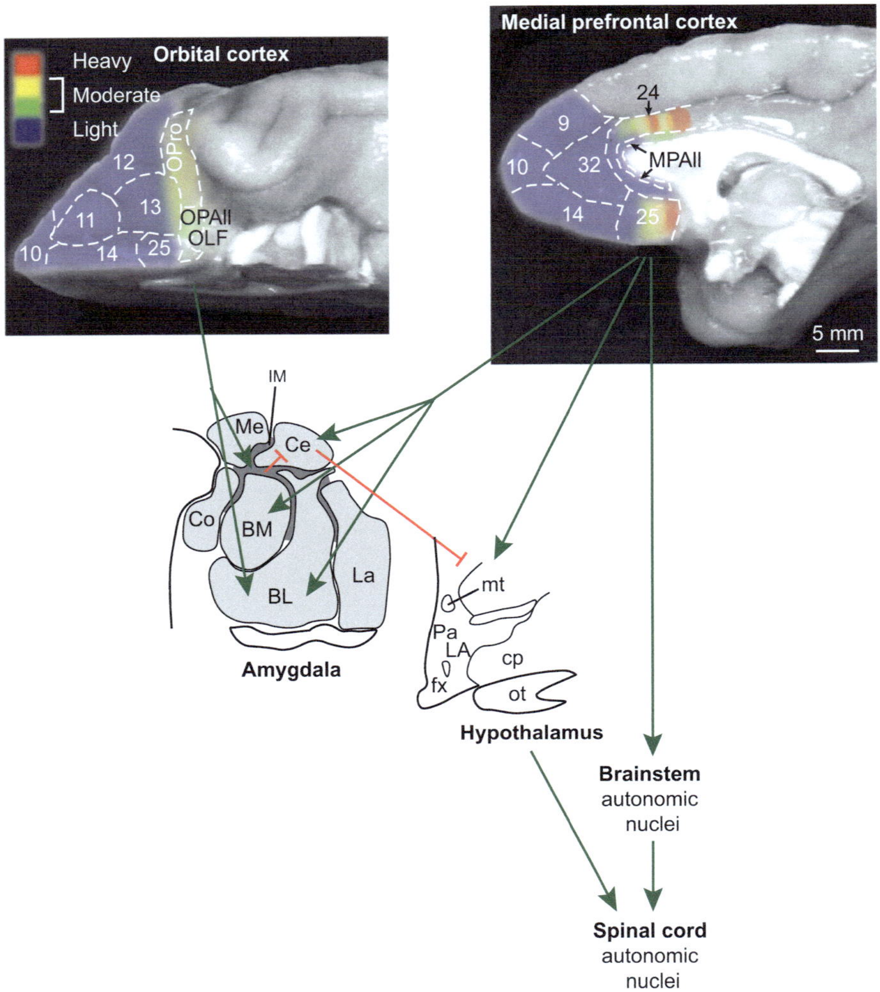


**Fig. 4.5** Pathways from the ACC and pOFC to the amygdala and autonomic structures. The ACC sends the strongest pathways to the amygdala, including its output central (*Ce*) nucleus, as well as to autonomic nuclei of the hypothalamus and brain stem that project to spinal autonomic structures. The pOFC also innervates the hypothalamus, but not as strongly as the ACC. The pOFC innervates the amygdala as well and especially the inhibitory intercalated masses (*IM*). However, input from the amygdala to pOFC exceeds its output to the amygdala (compare Fig. 4.1, *top*, with Fig. 4.5, *top left*). The ACC has the opposite relationship with the amygdala. *Abbreviations*: *BL* basolateral nucleus, *BM* basomedial nucleus (also known as accessory basal), *Ce* central nucleus, *Co* cortical nucleus, *cp* cerebral peduncle, *fx* fornix, *IM* intercalated masses, *La* lateral nucleus, *LA* lateral hypothalamic area, *Me* medial nucleus, *MPAll* medial periallocortex, *mt* mammillothalamic tract, *OLF* primary olfactory cortex, *OPAll* orbital periallocortex, *OPro* orbital proisocortex, *ot* optic tract, *Pa* paraventricular nucleus. Top panels were adapted from Ghashghaei et al. (2007)

and direct pathways of ACC to output nuclei of the amygdala and autonomic centers suggest that it is a cortical effector for emotions. There is an interesting and robust interaction between the ACC and the pOFC, suggesting collaborative processing between the two regions for the perception and expression of emotions (García-Cabezas and Barbas 2016).

### 4.3.2 Specialization of Prefrontal Eulaminate Areas

The above provides an overview of the organization of the cortical connections from sensory association areas to the two limbic regions of the prefrontal cortex and their connections with key subcortical structures associated with emotions. How do other parts of the prefrontal cortex compare in their connections? Lateral prefrontal areas are associated with cognitive processes, such as working memory – the temporary hold of information in mind long enough to accomplish a transient task, such as remembering directions while driving or the telephone number to dial (reviewed in Goldman-Rakic 1988; Fuster 1989; Funahashi and Kubota 1994; Petrides 1996; Miller 2000). How do the connections of lateral prefrontal cortices support cognitive processes? Lateral prefrontal areas differ from the pOFC in their connections with sensory association cortices in several ways. Unlike the origin of pathways directed to the pOFC, sensory pathways to lateral eulaminate areas originate from areas that are closer to the primary areas, though not the primary areas themselves (Pandya et al. 1988). For example, projections to the frontal eye fields (FEF), a region found at the confluence of the upper and lower limbs of the arcuate sulcus, originate in a large variety of visual association cortices, ranging from visual area 2 (V2), V4, V3, MT, and inferior temporal cortices (Barbas and Mesulam 1981; Barbas 1988; Schall et al. 1995) (Fig. 4.4c). The wealth of visual information to the FEF, which also has strong connections with parietal visuomotor cortices (Barbas and Mesulam 1981; Medalla and Barbas 2006), makes it possible to direct attention to relevant signals with precisely timed eye movements (Schiller 1998).

Unlike the pOFC, lateral prefrontal areas do not receive direct projections from olfactory areas. Lateral prefrontal areas receive pathways from unimodal sensory areas that arise mostly from visual and auditory association cortices. Similar to the projections from visual cortices, pathways from other sensory association cortices to lateral prefrontal areas, such as the auditory and the somatosensory, also originate from areas that are closer to the primary sensory areas, than those that project to pOFC. These features suggest that lateral prefrontal cortices receive more detailed information from earlier-processing sensory association cortices.

Lateral prefrontal areas also receive more limited direct projections from either cortical or subcortical limbic structures (reviewed in Barbas 2015). With regard to the example of the amygdala and the hypothalamus, all lateral prefrontal cortices receive some input from each of these structures. However, unlike either the pOFC or the ACC, pathways from the amygdala and the hypothalamus are sparser to lateral prefrontal cortices (Rempel-Clower and Barbas 1998). Moreover, lateral prefrontal areas have fewer, if any, projections to the amygdala and do not have direct projections to the hypothalamus (Ghashghaei et al. 2007; Barbas et al. 2003).

### *4.3.3 Regularity in the Laminar Pattern of Connections Between Limbic and Eulaminate Cortices*

Understanding the systematic variation of the cortex – traced from the limbic areas with the simplest laminar structure through a series of areas that culminate in eulaminate areas with the most elaborate laminar structure – is essential for summarizing the organization of connections. The systematic variation of cortical areas is also central to understanding the remarkably consistent laminar pattern of their interconnections (reviewed in Barbas 2015), as shown in Fig. 4.6. Thus, projections from limbic cortices originate in the deep layers (V and VI) when their targets are eulaminate areas, and their axons terminate mostly in the upper layers of prefrontal eulaminate areas. In the reverse direction, projections from eulaminate areas to limbic cortices originate in the upper layers (II–III), and their axons terminate in the middle-deep layers of limbic cortices (Barbas 1986; Barbas and Rempel-Clower 1997). In the sensory areas, projections originating from the upper layers of one area and terminating in the middle-deep layers of another area have been called feedforward. Pathways projecting from the deep layers of one area and terminating in the upper layers (especially layer I) of another area have been called feedback (reviewed in Felleman and Van Essen 1991). In the context of systematic variation of the cortex, feedforward pathways originate in areas with more complex laminar structure than the area of termination; feedback describes connections with the reverse relationship (Fig. 4.6). Importantly, this rule – which we call the structural model for connections – applies to the interconnections of the entire cortex (reviewed in Barbas 2015; Hilgetag et al. 2016). The distribution of connections within layers depends on the relative structural similarity of the linked areas (Barbas and Rempel-Clower 1997).

### *4.3.4 Thalamic Connections Differ According to the Systematic Variation of Prefrontal Areas*

The dorsal thalamus is connected with all areas of the cerebral cortex, including all of the prefrontal cortex (Jones 1985). Peripheral signals from the modalities of vision, audition, and somesthesis (including gustatory signals) project to thalamic relay nuclei, which then project to the primary areas of the cerebral cortex (Jones 1985). As a high-order association region, the prefrontal cortex does not receive direct input from the sensory relay nuclei of the thalamus. The principal thalamic

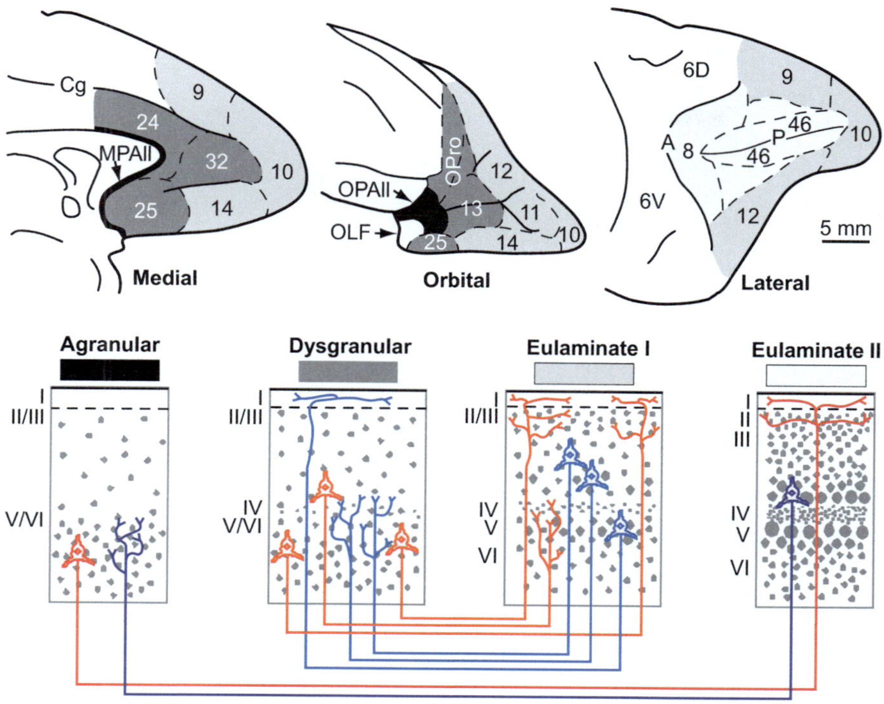


**Fig. 4.6** The laminar pattern of connections depends on the structural relationship of the linked areas. *Top*, different types of cortices include agranular (*black*) and dysgranular (*dark gray*) limbic areas and eulaminate areas (*light shades of gray*) in the prefrontal region. *Bottom*, the cartoons depict agranular areas with the simplest laminar structure and gradation to the best laminated areas (eulaminate II). Agranular areas project to eulaminate areas through their deep layers (*far left*, *red*), and their axons terminate in the upper layers of eulaminate areas (*far right*). In the reverse pathway projection, neurons from eulaminate areas (eulaminate II, *blue*) project to limbic areas from the upper layers, and their axons terminate in the middle-deep layers of limbic cortices (*far left*). The central panels show an intermediate pattern, seen between areas that have small differences in laminar structure (e.g., dysgranular and eulaminate I). The diagram is simplified and does not include all possible combinations of connections or gradations in laminar structure. *Abbreviations*: *A* arcuate sulcus, *Cg* cingulate sulcus, *MPAll* medial periallocortex, *OLF* primary olfactory cortex, *OPAll* orbital periallocortex, *OPro* orbital proisocortex, *P* principal sulcus

nucleus for the prefrontal cortex is the mediodorsal (MD). But other thalamic nuclei, including the medial pulvinar, ventral nuclei (especially the ventral anterior), anterior nuclei, and a series of midline and several intralaminar nuclei, are connected with the prefrontal cortex as well (e.g., Kievit and Kuypers 1977). Since prefrontal areas differ in structure, are the differences reflected in the connections with the thalamus?

The differences in the thalamic connections of prefrontal limbic and eulaminate cortices are striking. The pOFC and ACC, the two limbic sectors of the prefrontal cortex, have considerably more widespread thalamic connections than the lateral prefrontal cortices (Barbas 1997) (Fig. 4.7). The MD thalamic nucleus has three

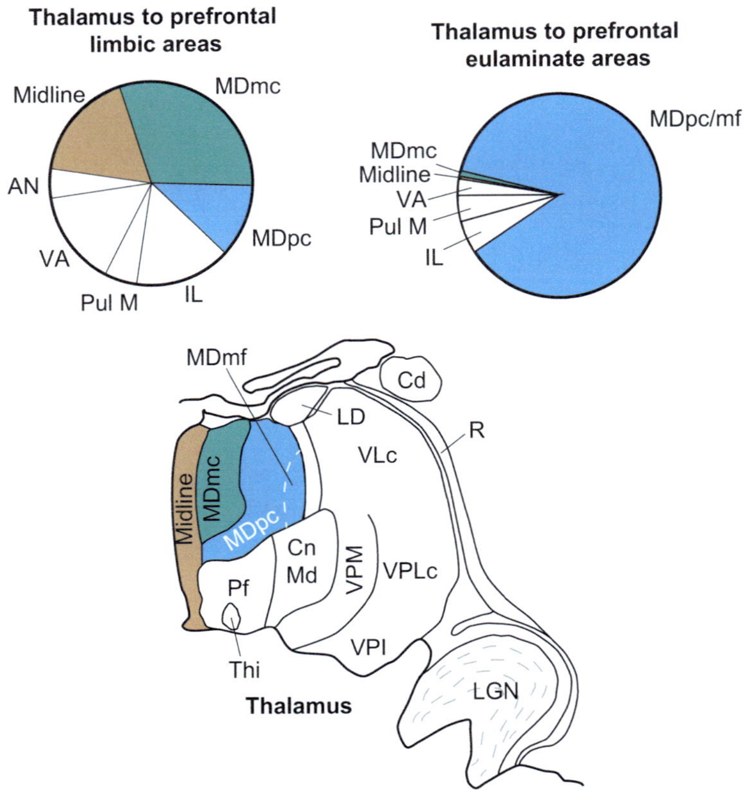


**Fig. 4.7** Differences in the thalamic connections of prefrontal limbic and eulaminate prefrontal areas. *Top, right*, lateral prefrontal areas receive the majority of their thalamic projections from neurons in the mediodorsal thalamic nucleus (MDpc and MDmf) and comparatively fewer projections from other thalamic nuclei. *Top, left*, limbic prefrontal areas receive more distributed projections from the thalamus, originating in MDmc, midline, anterior, intralaminar, medial pulvinar, and ventral anterior nuclei. *Bottom*, cross section through the thalamus shows some of the nuclei that are connected with prefrontal cortices. *Abbreviations*: *AN* anterior nuclei, *Cd* caudate, *CnMd* centromedian nucleus, *IL* intralaminar nuclei, *LD* lateral dorsal nucleus, *LGN* lateral geniculate nucleus, *MDmc* magnocellular division of the mediodorsal nucleus, *MDmf* multiform division of the mediodorsal nucleus, *MDpc* parvicellular division of the mediodorsal nucleus, *Pf* parafascicular nucleus, *Pul M* medial pulvinar nucleus, *R* reticular nucleus, *Thi* habenulo-interpeduncular tract, *VA* ventral anterior nucleus, *VLc* ventral lateral caudal nucleus, *VPI* ventral posterior inferior nucleus, *VPLc* ventral posterior lateral caudal nucleus, *VPM* ventral posterior medial nucleus. Thalamic nuclei are according to the map of Olszewski (1952)

major subdivisions, which include a medially situated magnocellular sector (MDmc), a centrally situated parvicellular sector (MDpc), and a lateral sector called multiform (MDmf). These sectors of MD are preferentially connected with prefrontal areas that show differences in laminar structure. Thus, the pOFC has the strongest association with MDmc, lateral prefrontal areas with MDpc, and the FEF with MDmf (Barbas and Mesulam 1981; Giguere and Goldman-Rakic 1988; Barbas et al. 1991; Siwek and Pandya 1991; Dermon and Barbas 1994). Lateral prefrontal cortices receive the large majority of their thalamic connections from MD, and only a small proportion from other thalamic nuclei, mostly from the ventral anterior and medial pulvinar. On the other hand, MD contributes only about half of the projection neurons directed to pOFC and a lower proportion of neurons directed to ACC. For both of these limbic prefrontal regions, significant proportions of thalamic projection neurons are found in midline, intralaminar, anterior, ventral, and medial pulvinar nuclei. Thus, the pOFC and ACC show a higher diversity in their thalamic connections compared with the eulaminate cortices (Barbas et al. 1991; Dermon and Barbas 1994). In this context, it is interesting that the olfactory bulb does not project to the thalamus but to the primary olfactory cortex (Shepherd 2007), which is limbic by structure and projects strongly to pOFC, as noted above.

#### Specialization of Direct and Indirect Pathways Through the Thalamus to Prefrontal Areas

As discussed above (Sect. 4.3.1), sensory input reaches the pOFC directly through cortical pathways, and indirectly through the amygdala, which innervates most densely the pOFC (Ghashghaei et al. 2007). The indirect sensory pathways through the amygdala are thought to convey information about emotional import. But is information about the internal state of emotions also conveyed to the thalamus? The amygdala, which is a chief processor of signals with affective significance (LeDoux 2003; Davis and Whalen 2001; Murray 2007), sends a strong projection to MDmc, which sends a strong projection to pOFC (Porrino et al. 1981; Timbie and Barbas 2014). Pathways from the amygdala thus reach pOFC directly and indirectly through MDmc. The amygdala thus may convey information about the internal (emotional) environment to MDmc. Recent findings show that pathways from the amygdala indeed innervate robustly MDmc sites and individual neurons that project to pOFC (Timbie and Barbas 2015). The presence of direct and indirect pathways from the amygdala suggests assurance of transmission of signals associated with emotional import to pOFC.

On the other hand, the lateral part of MD, which is connected with the FEF, receives signals from the superior colliculus (reviewed in Barbas 1995). This pattern of connection is consistent with the role of the FEF in goal-directed eye movements and search of the environment for cognitive operations. The thalamic connections of limbic and eulaminate prefrontal areas thus reflect their functional specialization.

## 4.4 Linkage of Areas Associated with Emotions and Cognition

The two limbic regions and the eulaminate prefrontal areas thus have specializations with regard to their connections with other cortices and subcortical structures. There are also prefrontal areas that are situated between the limbic areas, on one hand, and the best laminated areas in the caudal lateral prefrontal cortex, on the other hand. These areas, which are also eulaminate, have an intermediate position in both structure and pattern of connections. They include the anterior part of the orbital and medial prefrontal cortices, as well as the anterior part of the lateral prefrontal region (Barbas and Pandya 1989). The limbic areas and the eulaminate areas with the most elaborate structure are linked with each other directly but sparsely, or indirectly and profusely, through the intermediate eulaminate areas (Fig. 4.8). The connections between the two limbic prefrontal sectors with lateral prefrontal areas provide the basis for the linkage of cortical areas associated with emotions and cognition. This linkage is exemplified in functional studies that show influence of affective stimuli and motivation on cognitive functions (e.g., Roberts and Wallis 2000; Sakagami and Watanabe 2007; Watanabe 2007).

The linkage of limbic and eulaminate prefrontal cortices is essential for accomplishing even simple tasks in everyday life. Let us consider a simple example of preparing a meal at home. Lateral prefrontal areas must be engaged to keep track of the sequence of steps needed to follow a recipe. But it is also necessary to retrieve information from long-term memory: where the pots and pans are stored and where the ingredients are found. Pathways from the pOFC and the ACC may thus become activated, based on their robust connections with medial temporal memory-related

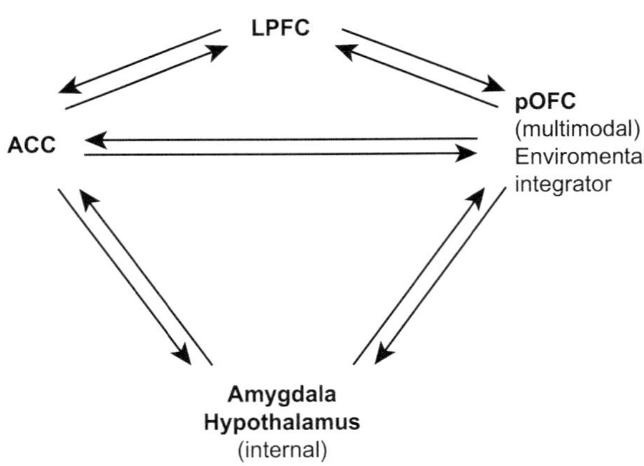


**Fig. 4.8** Convergence of pathways associated with emotions and cognition. The pOFC and ACC, which have the strongest connections with the amygdala and hypothalamus, are ultimately connected either directly (*top*) or through intermediate steps (not shown) with lateral prefrontal cortex (*LPFC*), ultimately linking areas associated with emotion and cognition. *Abbreviations*: *ACC* anterior cingulate cortex, *LPFC* lateral prefrontal cortex, *pOFC* posterior orbitofrontal cortex

areas (Bunce and Barbas 2011; Bunce et al. 2013), as well as the strong pathways they receive from the hippocampus. The hippocampus innervates robustly the ACC and to a lesser extent the pOFC (Rosene and Van Hoesen 1977; Barbas and Blatt 1995; Insausti and Munoz 2001; reviewed in Anderson et al. 2015). In contrast, none of the memory-related medial temporal areas provide significant direct projections to lateral prefrontal cortices. The connections between the two limbic prefrontal regions with lateral prefrontal areas thus allow access to information from memory for cognitive operations. In view of the rich connections of prefrontal limbic areas with subcortical limbic structures, their linkage with eulaminate cortices allows access of signals from vital autonomic processes to areas associated with cognition (Barbas et al. 2003).

### 4.4.1 Disruption of Linkage Between Areas Associated with Emotions and Cognition in Neurologic and Psychiatric Diseases

The orderly pattern of connections between areas associated with cognition and emotion appears to be disrupted in psychiatric and neurologic diseases. The symptoms in schizophrenia, for example, include distractibility, disordered thought process, and hallucinations, which frequently are auditory. Can these symptoms be mapped on the interactions of cortical areas associated with emotion and cognition? The ACC is a good candidate in this context because it is also associated with attention (e.g., Carter et al. 1999; Johnston et al. 2007), a process that is disrupted in schizophrenia. Among the most prominent features of ACC are its strong connections with the rest of the prefrontal cortex, including the pOFC, but also lateral prefrontal areas, including areas 9, 10, and 46 (Barbas et al. 1999).

Studies of pathways from ACC to lateral prefrontal areas at the synaptic level have made it possible to identify the extent of interaction of pathways with excitatory as well as specialized inhibitory neurons at the site of termination. Most synapses made by corticocortical pathways are on excitatory neurons (White 1989), but a significant proportion (~20%) are made with inhibitory neurons. When ACC fibers from area 32 innervate inhibitory neurons in the upper layers of lateral area 9, they form synapses preferentially with the specialized calbindin inhibitory neurons (Medalla and Barbas 2009). Physiologic and computational studies have shown that calbindin inhibitory neurons are synaptically suited to reduce noise and enhance signal in cognitive operations (Wang et al. 2004). Others have shown that there is a reduction in the number of pyramidal neurons in the deep layers of ACC in the brains of patients with schizophrenia (Benes et al. 2001). The pathway from ACC to lateral prefrontal cortices originates in the deep layers and terminates in the upper layers of lateral prefrontal areas. This pattern is predicted by the structural model for connections and confirmed in empirical studies in nonhuman primates (Barbas and Rempel-Clower 1997; Medalla and

Barbas 2009). Accordingly, the pathway from ACC to lateral prefrontal areas is expected to be weakened in schizophrenia, which may help explain the distractibility in this disease.

ACC area 32 projects to the frontal polar region (area 10) as well, where it forms large and efficient synapses with spines of excitatory neurons (Medalla and Barbas 2010, 2014). In humans, area 10 is active when one has to temporarily suspend a task to attend to another task and then return to the first task (Dreher et al. 2008). For example, if one interrupts preparation of a meal to answer the phone and then returns to the task, the cook must remember the point of interruption so as not to add an ingredient twice or forget to add an important ingredient. ACC area 32 projects to area 10 from the deep layers, and presumably this pathway would also be weakened in schizophrenia. Weakening of this pathway may help explain the disruption of the orderly sequencing of information within working memory, which may help explain the disordered thought process in schizophrenia. Finally, ACC area 32 has strong bidirectional connections with auditory association cortices (Vogt and Barbas 1988; Romanski and Goldman-Rakic 2002; Barbas et al. 1999). Since the ACC is hypoactive in schizophrenia, an imbalance in its pathways with auditory cortices may help explain the misattribution of thoughts to external voices in the disease.

The above example shows how a model based on the fundamental laminar structure of the cortex helps explain connections at the level of cortical layers. The model can help predict the pattern of connections in humans by the study of cortical structure in postmortem brains. Though less is known about circuits in other psychiatric diseases, the symptoms in several diseases suggest disconnection in pathways that link areas associated with cognition and emotion. In depression, for example, patients ruminate about negative events even in the presence of positive events in their lives (Mayberg 2007). Cognitive reasoning about the circumstances does not appear to influence the perception of events. In anxiety disorders, such as phobias or post-traumatic stress disorder, a balance also seems to be affected so that fear dominates behavior (reviewed in John et al. 2013).

## Consequences for Disconnection of Cognition from Emotion in Society

We started with a brief discussion surrounding the rejection or acceptance of the influence of emotions on cognition and vice versa. Why do some consider that influence of emotions on cognition negatively affects behavior? The answer is not clear, but perhaps it is based on thinking about emotions in their extreme expression. In extreme cases, there is a tilt in the balance so that strong emotions may indeed cloud judgment and disable cognitive control. There is evidence that in cases of extreme fear and anxiety, as in post-traumatic stress disorder, the prefrontal cortex fails to influence the amygdala, which seems to go into overdrive (Rauch et al. 2006; Lanius et al. 2006). In the throes of extreme rage, one may kill even a loved one. These situations argue strongly for the necessity for interaction of areas associated with cognition and emotion and the dire consequences of their disengagement. A balance of influence of emotion and cognition facilitates normal function. Constructs, such as fairness, judgment, and compassion, suggest the inexorable influence of emotion on cognition and vice versa, as the pathways in the brain signify.

## 4.5 Summary and Conclusion

There are both specializations and strong interactions in the connections of areas associated with emotions and cognition in the prefrontal cortex. The pOFC is connected with areas that receive signals from both the external (sensory) environment and from the internal environment of motives and drives. Pathways that transmit information to pOFC originate in high-order sensory association areas that provide a broad overview of the external environment. A sister limbic region, the ACC, appears to specialize in the expression of emotions through strong pathways to central autonomic structures. On the other hand, lateral prefrontal areas receive information from earlier-processing sensory association areas, suggesting that they provide more detailed information compared to what is sent to prefrontal limbic areas. The projections from the thalamus are widespread to prefrontal limbic areas and more focal to eulaminate areas. The pathways suggest a broad generalization by limbic areas and specialization by eulaminate areas. Ultimately, limbic and eulaminate prefrontal areas are interlinked according to principles that are based on their fundamental structure. Limbic cortices project through their deep layers, and their axons innervate extensively the upper layers (I and II) of eulaminate areas. The upper layers of the cortex include the apical dendrites of neurons from many of the neurons below. Widespread projections to the upper layers thus have a potential impact on the neurons below, suggesting that the limbic areas have a tonic influence on eulaminate cortices (Barbas 2000b). This linkage is an essential consequence of the structure of the cortex and likely its evolution. This linkage also suggests that there may be profound consequences when the bonds that tie areas associated with emotions and cognition are disrupted in psychiatric and neurologic diseases.

**Acknowledgments** This work is supported by grants from NIH (R01 MH057414, R01 NS024760 (HB)); M. Á. García-Cabezas was the recipient of a 2014 NARSAD Young Investigator Grant from the Brain & Behavior Research Foundation (grant number 22777, P&S Fund Investigator).

## References

Anderson MC, Bunce JG, Barbas H (2015) Prefrontal-hippocampal pathways underlying inhibitory control over memory. Neurobiol Learn Mem. doi:10.1016/j.nlm.2015.11.008

Armstrong E (1991) The limbic system and culture: an allometric analysis of the neocortex and limbic nuclei. Hum Nat 2:117–136

Barbas H (1986) Pattern in the laminar origin of corticocortical connections. J Comp Neurol 252:415–422
Barbas H (1988) Anatomic organization of basoventral and mediodorsal visual recipient prefrontal regions in the rhesus monkey. J Comp Neurol 276:313–342
Barbas H (1993) Organization of cortical afferent input to orbitofrontal areas in the rhesus monkey. Neuroscience 56:841–864
Barbas H (1995) Anatomic basis of cognitive-emotional interactions in the primate prefrontal cortex. Neurosci Biobehav Rev 19:499–510
Barbas H (1997) Two prefrontal limbic systems: their common and unique features. In: Sakata H, Mikami A, Fuster JM (eds) The association cortex: structure and function. Harwood Academic Publising, Amsterdam, pp 99–115
Barbas H (2000a) Complementary role of prefrontal cortical regions in cognition, memory and emotion in primates. Adv Neurol 84:87–110
Barbas H (2000b) Connections underlying the synthesis of cognition, memory, and emotion in primate prefrontal cortices. Brain Res Bull 52:319–330
Barbas H (2015) General cortical and special prefrontal connections: principles from structure to function. Annu Rev Neurosci 38:269–289
Barbas H, Blatt GJ (1995) Topographically specific hippocampal projections target functionally distinct prefrontal areas in the rhesus monkey. Hippocampus 5:511–533
Barbas H, García-Cabezas MA (2015) Motor cortex layer 4: less is more. Trends Neurosci 38 (5):259–261
Barbas H, Mesulam MM (1981) Organization of afferent input to subdivisions of area 8 in the rhesus monkey. J Comp Neurol 200:407–431
Barbas H, Pandya DN (1989) Architecture and intrinsic connections of the prefrontal cortex in the rhesus monkey. J Comp Neurol 286(3):353–375
Barbas H, Rempel-Clower N (1997) Cortical structure predicts the pattern of corticocortical connections. Cereb Cortex 7:635–646
Barbas H, Henion TH, Dermon CR (1991) Diverse thalamic projections to the prefrontal cortex in the rhesus monkey. J Comp Neurol 313:65–94
Barbas H, Ghashghaei H, Dombrowski SM, Rempel-Clower NL (1999) Medial prefrontal cortices are unified by common connections with superior temporal cortices and distinguished by input from memory-related areas in the rhesus monkey. J Comp Neurol 410:343–367
Barbas H, Ghashghaei H, Rempel-Clower N, Xiao D (2002) Anatomic basis of functional specialization in prefrontal cortices in primates. In: Grafman J (ed) Handbook of neuropsychology, vol 7: the frontal lobes, vol 2. Elsevier Science B.V, Amsterdam, pp 1–27
Barbas H, Saha S, Rempel-Clower N, Ghashghaei T (2003) Serial pathways from primate prefrontal cortex to autonomic areas may influence emotional expression. BMC Neurosci 4 (1):25
Barbas H, Hilgetag CC, Saha S, Dermon CR, Suski JL (2005) Parallel organization of contralateral and ipsilateral prefrontal cortical projections in the rhesus monkey. BMC Neurosci 6(1):32
Benes FM, Vincent SL, Todtenkopf M (2001) The density of pyramidal and nonpyramidal neurons in anterior cingulate cortex of schizophrenic and bipolar subjects. Biol Psychiatry 50 (6):395–406
Bereket A, Kiess W, Lustig RH, Muller HL, Goldstone AP, Weiss R, Yavuz Y, Hochberg Z (2012) Hypothalamic obesity in children. Obes Rev 13(9):780–798
Broca P (1878) Anatomie comparée des circonvolutions cérébrales: Le grand lobe limbique et la scissure limbique dans la série des mammifères. Rev D'anthropol 1:385–498
Brown JA, Woodworth HL, Leinninger GM (2015) To ingest or rest? Specialized roles of lateral hypothalamic area neurons in coordinating energy balance. Front Syst Neurosci 9:9. doi:10. 3389/fnsys.2015.00009
Bunce JG, Barbas H (2011) Prefrontal pathways target excitatory and inhibitory systems in memory-related medial temporal cortices. Neuroimage 55(4):1461–1474

Bunce JG, Zikopoulos B, Feinberg M, Barbas H (2013) Parallel prefrontal pathways reach distinct excitatory and inhibitory systems in memory-related rhinal cortices. J Comp Neurol 512 (18):4260–4283
Carmichael ST, Price JL (1995) Sensory and premotor connections of the orbital and medial prefrontal cortex of macaque monkeys. J Comp Neurol 363:642–664
Carmichael ST, Clugnet MC, Price JL (1994) Central olfactory connections in the macaque monkey. J Comp Neurol 346(3):403–434
Carter CS, Botvinick MM, Cohen JD (1999) The contribution of the anterior cingulate cortex to executive processes in cognition. Rev Neurosci 10(1):49–57
Cavada C, Company T, Tejedor J, Cruz-Rizzolo RJ, Reinoso-Suarez F (2000) The anatomical connections of the macaque monkey orbitofrontal cortex. A review. Cereb Cortex 10:220–242
Damasio AR (1994) Descarte's error: emotion, reason, and the human brain, vol 1. G. P. Putnam's Sons, New York
Darwin C (1872) The expression of the emotions in man and animals. J. Murray, London
Davis M, Whalen PJ (2001) The amygdala: vigilance and emotion. Mol Psychiatry 6(1):13–34
De Olmos J (1990) Amygdaloid nuclear gray complex. In: Paxinos G (ed) The human nervous system. Academic, San Diego, pp 583–710
Dermon CR, Barbas H (1994) Contralateral thalamic projections predominantly reach transitional cortices in the rhesus monkey. J Comp Neurol 344:508–531
Dreher JC, Koechlin E, Tierney M, Grafman J (2008) Damage to the fronto-polar cortex is associated with impaired multitasking. PLoS One 3(9):e3227
Felleman DJ, Van Essen DC (1991) Distributed hierarchical processing in the primate cerebral cortex. Cereb Cortex 1:1–47
Funahashi S, Kubota K (1994) Working memory and prefrontal cortex. Neurosci Res 21:1–11
Fuster JM (1989) The prefrontal cortex, vol 2. Raven Press, New York
García-Cabezas MA, Barbas H (2016) Anterior cingulate pathways may affect emotions through orbitofrontal cortex. Cereb Cortex. doi:10.1093/cercor/bhw284
García-Cabezas MA, Barbas H (2014) Area 4 has layer IV in adult primates. Eur J Neurosci 39:1824–1834
Ghashghaei HT, Barbas H (2002) Pathways for emotion: interactions of prefrontal and anterior temporal pathways in the amygdala of the rhesus monkey. Neuroscience 115:1261–1279
Ghashghaei HT, Hilgetag CC, Barbas H (2007) Sequence of information processing for emotions based on the anatomic dialogue between prefrontal cortex and amygdala. Neuroimage 34 (3):905–923
Giguere M, Goldman-Rakic PS (1988) Mediodorsal nucleus: areal, laminar, and tangential distribution of afferents and efferents in the frontal lobe of rhesus monkeys. J Comp Neurol 277:195–213
Goldman-Rakic PS (1988) Topography of cognition: parallel distributed networks in primate association cortex. Annu Rev Neurosci 11:137–156
Gross CG (1992) Representation of visual stimuli in inferior temporal cortex. Philos Theol R Soc B 335:3–10
Gross CG, Bender DB, Rocha-Miranda CE (1969) Visual receptive fields of neurons in inferotemporal cortex of the monkey. Science 166:1303–1306
Hilgetag CC, Medalla M, Beul SF, Barbas H (2016) The primate connectome in context: Principles of connections of the cortical visual system. NeuroImage 134:685–702
Insausti R, Munoz M (2001) Cortical projections of the non-entorhinal hippocampal formation in the cynomolgus monkey (*Macaca fascicularis*). Eur J Neurosci 14(3):435–451
John YJ, Bullock D, Zikopoulos B, Barbas H (2013) Anatomy and computational modeling of networks underlying cognitive-emotional interaction. Front Hum Neurosci 7:101. doi:10.3389/fnhum.2013.00101
Johnston JB (1923) Further contributions to the study of the evolution of the forebrain. J Comp Neurol 35:337–481

Johnston K, Levin HM, Koval MJ, Everling S (2007) Top-down control-signal dynamics in anterior cingulate and prefrontal cortex neurons following task switching. Neuron 53 (3):453–462
Jones EG (1985) The thalamus. Plenum Press, New York
Jongen-Relo AL, Amaral DG (1998) Evidence for a GABAergic projection from the central nucleus of the amygdala to the brainstem of the macaque monkey: a combined retrograde tracing and in situ hybridization study. Eur J Neurosci 10:2924–2933
Kievit J, Kuypers HGJM (1977) Organization of the thalamo-cortical connexions to the frontal lobe in the rhesus monkey. Exp Brain Res 29:299–322
Lanius RA, Bluhm R, Lanius U, Pain C (2006) A review of neuroimaging studies in PTSD: heterogeneity of response to symptom provocation. J Psychiatr Res 40(8):709–729
LeDoux JE (2000) Emotion circuits in the brain. Annu Rev Neurosci 23:155–184
LeDoux J (2003) The emotional brain, fear, and the amygdala. Cell Mol Neurobiol 23 (4–5):727–738
MacLean PD (1952) Some psychiatric implications of physiological studies on frontotemporal portion of limbic system (visceral brain). Electroencephalogr Clin Neurophysiol 4(4):407–418
Mayberg HS (2007) Defining the neural circuitry of depression: toward a new nosology with therapeutic implications. Biol Psychiatry 61(6):729–730
Medalla M, Barbas H (2006) Diversity of laminar connections linking periarcuate and lateral intraparietal areas depends on cortical structure. Eur J Neurosci 23(1):161–179
Medalla M, Barbas H (2009) Synapses with inhibitory neurons differentiate anterior cingulate from dorsolateral prefrontal pathways associated with cognitive control. Neuron 61 (4):609–620
Medalla M, Barbas H (2010) Anterior cingulate synapses in prefrontal areas 10 and 46 suggest differential influence in cognitive control. J Neurosci 30(48):16068–16081
Medalla M, Barbas H (2014) Specialized prefrontal "auditory fields": organization of primate prefrontal-temporal pathways. Front Neurosci 8:77. doi:10.3389/fnins.2014.00077
Milad MR, Quirk GJ, Pitman RK, Orr SP, Fischl B, Rauch SL (2007) A role for the human dorsal anterior cingulate cortex in fear expression. Biol Psychiatry 62(10):1191–1194
Miller EK (2000) The prefrontal cortex and cognitive control. Nat Rev Neurosci 1:59–65
Morecraft RJ, Geula C, Mesulam MM (1992) Cytoarchitecture and neural afferents of orbitofrontal cortex in the brain of the monkey. J Comp Neurol 323:341–358
Mosher CP, Zimmerman PE, Gothard KM (2010) Response characteristics of basolateral and centromedial neurons in the primate amygdala. J Neurosci 30(48):16197–16207. doi:10.1523/JNEUROSCI.3225-10.2010
Murray EA (2007) The amygdala, reward and emotion. Trends Cogn Sci 11(11):489–497
Nauta WJH (1971) The problem of the frontal lobe: a reinterpretation. J Psychiatr Res 8:167–187
Nauta WJH (1979) Expanding borders of the limbic system concept. In: Rasmussen T, Marino R (eds) Functional neurosurgery. Raven Press, New York, pp 7–23
Nishijo H, Hori E, Tazumi T, Ono T (2008) Neural correlates to both emotion and cognitive functions in the monkey amygdala. Behav Brain Res 188(1):14–23
Olszewski J (1952) The thalamus of the *Macaca mulatta*. An atlas for use with the stereotaxic instrument. Karger, Basel
Pandya DN, Seltzer B, Barbas H (1988) Input-output organization of the primate cerebral cortex. In: Steklis HD, Erwin J (eds) Comparative primate biology, Neurosciences, vol 4. Alan R. Liss, New York, pp 39–80
Paré D, Smith Y (1993) The intercalated cell masses project to the central and medial nuclei of the amygdala in cats. Neuroscience 57:1077–1090
Paré D, Quirk GJ, LeDoux JE (2004) New vistas on amygdala networks in conditioned fear. J Neurophysiol 92(1):1–9
Paton JJ, Belova MA, Morrison SE, Salzman CD (2006) The primate amygdala represents the positive and negative value of visual stimuli during learning. Nature 439(7078):865–870

Pessoa L (2013) The cognitive-emotional brain: from interactions to integration. The MIT Press, Cambridge, MA
Petrides M (1996) Lateral frontal cortical contribution to memory. Semin Neurosci 8:57–63
Petrovich GD, Gallagher M (2007) Control of food consumption by learned cues: a forebrain-hypothalamic network. Physiol Behav 91(4):397–403
Plato (1892/impression of 1931) Phaedrus. In: The dialogues of Plato translated into English with analysis and introductions, vol I, 3rd edn. Oxford University Press, London, pp 246–254
Poremba A, Mishkin M (2007) Exploring the extent and function of higher-order auditory cortex in rhesus monkeys. Hear Res 229(1–2):14–23
Porrino LJ, Crane AM, Goldman-Rakic PS (1981) Direct and indirect pathways from the amygdala to the frontal lobe in rhesus monkeys. J Comp Neurol 198:121–136
Rauch SL, Shin LM, Phelps EA (2006) Neurocircuitry models of posttraumatic stress disorder and extinction: human neuroimaging research – past, present, and future. Biol Psychiatry 60 (4):376–382
Rempel-Clower NL, Barbas H (1998) Topographic organization of connections between the hypothalamus and prefrontal cortex in the rhesus monkey. J Comp Neurol 398:393–419
Roberts AC, Wallis JD (2000) Inhibitory control and affective processing in the prefrontal cortex: neuropsychological studies in the common marmoset. Cereb Cortex 10:252–262
Rockel AJ, Hiorns RW, Powell TP (1980) The basic uniformity in structure of the neocortex. Brain 103(2):221–244
Romanski LM, Goldman-Rakic PS (2002) An auditory domain in primate prefrontal cortex. Nat Neurosci 5(1):15–16
Rosene DL, Van Hoesen GW (1977) Hippocampal efferents reach widespread areas of cerebral cortex and amygdala in the rhesus monkey. Science 198:315–317
Saha S, Batten TF, Henderson Z (2000) A GABAergic projection from the central nucleus of the amygdala to the nucleus of the solitary tract: a combined anterograde tracing and electron microscopic immunohistochemical study. Neuroscience 99(4):613–626
Sakagami M, Watanabe M (2007) Integration of cognitive and motivational information in the primate lateral prefrontal cortex. Ann N Y Acad Sci 1104:89–107
Sanides F (1970) Functional architecture of motor and sensory cortices in primates in the light of a new concept of neocortex evolution. In: Noback CR, Montagna W (eds) The primate brain: advances in primatology. Appleton-Century-Crofts Educational Division/Meredith Corporation, New York, pp 137–208
Schall JD, Morel A, King DJ, Bullier J (1995) Topography of visual cortex connections with frontal eye field in macaque: convergence and segregation of processing streams. J Neurosci 15:4464–4487
Schiller PH (1998) The neural control of visually guided eye movements. In: Richards JE (ed) Cognitive neuroscience of attention. Lawrence Erlbaum Assoc. Publ, New Jersey, pp 3–50
Shepherd GM (2007) Perspectives on olfactory processing, conscious perception, and orbitofrontal cortex. Ann N Y Acad Sci 1121:87–101
Shepherd GM (2012) Neurogastronomy: how the brain creates flavor and why it matters. Columbia University Press, New York
Siwek DF, Pandya DN (1991) Prefrontal projections to the mediodorsal nucleus of the thalamus in the rhesus monkey. J Comp Neurol 312:509–524
Timbie C, Barbas H (2014) Specialized pathways from the primate amygdala to posterior orbitofrontal cortex. J Neurosci 34(24):8106–8118
Timbie C, Barbas H (2015) Pathways for emotions: specializations in the amygdalar, mediodorsal thalamic, and posterior orbitofrontal network. J Neurosci 35(34):11976–11987
Vogt BA, Barbas H (1988) Structure and connections of the cingulate vocalization region in the rhesus monkey. In: Newman JD (ed) The physiological control of mammalian vocalization. Plenum Publ. Corp., New York, pp 203–225

Wang XJ, Tegner J, Constantinidis C, Goldman-Rakic PS (2004) Division of labor among distinct subtypes of inhibitory neurons in a cortical microcircuit of working memory. Proc Natl Acad Sci U S A 101(5):1368–1373

Watanabe M (2007) Role of anticipated reward in cognitive behavioral control. Curr Opin Neurobiol 17(2):213–219

White EL (1989) Cortical circuits. Synaptic organization of the cerebral cortex. Structure, function and theory. Birkhäuser, Boston

Yakovlev PI (1948) Motility, behavior and the brain: stereodynamic organization and neurocoordinates of behavior. J Nerv Ment Dis 107:313–335

Yakovlev PI (1959) Pathoarchitectonic studies of cerebral malformations. III. Arrhinencephalies (holotelencephalies). J Neuropathol Exp Neurol 18(1):22–55

Zhang W, Schneider DM, Belova MA, Morrison SE, Paton JJ, Salzman CD (2013) Functional circuits and anatomical distribution of response properties in the primate amygdala. J Neurosci 33(2):722–733

# Chapter 5
# Interaction of Dopamine and Glutamate Release in the Primate Prefrontal Cortex in Relation to Working Memory and Reward

**Tohru Kodama and Masataka Watanabe**

**Abstract** The prefrontal cortex (PFC) aggregates information widely from other brain areas. The PFC plays the role of a command center that determines actions related to higher cognitive functions, such as working memory (WM), planning, decision-making, and behavioral inhibition, for the sake of better adaptation to the outside environment. Activities related to the higher cognitive function observed in the PFC are regulated by several neurotransmitters such as dopamine (DA), glutamate, serotonin, norepinephrine, and gamma-aminobutyric acid (GABA). Abnormalities of neurotransmitters, i.e., when these substances are out of the suitable range, impair the human cognitive performance and sometimes cause psychiatric disorders. While there are many reports about changes in neurotransmitters in the PFC of the human and rodent, there are very few studies on the PFC of the nonhuman primate. In this chapter, we discuss the function of the PFC in relation to changes in neurotransmitters, especially DA and glutamate, and their interactions, referring to our studies in which we investigated changes in neurotransmitters in the primate PFC in relation to cognitive task performance and reward.

**Keywords** Dopamine (DA) • Glutamate • Microdialysis • Primate • Prefrontal cortex (PFC) • Working memory (WM) • Reward

## 5.1 Introduction

The prefrontal cortex (PFC) aggregates information widely from other brain areas and plays the role of a command center that determines actions related to higher cognitive functions, such as working memory (WM), planning, decision-making, and behavioral inhibition, for the sake of better adaptation to the outside environment.

T. Kodama (✉) • M. Watanabe
Department of Physiological Psychology, Tokyo Metropolitan Institute of Medical Science, Tokyo 156-8506, Japan
e-mail: kodama-tr@igakuken.or.jp



M. Watanabe (ed.), *The Prefrontal Cortex as an Executive, Emotional, and Social Brain*, DOI 10.1007/978-4-431-56508-6_5

As noninvasive neuroimaging studies in the human, recordings of neuronal activities (unit discharges) in animals have been leading brain science in the field of higher cognitive function. The unit discharge pattern by itself has a significant meaning as a memory- and reward-related function. For example, the existence of "grandmother neurons" or "mirror neurons," which discharge when the human or animal recognizes a special pattern, like the face of a grandmother or when it feels others' behaviors as if it acts itself, provides direct evidence that these neurons are related to a specific brain function. Event-related increases or decreases in neuronal discharge are also very important for understanding the function of the PFC. For example, "cabbage neurons" reported in the primate PFC (Watanabe 1996), which discharge at a higher rate during the memory retention period and when expecting a preferable reward, demonstrate that the PFC plays a role in the integration of memory and reward expectancy.

On the other hand, compared to the outputs to other brain regions from the PFC, inputs from other brain regions that generate event-related neuronal activities in the PFC are poorly clarified. Information from other areas of the brain reaches the dendrite or soma of PFC neurons as action potentials, which are translated to the release of neurotransmitters. The neurotransmitters are aggregated to depolarize or hyperpolarize membrane potentials and are finally converted again to neuronal activities as an output signal. Therefore, the field of the synaptic gap is the place where the information of neuronal inputs are summed and interact with each other, contributing to a greatly complicated transition of information. As described later, the neurotransmitters work synergistically, changing the receptor sensitivity and regulating the release of neurotransmitters.

Activities related to the higher brain functions observed in the PFC are regulated by several neurotransmitters, such as dopamine (DA), glutamate, serotonin, norepinephrine, and gamma-aminobutyric acid (GABA). When these substances are out of the suitable range (deficient or too abundant), abnormalities of neurotransmitters impair cognitive operations, such as WM, planning, and decision-making, and sometimes cause affective disorders. For the treatment of these disorders, research about changes in these neurotransmitters would provide us with important information. While there are many reports about changes in neurotransmitters in the PFC using human positron emission tomography (PET) as well as rodent microdialysis studies, there are very few studies on the PFC of the nonhuman primate, which is the animal closest to humans available for invasive studies. Importantly, because of technical difficulties associated with the low level of DA concentration in the PFC, there has been no study to examine DA concentrations in the human PFC. In the rat where the DA concentration in the PFC can be studied, the volume of the PFC is quite small compared with that of the primate, and the rat PFC is indicated to have no comparable part of the primate lateral PFC (LPFC) that is essential for cognition and attention (Preuss 1995). In this chapter, we discuss the function of the PFC in relation to changes in neurotransmitters, especially DA and glutamate, and their interactions, referring to our microdialysis studies in which we investigated changes in neurotransmitters in the primate PFC in relation to cognitive task performance and reward (Watanabe et al. 1997; Kodama et al. 2002a, b, 2014, 2015).

Microdialysis which appears frequently in this chapter is based on the principle of dialysis. The semipermeable membrane separates brain interstitial fluid from the perfusing solution. If there is a difference in the concentration of any substance to which the membrane is permeable, the concentration gradient causes diffusion of the substance from the interstitial space into the dialysate (passing the dialysis membrane). Together with improvements in the detection system for neurotransmitters, the present microdialysis method has enabled us to measure various neurotransmitters in a small quantity, in vivo and in situ. The smallest probe now utilized is 0.22 mm in diameter and 0.5 mm in length. Taking the perfusion area of 0.5 mm into consideration, the research target must be bigger than 0.5 $mm^3$. This observation size is still quite large compared to the size of a neuron but is comparable to that of the local field potential (LFP). Thus, this method is advantageous when we are interested in the interaction between and among neurotransmitters.

DA released by exocytosis binds to DA receptors, which are located relatively distant from the site of release, to modulate the physiological function of the target neurons. All the DA receptors are G protein coupled and are responsible for the slow signal transfer or modification of neuronal activities. It is well known that most DA receptors are expressed outside the synapse and that a considerable proportion of DA receptors do not form a distinct synaptic structure. As the signal transduction carried by DA is based on diffusion (volume transmission), DA is thought to affect widely and relatively distant areas. Therefore, the microdialysis method is highly suitable for investigating the wide and slow integration of various neural inputs.

## 5.2 Important Roles of DA in the Prefrontal Cortex

DA in the PFC is known to be the most important neurotransmitter that mediates memory, reward, attention, and so on. The dysregulation of DA leads to a decline in cognitive functions, especially memory, attention, and problem-solving, and furthermore it is linked to psychiatric diseases, such as schizophrenia, and to developmental disorders, such as attention deficit hyperactivity disorder (ADHD). Antipsychotics mainly inhibit DA at the receptor level to improve the symptoms of psychosis. Adequate levels of DA in the brain, especially in the PFC, help to improve WM. However, this is on a delicate balance, and as DA levels increase or decrease to abnormal levels, cognitive functions are impaired (Williams and Castner 2000). DA is necessary for focusing and directing one's attention. In other words, DA may be important for determining what should be kept in the WM.

DA neurons are located in the substantia nigra (A8, A9), ventral tegmental area (VTA) (A10), posterior hypothalamus (A11), arcuate nucleus (A12), zona incerta (A13), and periventricular nucleus (A14). Of these areas, VTA DA neurons mainly project to the PFC via the neocortical pathway, while another smaller group projects to the nucleus accumbens (NAC) via the mesolimbic pathway, as well as sending minor projections to the amygdala, cingulate gyrus, hippocampus, and olfactory bulb (Björklund and Dunnett 2007; Malenka et al. 2009). Most DA-sensitive neurons are located in the deep (fifth and sixth) layers of the PFC, and half of them are excitatory.

The activity of DA neurons has both phasic and tonic properties. DA neurons respond to external stimuli, such as reward, in a phasic way, or show tonic activities in relation to internal or external environmental situations. It has hitherto been suggested that microdialysis estimates the overflow of DA accompanied by tonic activity, whereas voltammetry estimates that released by phasic activity. Recently, Di Chiara (2016) reported that DA release, as estimated by microdialysis, is largely accounted for by the burst firing of DA neurons. The amount of DA release accompanied by burst discharges is much larger than that by tonic ones. However, the DA released from the axon terminals by phasic discharges is retaken up by dopamine transporters (DAT) and quickly removed from the synaptic cleft in the striatum. On the other hand, as the density of DAT is low in the PFC compared to that in the striatum, a large portion of DA released from the axon terminal is overflowed outside the synaptic cleft. Taken together, DA release in the PFC that has been detected by microdialysis is considered to be related to the burst activity of DA neurons.

As described so far, the appropriate amount of DA released in the PFC is critical for normal brain function. Not only a deficiency of DA but also an abundance of DA causes cognitive malfunction (Williams and Castner 2000). It has been reported that there is an inverted U-shaped relationship between the performance of cognitive task and amount of DA in the PFC. Keeping the DA concentration at "an optimal level" is essential for the efficient work of the PFC. Therefore, information about changes in DA release is important to discuss higher cognitive functions, for example, memory, attention, and motivation.

There are five subtypes of DA receptor: $D_1$, $D_2$, $D_3$, $D_4$, and $D_5$. $D_1$ and $D_4$ receptors are responsible for the cognitive operation of DA in the PFC. Reduced DA concentrations and abnormally high $D_2$ receptor function in the PFC are supposed to be related to the symptoms of attention deficit hyperactivity disorder (ADHD), and low $D_2$ function, to be related to social anxiety or social phobia. In schizophrenia patients, it is reported that the binding of the $D_1$ receptor, which is related to negative symptoms, and the binding of the $D_2$ receptor, which is related to positive symptoms, decrease in the PFC and in the anterior cingulate cortex, respectively. (Okubo et al. 1997)

### 5.2.1 DA Changes in the Dorsolateral PFC (DLPFC)

**DA Changes Related to WM in the DLPFC**

In previous studies, the importance of PFC DA for WM was demonstrated by the effect of altered DA neurotransmission (Arnsten et al. 1994; Brozoski et al. 1979; Diamond 1996; Diamond et al. 1994; Luciana and Collins 1997; Murphy et al. 1996a, b; Sawaguchi and Goldman-Rakic 1991). In line with these studies, we provided direct evidence for an increase of DA in the dorsolateral PFC (DLPFC) during a typical WM task, a delayed alternation task. We examined DA changes in the three PFC areas (DLPFC; orbitofrontal cortex, OF; and arcuate cortex, ARC) and one non-PFC frontal area (premotor cortex, PM) (Fig. 5.1 left) (Watanabe et al. 1997). Basal levels for the DLPFC, PM, OF, and ARC areas were $0.098 \pm 0.013$, $0.109 \pm 0.017$, $0.116 \pm 0.021$, and $0.147 \pm 0.027$ fmol/μl (means ± SEM), respectively, and these differences were not statistically significant, although it is reported that there is a medial-lateral gradient of decreasing fiber density of DA in the dorsal cortical regions (Williams and Goldman-Rakic 1993). We revealed a significant difference in the extracellular DA concentration between the delayed alternation (WM) and sensory-guided control (non-WM) tasks in the primate DLPFC (17% increase), but not in the other frontal areas investigated (Watanabe et al. 1997). Compared with the basal value, significant increases in DA concentration were observed during the WM task in both the DLPFC and PM areas, but during the sensory-guided task, only in the PM area (Fig. 5.2a).

Furthermore, we compared changes in the DA concentration in detail among the DLPFC subareas, i.e., the dorsal gyrus (DLd), principalis (PS), and ventral gyrus (DLv) (Fig. 5.1 right). The DA concentration was significantly higher during the WM task than during the sensory-guided task in the DLd and DLv subareas, but not in the PS subarea (Fig. 5.2c).

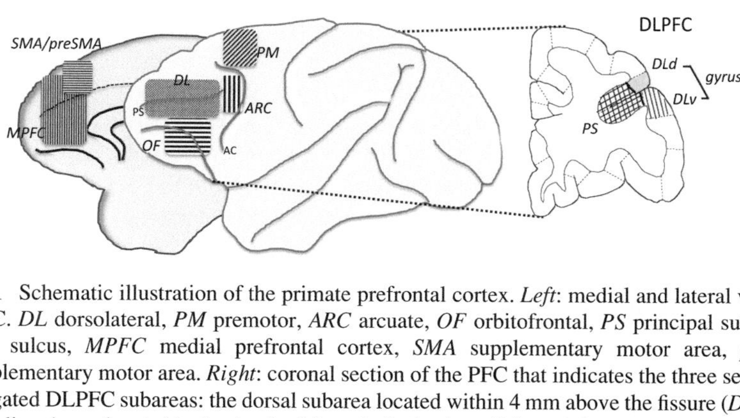


**Fig. 5.1** Schematic illustration of the primate prefrontal cortex. *Left*: medial and lateral views of the PFC. *DL* dorsolateral, *PM* premotor, *ARC* arcuate, *OF* orbitofrontal, *PS* principal sulcus, *AS* arcuate sulcus, *MPFC* medial prefrontal cortex, *SMA* supplementary motor area, *preSMA* presupplementary motor area. *Right*: coronal section of the PFC that indicates the three separately investigated DLPFC subareas: the dorsal subarea located within 4 mm above the fissure (*DLd*), the principalis subarea located in the depth of the principal sulcus (*PS*), and the ventral subarea located within 4 mm below the fissure (*DLv*)

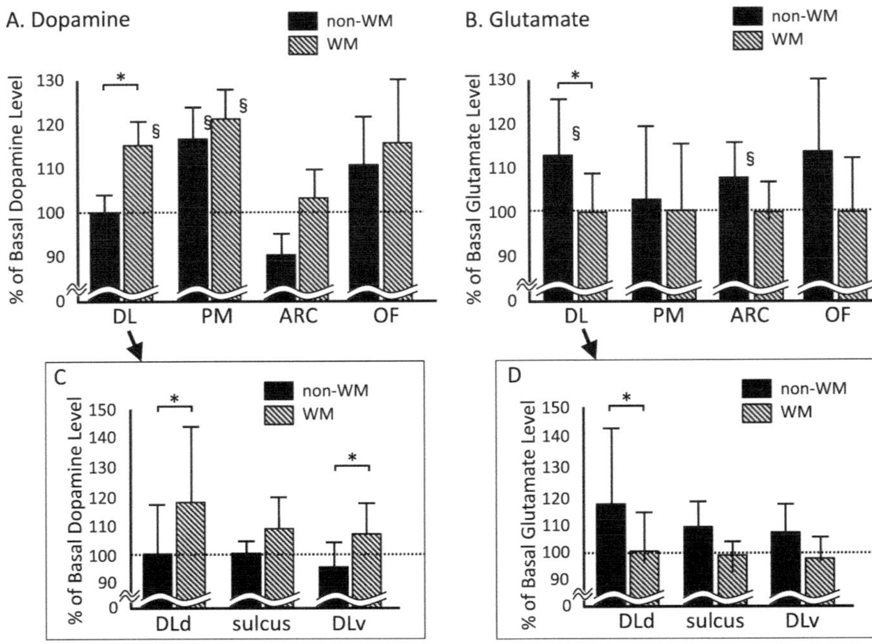


**Fig. 5.2** WM-related dopamine and glutamate changes in the frontal cortex. (**a**) Mean dopamine (DA) concentration (± SEM) expressed as a percent of the basal rest level during the WM (delayed alternation) task and non-WM (sensory-guided) task in four frontal areas, *DL*, *PM*, *ARC*, and *OF*. (**b**) Mean glutamate concentration (± SEM) expressed as a percent of the basal rest level during the WM task and non-WM task in four frontal areas, *DL*, *PM*, *ARC*, and *OF*. (**c**) Mean DA concentration (± SEM) expressed as a percent of the basal rest level during the WM task and non-WM task in the three DLPFC subareas, *DLd*, *PS*, and *DLv*. (**d**) Mean glutamate concentration (± SEM) expressed as a percent of the basal rest level during the WM task and non-WM task in the three DLPFC subareas, *DLd*, *PS*, and *DLv*. For (**a**–**d**), *filled bars* indicate data for the non-WM task (sensory-guided task), and *shaded bars* indicate data for the WM task (delayed alternation task). Abbreviations: *DL* dorsolateral prefrontal cortex, *PM* premotor cortex, *ARC* arcuate cortex, *OF* orbitofrontal cortex, *DLd* the dorsal subarea of the DLPFC, *PS* the principalis subarea of the DLPFC, *DLv* the ventral subarea of the DLPFC. *Asterisk* (*) indicates a statistically significant difference, and *section mark* (§) indicates a statistically significant difference from the basal resting level

Different from DLPFC neurons, midbrain DA neurons that innervate the DLPFC do not show sustained activities during the delay period in the delayed alternation task (Ljungberg et al. 1991). Although DA neurons do not respond to the predicted rewards in reaction time tasks (Schultz 1992), they are activated by the reward in delayed alternation, probably because the instructional components of reward delivery arouse the attention of the animal to guide its goal-directed behavior (Schultz 1992). This activation in turn may contribute to the increase in DA levels in the DLPFC. In this regard, the difference in attentional demand or that of task difficulty, besides that of the WM requirement between WM and sensory-guided (non-WM) tasks, may also have contributed to the observed difference in DA concentration between these two tasks.

In the primate PFC, deficient DA is detrimental for WM task performances (Brozoski et al. 1979; Sawaguchi and Goldman-Rakic 1991; Arnsten et al. 1994; Murphy et al. 1996a, b). Also, Sawaguchi et al. (1990a, b) indicated that the iontophoretic application of DA in the DLPFC induces activation in WM-related neuronal activity as well as inducing an improvement of the S/N ratio to background activity in DLPFC neurons. The increased DA concentration observed in our study may thus have worked to keep the spatial WM online during the delay period, by activating WM-related neuronal activity and by improving the S/N ratio.

## DA Changes Related to Reward Expectancy in the DLPFC

It has been well documented that WM-related DLPFC neuronal activity is enhanced when a preferred, compared with a less-preferred, reward is used (Watanabe 1996; Watanabe et al. 2002). Also, microinjection of an appropriate amount of DA or DA $D_1$ receptor agonist to the PFC enhances WM-related neuronal activity (Mantini et al. 2011; Cools and D'Esposito 2011). Thus, it was expected that there would be a higher DA release in the primate DLPFC when a more preferred, compared with a less-preferred, reward is delivered during a WM task.

Contrary to our expectation, compared with the DA concentration during the basal resting period, the concentration was significantly higher when water (less preferred), but not when juice (more preferred), was given as a reward during the WM task performance in the DLPFC. Furthermore, the DA concentration was significantly higher in the water than in the juice reward condition. Thus, DA release was higher in the DLPFC when a less rather than a more preferred reward was used during a WM task (Kodama et al. 2014) (Fig. 5.3a). It was also found that the increase in DA release related to rewards was higher in the gyrus than in the sulcus of the DLPFC (Fig. 5.3a), as a WM-related DA increase was higher in the gyrus than in the sulcus of the DLPFC (Watanabe et al. 1997).

Behaviorally, the monkey performed the task almost without error irrespective of the difference in the reward. However, the reaction time was significantly shorter in juice than in water reward trials, indicating that the monkey was more motivated to perform the task in juice reward trials. Although DA release in the PFC is indicated to be influenced by changes in the deprivation state of animals and incentive value of the reward (Kodama et al. 2002a), the DA concentration may not directly be concerned with the motivational level itself.

No change in DA release was observed during the sensory-guided task compared with during the basal resting period, although the same reward was used as during the WM task (Watanabe et al. 1997). It is indicated that the magnitude of DA release in the rat PFC predicts the accuracy of memory on a delayed response task (Phillips et al. 2004). It is further proposed that PFC DA plays a key role in overcoming response costs and enabling high-effort behaviors (Harrison et al. 2007; Garrity et al. 2007). It is speculated that when a monkey is less motivated to

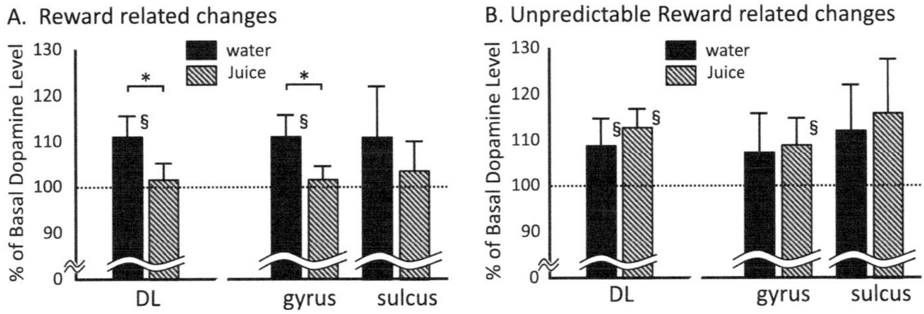


**Fig. 5.3** Reward-related dopamine changes in the dorsolateral prefrontal cortex. (**a**) Mean DA concentration (± SEM) expressed as a percent of the basal rest level during the WM task with a preferred (*diagonal bar*) and less-preferred (*black bar*) reward. (**b**) Mean DA concentration (± SEM) expressed as a percent of the basal rest level during the unpredictable reward delivery period with a preferred (*diagonal bar*) and less-preferred (*black bar*) reward. Abbreviations: *DL* dorsolateral prefrontal cortex. *Asterisk* (*) indicates a statistically significant difference, and *section mark* (§) indicates a statistically significant difference from the basal resting level

perform a task during the less-preferred reward condition, more psychological effort or more cognitive resources may be required to overcome the unfavorable situation of working for a less-preferred outcome. Such a cognitive load may have induced more DA release in the DLPFC, which in turn may have enabled the animal to overcome the unfavorable situation. The higher DA release in the DLPFC with a less-preferred reward may be beneficial for the monkey to cope with the mildly stressful and unfavorable situation for proficient WM task performance.

## DA Changes Related to Unpredictable Reward in the DLPFC

We examined also changes in DA release induced by unpredictable reward delivered after the daily WM task performance was over in the DLPFC (Kodama et al. 2014). There was a significant increase in the DA concentration during the unpredictable reward delivery period compared with that during the basal resting period irrespective of the difference in reward. When the DA concentration was examined separately for the gyrus and sulcus, a significant increase in the DA concentration was found only in the gyrus during the preferred reward delivery period (Fig. 5.3b).

Interestingly, a significant increase in DA release caused by unpredictable reward delivery was observed even though each monkey's motivation for liquid was not high after obtaining a certain amount of juice and water as a reward by performing the WM task. Furthermore, there was no significant difference in DA release between the two kinds of unpredictable reward even though the incentive values of these two kinds of reward were different. PFC DA activation is proposed to be over and above that provided by the passive receipt of rewards (Robbins and Arnsten 2009). As the higher DA release associated with the less-preferred reward during the WM task may not be caused by motivational factors, but by the difference in cognitive factors between the two different kinds of reward conditions, the significant increase in DA release by unpredictable reward delivery may be caused not by the delivery of the reward itself but by the cognitive demand for the monkey to cope with the uncertainty in reward delivery.

In summary, the roles of DA in motivational operations appear to be not as significant as those in cognitive operations in the DLPFC. Thus, DLPFC DA may be predominantly concerned with cognitive operations and may play a significant role in overcoming unfavorable situations.

## 5.2.2 DA Changes in the Medial PFC

### DA Changes Related to WM in the Medial PFC

The role of the medial PFC (MPFC) is reported to be concerned with memory (Takashima et al. 2006), reward-guided learning (Rushworth et al. 2001), decision-making (Botvinick et al. 2004), error detection (Holroyd et al. 2002), and executive control (Posner et al. 2007; Ridderinkhof et al. 2004). We were interested in examining neurotransmitter release in the MPFC during WM task performance compared with that during rest. We also examined neurotransmitter release in the medial motor areas (supplementary motor areas, SMAs, and presupplementary motor areas, preSMAs) as non-PFC control areas. The mean DA concentration during the basal resting period was $0.247 \pm 0.077$ fmol/μl (means $\pm$ SEM) in the MPFC and $0.121 \pm 0.035$ fmol/μl (means $\pm$ SEM) in the SMA/preSMA. We found a significant decrease in DA release in the MPFC and a significant increase in DA release in the SMA/preSMA during the WM task compared with that during rest. Furthermore, we observed significant differences in DA release during the WM task between the MPFC and SMA/preSMA (Fig. 5.4a).

### DA Changes Related to Unpredictable Reward Delivery in the MPFC

We also examined DA changes in relation to unpredictable reward delivery in the MPFC and SMA/preSMA. We observed a significant decrease in DA release in the MPFC and a significant increase in DA release in the SMA/preSMA during the unpredictable reward delivery period (Fig. 5.4b), compared with that during rest. There was a significant difference in the DA change between the MPFC and SMA/preSMA during the unpredictable reward delivery period. As indicated in Fig. 5.3b, we observed an increase in DA in the DLPFC with unpredictable reward delivery compared with that during the basal resting period. Because monkeys could not predict the time of reward delivery, they were considered to be continuously attentive to reward delivery. The higher level of attention caused by

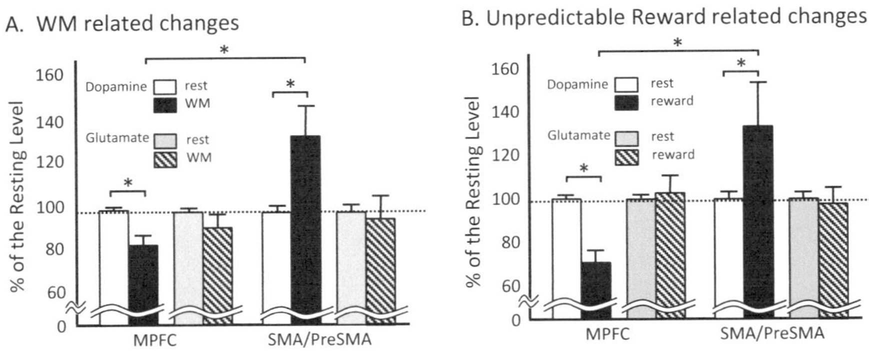


**Fig. 5.4** Working memory-related and unpredictable reward-related dopamine and glutamate changes in the medial prefrontal cortex. (**a**) Mean DA and glutamate concentration (± SEM) expressed as a percent of the basal rest level in the MPFC and SMA/preSMA during the resting period (*white* for DA and *gray* for glutamate) and during the WM task (*black* for DA and diagonal for glutamate). (**b**) Mean DA and glutamate concentration (± SEM) expressed as a percent of the basal rest level in the MPFC and the SMA/preSMA during the resting period (*white* for DA and *gray* for glutamate) and during the unpredictable reward delivery period (*black* for DA and diagonal for glutamate). For both (**a**, **b**), an *asterisk* (*) indicates a statistically significant difference

unpredictable reward delivery may have induced an increase and a decrease in DA release in the DLPFC and MPFC, respectively.

## DA Changes Related to the Default Mode of Brain Activity

Human functional brain imaging studies, such as PET and functional magnetic resonance imaging (fMRI), have well documented that there are areas in the brain that show task-induced activity decreases (default mode of brain activity)(Buckner et al. 2008; Gusnard and Raichle 2001) (see also Chapter 12. by M. Watanabe). These areas consist of mainly medial prefrontal and medial parietal areas (default system), and the MPFC constitutes the anterior default system. Our study showing WM-related as well as unpredictable reward-delivery-related decreases in DA release in the MPFC (Kodama et al. 2015) (Fig. 5.4) indicates increased DA release in the monkey anterior default system during rest compared with that during a cognitively demanding or unpredictable situation. An increase in regional cerebral blood flow (rCBF) in the monkey DLPFC has been reported during the WM task (Inoue et al. 2004), and increased DA release in the DLPFC is considered to play an important role in WM task performance (Watanabe et al. 1997). Human neuroimaging studies have indicated that the default mode of brain activity may be concerned with internal thought processes (Buckner et al. 2008). Thus, the increase in both rCBF (Kojima et al. 2009) and DA release (Kodama et al. 2015) in the monkey MPFC during rest compared with that during a WM task may indicate that the mental operations, such as internal thoughts, performed in the MPFC during rest may be supported by an increased DA release.

Conversely, in the SMA/preSMA, which is not a part of the monkey default system, there were increases in DA release during the WM task and unpredictable reward delivery periods compared with that during rest. These results are similar to those obtained in our previous study in the lateral premotor area (see Fig. 5.2), which is another non-default frontal area (Watanabe et al. 1997).

## 5.3 Important Roles of Glutamate in the Prefrontal Cortex

Glutamate is the major neurotransmitter mediating information in the mammalian brain, and glutamatergic neurotransmission in the PFC is related to cognitive functions (Ghoneim et al. 1985; Krystal et al. 1994; Malhotra et al. 1996). Glutamate in the PFC plays an important role in WM task performance both in the rat (Moghaddam et al. 1997; Romanides et al. 1999; Verma and Moghaddam 1996) and in the monkey (Dudkin et al. 1996). On the other hand, the noncompetitive N-methyl-D-aspartate (NMDA) antagonists, ketamine and MK-801, dose-dependently impair WM task performance in the rat (Verma and Moghaddam 1996). Decreased glutamatergic neurotransmission plays a role in the pathophysiology of psychosis (Kim et al. 1980; Riederer et al. 1992). NMDA receptor hypofunction may be related to the underlying pathological process in schizophrenia in combination with DA system dysfunction (Olney and Farber 1995).

Glutamate plays a role not only in the local PFC circuit but also in the excitatory connectivity of the PFC with the striatum, thalamus, and limbic structures. The integration of glutamatergic inputs from the cortical and subcortical areas is essential for goal-directed behavior (Miller 2000). The complexity of neuronal pathways involved, combined with the multifarious effects glutamate could mediate via pre-and postsynaptic interactions with various receptor subtypes, have led to controversies regarding the exact role glutamate plays in psychiatric diseases. Abnormalities of glutamate neurotransmission can cause cognitive deficits and negative symptoms. Also autoregulation of glutamate neurons can reduce DA transmission too much, which causes cognitive deficits and makes negative symptoms worse. The DA excess derived secondarily by glutamate excess is believed to cause positive symptoms. We thus examined changes in extracellular glutamate in the primate PFC in relation to WM task performance.

### 5.3.1 Glutamate Changes in Relation to Sensory Processing in the DLPFC

Glutamate concentrations were examined in the same study where WM-related DA changes were investigated in the four frontal areas: DLPFC, PM, OF, and ARC areas. Basal levels for the DLPFC, PM, OF, and ARC areas were 2.03 ± 0.18, 2.21 ± 0.34, 1.29 ± 0.15, and 2.03 ± 0.15 pmol/μl (means ± SEM), respectively. There was no significant difference in the basal glutamate level among these four frontal areas. There was a significant difference in the glutamate concentration between the sensory-guided and WM tasks only in the DLPFC. Compared to the basal resting level, there was almost no change in glutamate concentration during the WM task in all the four frontal areas, whereas the concentration during the non-WM task (sensory-guided control task) was significantly higher in the DLPFC and ARC areas (Fig. 5.2b).

Taken together, significant increases in the glutamate concentration compared with the basal resting level were observed in the DLPFC and ARC areas during the sensory-guided task, but no change was observed in any area examined during the WM task.

Since the DLPFC appears to be essential for WM tasks (Mishkin 1957), and WM-related DA increases have been observed in the gyrus but not in the depth area of the principal sulcus in the DLPFC (Watanabe et al. 1997), glutamate levels were further compared among the three DLPFC subareas: dorsal (DLd), principalis (PS), and ventral (DLv). There was a significant difference in the glutamate concentration between the sensory-guided and WM tasks in the DLv area but not in the PS and DLd areas (Fig. 5.2d).

Because spatial information was not explicitly indicated to the animal but was determined by the animal's previous response to the right or left in this WM task (delayed alternation), the animal had to retain the spatial information during the delay period. On the other hand, during the non-WM, sensory-guided control task, the animal had to pay more attention to the visual cue that explicitly indicated the correct response to the animal. Interestingly, Oye et al. (1992) suggested that the impairment in human memory tasks induced by the application of an NMDA antagonist is not an impairment of memory but rather is produced by a general attenuation of sensory input. If this interpretation is correct and if glutamate in the PFC is not so involved in memory processes but is more concerned with sensory processing, an increased glutamate concentration concomitant with the lack of change in DA concentration during the sensory-guided control task in this experiment may have facilitated processing of visual information more effectively, possibly by directing more attention to the cue stimulus. In the same vein, the absence of an increase in glutamate during the WM task is speculated to be beneficial for the retention of important task information by not allowing too much irrelevant sensory input to enter the PFC.

According to Funahashi et al. (1990), population activities of primate DLPFC neurons appear to be much higher during the cue period than during the delay period in an oculomotor delayed response task, which is another kind of spatial WM task. The increased glutamate concentration observed during the non-WM task may be associated with higher DLPFC neuronal activities related to the visual cue stimulus indicating the correct response side to the animal.

In summary, the results are considered to indicate the importance of glutamate, not in retaining information in WM but in processing sensory information in the primate DLPFC and ARC areas. As there appears to be an optimal level of DA release in the PFC for WM task performance (Arnsten et al. 1994; Murphy et al. 1996a, b), there may be an optimal level of glutamate in the PFC for the cognitive task, whose level may be higher for task situations that demand sensory processing than for WM tasks.

### 5.3.2 *Glutamate Changes in Relation to WM and Unpredictable Reward in the MPFC*

Glutamate concentrations were also examined in the MPFC and medial frontal motor areas (SMA and preSMA) in relation to WM task performance (Kodama et al. 2015). The mean glutamate concentration during the basal resting period was $2.32 \pm 0.43$ pmol/μl (means ± SEM) in the MPFC and $4.75 \pm 1.36$ pmol/μl in the SMA/preSMA. There was no significant change in glutamate release during the WM task compared with during rest in both the MPFC and SMA/preSMA (Fig. 5.4a). There was also no significant change in both the MPFC and SMA/preSMA during the unpredictable reward delivery period compared with during rest (Fig. 5.4b).

## 5.4 Interactions of DA and Glutamate in the PFC

In the rat, it is suggested that abnormal basal or stimulus-activated DA neurotransmission in the PFC may be caused by glutamatergic dysregulation (Takahashi and Moghaddam 1998). It is further suggested that attenuation of glutamatergic neurotransmission at the NMDA receptor may cause activation of DA neurotransmission and concomitantly impair PFC-dependent cognitive functions (Verma and Moghaddam 1996). Here we discuss possible mechanisms of interaction between glutamate and DA in the primate PFC (Fig. 5.5).

There are two possible mechanisms. The first possible mechanism to control glutamate and DA release in the PFC is through feedback circuits from the PFC to the thalamus and the VTA. Neurons in the rat PFC receive inputs from the other areas of the cortex or from the thalamus. They receive inputs directly to the synapse of the pyramidal cells at the fourth layer or receive inputs by synapse contacting the

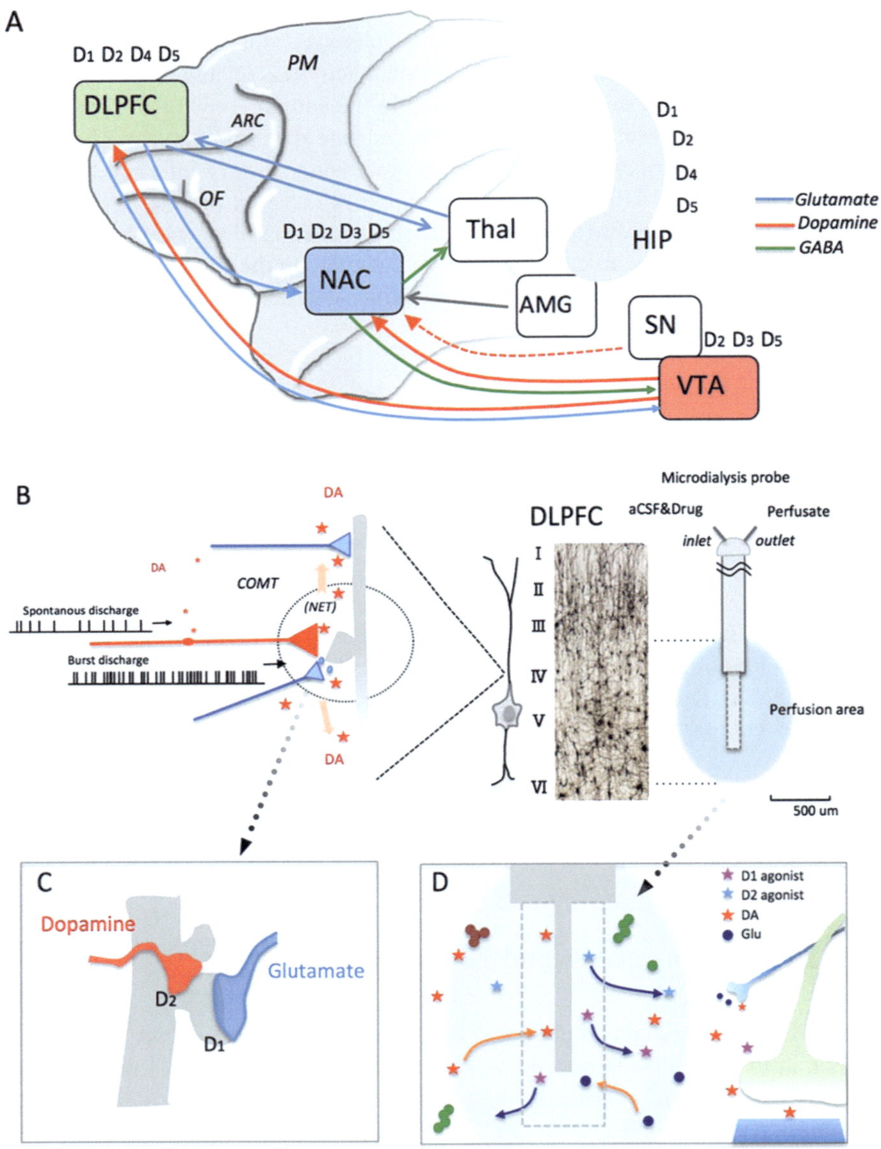


**Fig. 5.5** Circuit inside the PFC and the loop between the PFC and subcortical structures. Schematic illustration of the interaction between glutamate and DA inside PFC and among the PFC and the subcortical structures: (**a**) Major DA inputs to the PFC come from the VTA, and PFC neurons send the glutamatergic output to the VTA. PFC neurons also send glutamatergic projection to the nucleus accumbens (*NAC*), and the NAC regulates the VTA via GABAergic projection. There is a loop regulation between the PFC and VTA, glutamatergic-dopaminergic interaction. (**b**) *Left*: DA receptors are expressed outside the synapse and do not form a distinct synaptic structure. DA signal transduction is based on diffusion. *Right*: cortical lamination in DLPFC (Golgi staining) and approximate size of a microdialysis probe and estimated perfusion area. (**c**) Glutamate and DA may interact directly in the PFC. DA terminals synapse onto the dendritic spines, which also received converging synaptic glutamate inputs. This synaptic "triad" has been postulated to represent the anatomical substrate for the local modulation of excitatory inputs by DA. (**d**)

dendrite at the first layer. And they send the outputs to the thalamus from the sixth layers or send the outputs inside the cortex from the second/third layers (Fig. 5.5a).

It is well known that there is a closed loop in the information flow from the PFC to the basal ganglia, to the thalamus, and back to the PFC (Alexander et al. 1986). Also, there are reciprocal fiber connections between the PFC and VTA and between the PFC and limbic system (Goldman-Rakic 1987). The VTA also sends outputs to the striatum and limbic system (Oades and Halliday 1987), which in turn send information to the PFC (Goldman-Rakic 1987). Major DA inputs to the PFC come from the VTA, and PFC neurons send the glutamatergic output to the VTA. PFC neurons also send glutamatergic projection to the nucleus accumbens (NAC), and then the NAC regulates the VTA via GABAergic projection. This may be too simplified, but, roughly speaking, there is a loop regulation between the PFC and VTA, i.e., glutamatergic-dopaminergic interaction (Fig. 5.5a).

The second possible mechanism is the direct interaction between glutamate and DA in the PFC. DA released in the PFC binds to the DA receptors, which are located relatively distant from the site of release, to modulate neuronal transductions. DA receptors are expressed outside the synapse, and a considerable proportion of DA receptors do not form a distinct synaptic structure. Released dopamine is mainly removed by the norepinephrine transporter (NET) or catechol-O-methyl-transferase (COMT) in place of DAT in the PFC. DA inputs from VTA not only transfer to the local synaptic gap but distribute widely within the PFC to control the glutamate inputs (Fig. 5.5b). Thus, DA is thought to affect widely on a relatively longtime scale. By this special circumstance of the PFC, DA works as a modulator of the wide and slow integration of various neural inputs, especially that of glutamate (Fig. 5.5b).

It has also been shown that DA terminals in the primate PFC synapse onto the dendritic spines, which also received converging synaptic input from excitatory, glutamatergic afferents (Smiley and Goldman- Rakic 1993; Smiley et al. 1992). This synaptic "triad" has been postulated to represent the anatomical substrate for the local modulation of excitatory inputs by DA (Goldman-Rakic and Selemon 1997). In the primate PFC, it is suggested that DA regulates the excitability of the glutamatergic pyramidal neuron (Goldman-Rakic 1999) (Fig. 5.5c).

Malfunction of the $D_1$ receptor in the PFC and $D_2$ receptor in the anterior cingulate cortex is considered to be related to schizophrenia (Okubo et al. 1997). On the other hand, the cause of schizophrenia has also been speculated to be due to abnormalities of the NMDA receptor. Glutamate neurons in the PFC have an inhibitory control on DA release. The decrease of glutamate activities results in a DA increase, and this increase causes disinhibition in the limbic system and basal

**Fig. 5.5** (continued) Schematic illustration of the reverse microdialysis principle. Reverse dialysis is one of the methods that can be used to investigate changes in neurotransmitters by the perfusion of chemicals. When the substance is added to the perfusion liquid at a higher concentration than the interstitial fluid, the substance diffuses out of the membrane and into the tissue around the membrane

ganglia. This hyperexcitability would induce the strong anxiety and hypersensitivities observed in schizophrenia (Spencer et al. 2004).

### *5.4.1 Interaction of Glutamate and DA During Task Performance in the DLPFC*

As described, we observed differential changes between glutamate and DA depending on the task: there was an increase in glutamate but no change in DA during the sensory-guided task, whereas there was an increase in DA but no change in glutamate during the WM task (Fig. 5.6). There was also a significant difference in the percent change from the basal level between the two transmitters (Kodama et al. 2002a). It is thus suggested that in the primate DLPFC, increased glutamate tone without a DA increase facilitates sensory-guided task performance, while increased DA tone without a glutamate increase is beneficial for WM task performance, although previous studies on rats using direct stimulation of the PFC by chemical substances indicated that reduced glutamate tone and increased DA tone disrupt

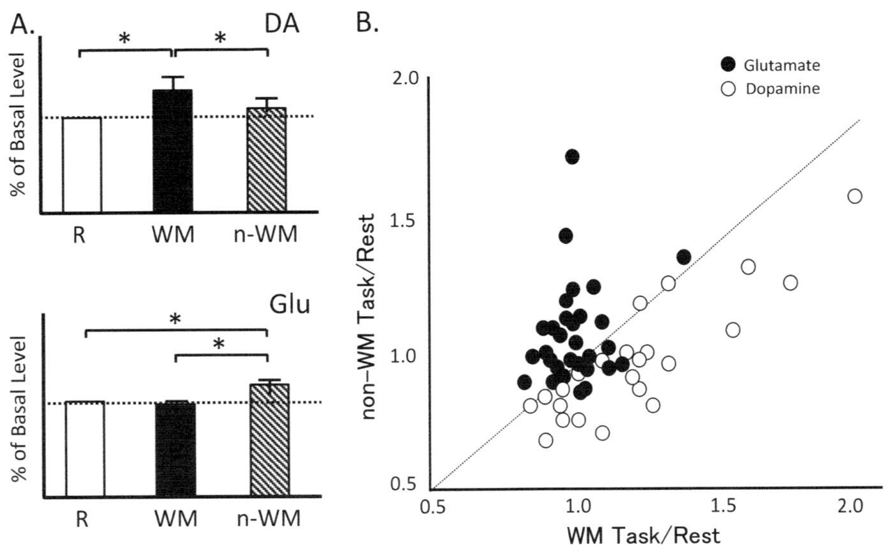


**Fig. 5.6** Dopamine-glutamate interaction in the dorsolateral prefrontal cortex. (**a**) Mean dopamine and glutamate concentration (± SEM) expressed as a percent of the basal rest level in relation to WM and non-WM tasks in the DLPFC. *Asterisk* (*) indicates a significant difference. (**b**) Relationship between the DA and glutamate concentration in relation to the WM task. Concentrations are expressed as percents of the basal resting level for DA (*unfilled circle*) and glutamate (*filled circle*) during the WM task (abscissa) and non-WM task (ordinate). Abbreviations: *R* rest period, *WM* working memory task (delayed alternation task), *n-WM* nonworking memory task. *Asterisk* (*) indicates a statistically significant difference

WM in a spatial delayed alternation task (Romanides et al. 1999; Verma and Moghaddam 1996).

The interesting point to be noted is the segregation in the changes of glutamate and DA in the PFC. Significant changes in DA in relation to the WM task were observed only in a limited area of the PFC, i.e., the DLPFC. Significant changes in glutamate in relation to the sensory-guided task were observed only in the DLPFC and arcuate PFC and not in the OF and PM areas.

### 5.4.2 *Interaction of Glutamate and DA During Task Performance in the MPFC*

Compared with the basal resting level, we found a significant decrease in DA release without an associated change in glutamate release in the MPFC, whereas we found a significant increase in DA release without an associated change in glutamate release in the SMS/preSMA during the WM task as well as during the unpredictable reward delivery period (Kodama et al. 2015) (Fig. 5.4). Furthermore, there were significant differences in the percent change between DA and glutamate release in the MPFC during the unpredictable reward delivery period and in the SMS/preSMA during both the WM task and the unpredictable reward delivery periods (Fig. 5.4) (Kodama et al. 2015). Therefore, there was a similar inhibitory relationship between the DA and glutamate release in the MPFC as previously observed in the DLPFC.

### 5.4.3 *Inhibitory Interaction Between Glutamate and DA*

The glutamate increase without a change in DA during the sensory-guided task and the DA increase without a change in glutamate during the WM task suggest that there are some inhibitory interactions between these two transmitters in relation to cognitive tasks. Indeed, previous studies on rats indicated interactions between glutamate and DA in the PFC. Takahashi and Moghaddam (1998) showed that a glutamate AMPA antagonist produced a reduction of nearly 40% in the DA level and activation of AMPA receptors enhanced DA release, whereas the NMDA antagonist AP5 increased the release of DA in the PFC. Similarly, it has been shown that stimulation of NMDA receptors in the rat PFC reduces the basal release of DA, and a blockade of NMDA receptors increases DA release in this region (Feenstra et al. 1995; Wedzony et al. 1993; Nishijima et al. 1994; Verma and Moghaddam 1996; Hata et al. 1990; Del Arco and Mora 1999). Jedema and Moghaddam (1996) showed that local infusion of AMPA or kainate produced a significant increase in extracellular levels of DA, and AMPA/kainate receptor antagonists blocked this increase, while NMDA infusion either did not increase

DA or even decreased it. These data suggest that there is an inhibitory control of DA release mediated by NMDA receptors and facilitatory control by AMPA/kainate receptors localized in the PFC.

However, DA also modulates glutamate transmission in the PFC. Cepeda et al. (1992) reported that the DA release regulated glutamate release in a slice of human PFC. DA agonists, amphetamine (Del Arco et al. 1998) and apomorphine (Porras et al. 1997), increase the extracellular concentration of glutamate in the rat PFC. On the other hand, Abekawa et al. (2000) showed that application of the $D_1$ selective DA agonist, SKF38393, reduced glutamate concentration in the rat PFC.

## 5.5 Pharmacological Approaches to Investigate DA-Glutamate Interaction

Much knowledge has been accumulated by pharmacological stimulation experiments using the "reverse dialysis method." In this method, when the substance is added to a perfusion liquid at a higher concentration than the interstitial fluid, the substance diffuses out of the membrane and into the tissue around the membrane (Fig. 5.5d). While the iontophoresis technique is the often used method to investigate neural responses with chemical application (postsynaptic effect), the reverse dialysis method is the best method to investigate presynaptic neurotransmitter changes with chemical application (presynaptic effect). Reverse dialysis is an easier and much more stable method to apply chemicals to the neurons than microinjection. Using iontophoresis, there are a couple of important reports about DA effects on the WM task in the primate (Sawaguchi et al. 1990a, b). However, there are few reports about presynaptic neurotransmitter changes in the primate. Here we show some unpublished preliminary data regarding changes in glutamate release by application of DA and its analogue using the reverse dialysis technique.

### 5.5.1 *Effects of DA on the Extracellular Glutamate Level in the Monkey DLPFC*

The application of DA (1 mM) decreased the extracellular glutamate level to 76% of the control. Figure 5.7a shows the mean glutamate level during the control period and during the application of DA (1 mM). The decrease of the extracellular glutamate level recovered within 20 min after the removal of DA.

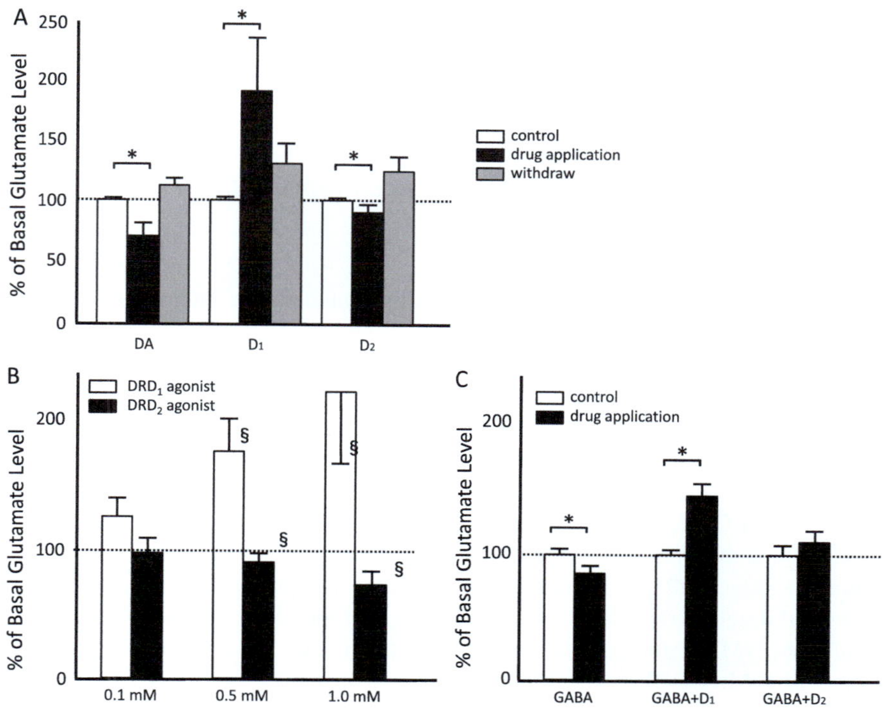


**Fig. 5.7** Neurotransmitters' interaction in the dorsolateral prefrontal cortex. (**a**) Mean glutamate concentration (± SEM) expressed as a percent of the control level, during the control period as well as during the application of DA, and dopamine receptor-related agonists ($DRD_1$ or $DRD_2$). DA, $DRD_1$, and $DRD_2$ were applied for 5 min., each for three times. (**b**) Mean glutamate concentration (± SEM) expressed as a percent of the control level during the single 5 min application of $DRD_1$ and $DRD_2$ agonists at various concentrations (0.1, 0.5, 1.0 mM). (**c**) Mean glutamate concentration (± SEM) expressed as a percent of the control level during the single 5 min application of GABA, GABA + $DRD_1$ agonists, and GABA + $DRD_2$ agonists. *Asterisk* (*) indicates a statistically significant difference, and *section mark* (§) indicates a statistically significant difference from the resting level

### 5.5.2 *Effects of DA Receptor Agonist on the Extracellular Glutamate Level in the Monkey DLPFC*

Application of the $D_1$ agonist SKF38393 (1.0 mM) increased the mean glutamate level, and application of the $D_2$ receptor agonist quinelorane (1.0 mM) decreased the mean glutamate level (Fig. 5.7a). These changes in the extracellular glutamate level recovered within 20 min during the removal of the DA agonists. Figure 5.7b shows the extracellular glutamate level during the single 5-min application of the $D_1$ and $D_2$ agonists at various concentrations (0.1, 0.5, 1.0 mM). We observed a significant change in the glutamate level during the application of the $D_1$ agonist at 0.5 mM and 1 mM and the application of the $D_2$ agonist at 0.5 mM and 1 mM but did not find a significant change in the glutamate level during the application of 0.1 mM of the $D_1$ and $D_2$ agonists.

### 5.5.3 *Effects of GABA on the Extracellular Glutamate Level in the Monkey DLPFC*

There could be other possible mechanisms that mediate the interaction between glutamate and DA in the DLPFC. Figure 5.7c shows the extracellular glutamate level during the application of GABA (1 mM), during the application of GABA + the $D_1$ receptor agonist SKF38393 (1 mM), and during the application of GABA + the $D_2$ receptor agonist quinelorane (1 mM). We observed significant changes in the glutamate level during the application of GABA and GABA + $D_1$ agonist compared with during the control period but did not find a significant change during the application of GABA + $D_2$ agonist.

There are many GABAergic inhibitory interneurons in the PFC (Smiley and Goldman-Rakic 1993). The interaction between the glutamate and DA may be attained through these GABA interneurons in addition to the direct DA-glutamate regulation. NMDA antagonists are reported to decrease the extracellular concentration of GABA in the rat PFC (Yonezawa et al. 1998), while increased endogenous extracellular glutamate increases extracellular GABA (Del Arcos and Mora 1999). Thus, glutamate has facilitatory effects on GABA. Santiago et al. (1993) reported that GABA-receptor agonists inhibit, whereas its antagonists facilitate, the release of DA in the rat PFC. Thus, stimulation of GABA release by NMDA, or attenuation of GABA release by NMDA receptor antagonists could, in turn, inhibit or increase DA release in the PFC, respectively (Jedema and Moghaddam 1996). It is suggested that in the rat PFC, GABA could modulate DA release through GABAergic receptor sites localized in DA terminals (Jedema and Moghaddam 1996). Further studies are needed to investigate the role of GABA in the modulation of DA release in the primate.

## 5.6 The Role of Other Neurotransmitters in the PFC

Both the excitatory inputs (glutamate and acetylcholine) and inhibitory inputs (GABA, serotonin, and norepinephrine) play a regulatory role in cognition. And, most importantly, these substances work while mutually affecting each other. Sensitivities of DLPFC neurons to these transmitters are reported to be not uniform among cortical layers (Sawaguchi and Matsumura 1985), and thus, each transmitter may regulate neuronal activities of particular layers of the monkey DLPFC. Their well-regulated releases are considered necessary for keeping the cognitive activity in good condition.

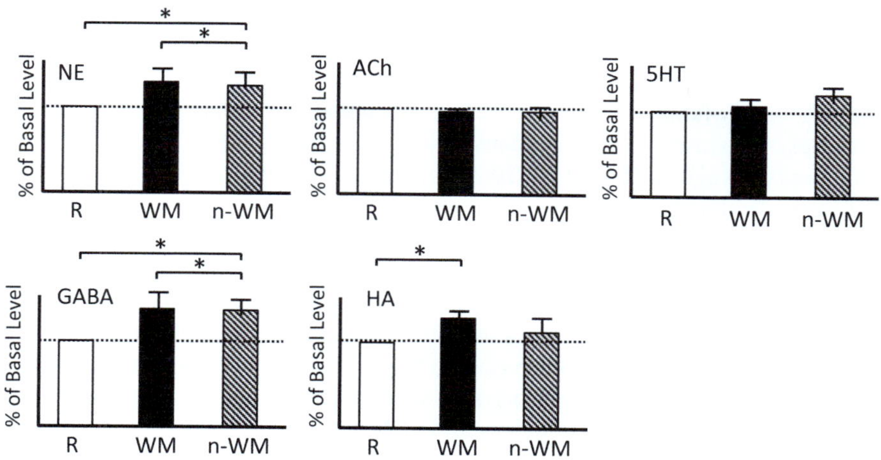


**Fig. 5.8** Other neurotransmitters' changes in the dorsolateral prefrontal cortex. Mean concentration (± SEM) expressed as a percent of the basal rest level for noradrenalin (*NE*), acetylcholine (*ACh*), serotonin (*5HT*), gamma-aminobutyric acid (*GABA*), and histamine (*HA*). The levels of NE, GABA, and histamine were significantly higher during the WM task than during the resting period. There were no task-related changes in the level of serotonin or acetylcholine. Abbreviations: *R* rest period, *WM* working memory task (delayed alternation task), *n-WM* nonworking memory task. *Asterisk* (*) indicates a statistically significant difference

Noradrenergic projections are diffuse through the cortex, and norepinephrine (NE) concentrations exceed those of DA in the rat PFC. NE works under the regulation of an inverted U-curve as does DA. The majority of PFC NE-sensitive neurons are reported to be NE-inhibited neurons and are located in layers III and IV. NE works as a global modulator of vigilance level and also plays an important role for a specific function of the DLPFC, such as WM. In our study, levels of NE were significantly higher during the WM and sensory-guided tasks than during the resting period (Kodama et al. 2002b) (Fig. 5.8) in the DLPFC.

The acetylcholine (ACh) projects from the basal forebrain and brainstem to the entire cortex, especially to the PFC, and is another important excitatory input related to cognitive function such as Alzheimer's disease. In the monkey DLPFC, half of iontophoretically tested neurons are ACh sensitive: one fourth are excited and located mainly in layers III and V, and one thirds are inhibited and located in layers III and IV (Sawaguchi and Matsumura 1985). ACh may be involved in the excitatory process in the DLPFC, playing an important role in the cognitive functions of attention and discrimination by keeping the vigilance level high (Arnsten and Robbins 2002). However, unlike DA or NE, there were no task-related changes in the levels of ACh in the DLPFC in our study (Kodama et al. 2002b) (Fig. 5.8).

Serotonin in the PFC is reported to function as a key in reversal learning (Robbins and Roberts 2007). Serotonin interacts with other neurotransmitters in the PFC, as DA and NE do. The OF exerts some of its control on emotion and impulsivity through serotonin neurons. However, in the DLPFC, there were no task-related changes in the level of serotonin in our study (Kodama et al. 2002b) (Fig. 5.8).

GABA is the prime inhibitory neurotransmitter. In the PFC, GABA supports local inhibitory functions, including lateral inhibition that enhances the saliency of excitatory responses. GABA-mediated inhibition is also critical for the cognitive processes (Sawaguchi et al. 1988, 1989; Wilson et al. 1994; Rao et al. 1999, 2000). For example, during tasks that require spatial WM, fast-spiking GABA neurons in the monkey DLPFC display delay-period activity that is selective for memoranda in specific spatial locations (Wilson et al. 1994; Rao et al. 1999, 2000). In our study, levels of GABA during the WM task were significantly higher than those during rest and during the non-WM task in the DLPFC (Kodama et al. 2002b) (Fig. 5.8).

Histamine is thought to be involved in the inflammatory response and has a central role as a mediator of pruritus. However, increasing evidence supports that it is involved in many regulatory functions: in CNS, histamine works as a sleep-wake regulator to maintain consciousness (Thakkar 2011) and plays an important role in memory and learning (Passani et al. 2000). Our data also showed that histamine levels during the WM task were significantly higher than those during rest (Fig. 5.8).

## 5.7 Conclusion

The PFC is the place where cognitive information is integrated. Various neurotransmitters, such as DA, glutamate, serotonin, acetylcholine, norepinephrine, histamine, and GABA, interact with each other to produce optimal conditions for higher brain function. Among these neurotransmitters, DA plays a particularly important role by modifying glutamate inputs to the PFC. DA-glutamate interaction in the PFC is produced by two patterns: one is the local synaptic interaction, and the other is the loop among the PFC and subcortical structures (VTA, thalamus, and basal ganglia) (Fig. 5.5). Local dopamine-glutamate interaction focused here is caused by the characteristic structure of the PFC: the weak function of the dopamine transporter DAT compared to the striatum. In the PFC, released DA is mainly removed by the NET or COMT in place of DAT. DA inputs from VTA not only transfer to the local synaptic gap but also distribute widely within the PFC to control the glutamate inputs. This relatively loose control is important for retrieving necessary information from a large amount of cognitive information coming to the PFC. This is also important for the allocation of limited resources in performing cognitive activities.

## References

Abekawa T, Ohmori T, Ito K, Koyama T (2000) D1 dopamine receptor activation reduces extracellular glutamate and GABA concentrations in the medial prefrontal cortex. Brain Res 867: 250–254

Alexander GE, DeLong MR, Strick PL (1986) Parallel organization of functionally segregated circuits linking basal ganglia and cortex. Annu Rev Neurosci 9:357–381

Arnsten AF, Robbins TW (2002) Neurochemical modulation of prefrontal cortical function in humans and animals. In: Stuss DT, Knight RT (eds) Principles of frontal lobe function. Oxford University Press, Oxford, pp 51–84

Arnsten AFT, Cai JX, Murphy BL, Goldman-Rakic PS (1994) Dopamine D1 receptor mechanisms in the cognitive performance of young adult and aged monkeys. Psychopharmacology 116: 143–151

Björklund A, Dunnett SB (2007) Dopamine neuron systems in the brain: an update. Trends Neurosci 30(5):194–202

Botvinick MM, Cohen JD, Carter CS (2004) Conflict monitoring and anterior cingulate cortex: an update. Trends Cogn Sci 8:539–546

Brozoski TJ, Brown RM, Rosvold HE, Goldman PS (1979) Cognitive deficit caused by regional depletion of dopamine in prefrontal cortex of rhesus monkey. Science 205(4409):929–932

Buckner RL, Andrews-Hanna JR, Schacter DL (2008) The brain's default network: anatomy, function, and relevance to disease. Ann N Y Acad Sci 1124:1–38

Cepeda C, Radisavljevic Z, Peacock W, Levine MS, Buchwald NA (1992) Differential modulation by dopamine of responses evoked by excitatory amino acids in human cortex. Synapse 11: 330–341

Cools R, D'Esposito M (2011) Inverted-U-shaped dopamine actions on human working memory and cognitive control. Biol Psychiatry 69:e113–e125

Del Arco A, Mora F (1999) Effects of endogenous glutamate on extracellular concentrations of GABA, dopamine and dopamine metabolites in the prefrontal cortex of the freely moving rat: involvement of NMDA and AMPA/KA receptors. Neurochem Res 24:1027–1035

Del Arco A, Martinez R, Mora F (1998) Amphetamine increases extracellular concentrations of glutamate in the prefrontal cortex of the awake rat: a microdialysis study. Neurochem Res 23: 1153–1158

Di Chiara G (2016) https://ca.pittcon.org/Technical+Program/tpabstra16.nsf/Agenda+Time+Slots+Web/15BE89CBF2D44B5485257EC5004D3086?Opendocument/

Diamond A (1996) Evidence for the importance of dopamine for prefrontal cortex functions early in life. Philos Trans R Soc Lond Ser B Biol Sci 351:1483–1494

Diamond A, Ciaramitaro V, Donner E, Djali S, Robinson MB (1994) An animal model of early-treated PKU

Dudkin KN, Kruchinin VK, Chueva IV (1996) Neurophysiological correlates of improvements in cognitive characteristics in monkeys during modification of NMDA-ergic structures of the prefrontal cortex. Neurosci Behav Physiol 26:545–551

Feenstra MGP, Van der Weij W, Botterblom MHA (1995) Concentration-dependent dual action of locally applied N-methyl-D-aspartate on extracellular dopamine in the rat prefrontal cortex in vivo. Neurosci Lett 201:175–178

Funahashi S, Bruce CJ, Goldman-Rakic PS (1990) Visuospatial coding in primate prefrontal neurons revealed by oculomotor paradigms. J Neurophysiol 63:814–831

Garrity AG, Pearlson GD, McKiernan K, Lloyd D, Kiehl KA, Calhoun VD (2007) Aberrant "default mode" functional connectivity in schizophrenia. Am J Psychiatry 164:450–457

Ghoneim MM, Hinrichs JV, Mewaldt SP, Petersen RC (1985) Ketamine: behavioral effects of subanesthetic doses. J Clin Psychopharmacol 5:70–77

Goldman-Rakic PS (1987) Circuitry of primate prefrontal cortex and regulation of behavior by representational memory. In: Plum F, Mountcastle V (eds) Handbook of physiology, the nervous system, Higher function of the brain, vol 5. Bethesda, American Physiological Society, pp 373–417

Goldman-Rakic PS (1999) The physiological approach: functional architecture of working memory and disordered cognition in schizophrenia. Biol Psychiatry 46:650–661

Goldman-Rakic PS, Selemon LD (1997) Function and anatomical aspects of prefrontal pathology in schizophrenia. Schizophr Bull 23:437–458

Gusnard DA, Raichle ME (2001) Searching for a baseline: functional imaging and the resting human brain. Nat Rev Neurosci 2:685–694

Harrison BJ, Yucel M, Pujol J, Pantelis C (2007) Task-induced deactivation of midline cortical regions in schizophrenia assessed with fMRI. Schizophr Res 91:82–86

Hata N, Nishikawa T, Umino A, Takahashi K (1990) Evidence for involvement of N-methyl-D-aspartate receptor in tonic inhibitory control of dopaminergic transmission in rat medial frontal cortex. Neurosci Lett 120:101–104

Holroyd CB, Coles MG, Nieuwenhuis S (2002) Medial prefrontal cortex and error potentials. Science 296:1610–1611

Inoue M, Mikami A, Ando I, Tsukada H (2004) Functional brain mapping of the macaque related to spatial working memory as revealed by PET. Cereb Cortex 14:106–119

Jedema HP, Moghaddam B (1996) Characterization of excitatory amino acid modulation of dopamine release in the prefrontal cortex of conscious rats. J Neurochem 66:1448–1453

Kim JS, Kornhuber HH, Schmid-Burgk W, Holzmuller B (1980) Low cerebrospinal fluid glutamate in schizophrenia patients and a new hypothesis on schizophrenia. Neurosci Lett 20: 379–382

Kodama T, Honda Y, Watanabe M, Hikosaka K (2002a) Release of neurotransmitters in the monkey frontal cortex is related to level of attention. Psychiatry Clin Neurosci 56(3):341–342

Kodama T, Hikosaka K, Watanabe M (2002b) Differential changes in glutamate concentration in the primate prefrontal cortex during delayed spatial alteration and sensory-guided tasks. Exp Brain Res 145(2):133–141

Kodama T, Hikosaka K, Honda Y, Kojima T, Watanabe M (2014) Higher dopamine release induced by less rather than more preferred reward during a working memory task in the primate prefrontal cortex. Behav Brain Res 266:104–107

Kodama T, Hikosaka K, Honda Y, Kojima T, Tsutsui K, Watanabe M (2015, November) Dopamine and glutamate release in the anterior default system during rest: a monkey microdialysis study. Behav Brain Res 294:194–197

Kojima T, Onoe H, Hikosaka K, Tsutsui K, Tsukada H, Watanabe M (2009) Default mode of brain activity demonstrated by positron emission tomography imaging in awake monkeys: higher rest-related than working memory-related activity in medial cortical areas. J Neurosci 29(46):14463–14471

Krystal JH, Karper LP, Seibyl JP, Freeman GK, Delaney R, Bremner J, Heninger GR, Bowers MB Jr, Charney DS (1994) Subanesthetic effects of the noncompetitive NMDA antagonist, ketamine, in humans. Arch Gen Psychiatry 51:199–214

Ljungberg T1, Apicella P, Schultz W (1991) Responses of monkey midbrain dopamine neurons during delayed alternation performance. Brain Res 567(2):337–341

Luciana M, Collins PF (1997) Dopaminergic modulation of working memory for spatial but not object cues in normal humans. J Cogn Neurosci 9:330–347

Malenka RC, Nestler EJ, Hyman SE (2009) Chapter 6: widely projecting systems: monoamines, acetylcholine, and orexin. In: Sydor A, Brown RY (eds) Molecular neuropharmacology: a foundation for clinical neuroscience, 2nd edn. McGraw-Hill Medical, New York, pp 147–148

Malhotra AK, Pinals DA, Weingartner H, Sirocco K, Missar CD, Pickar D, Breier A (1996) NMDA receptor function and human cognition: the effects of ketamine in healthy volunteers. Neuropsychopharmacology 14:301–307

Mantini D, Gerits A, Nelissen K, Durand JB, Joly O, Simone L, Sawamura H, Wardak C, Orban GA, Buckner RL, Vanduffel W (2011) Default mode of brain function in monkeys. J Neurosci 31: 12954–12962

Miller EK (2000) The prefrontal cortex and cognitive control. Nat Rev Neurosci 1:59–65

Mishkin M (1957) Effects of small frontal lesions on delayed alternation in monkey. J Neurophysiol 20:615–622

Moghaddam B, Adams B, Verma A, Daly D (1997) Activation of glutamatergic neurotransmission by ketamine: a novel step in the pathway from NMDA receptor blockade to dopaminergic and cognitive disruptions associated with the prefrontal cortex. J Neurosci 17:2921–2927

Murphy BL, Arnsten AF, Goldman-Rakic PS, Roth RH (1996a) Increased dopamine turnover in the prefrontal cortex impairs spatial working memory performance in rats and monkeys. Proc Natl Acad Sci U S A 93(3):1325–1329

Murphy BL, Arnsten AF, Jentsch JD, Roth RH (1996b) Dopamine and spatial working memory in rats and monkeys: pharmacological reversal of stress-induced impairment. J Neurosci 16: 7768–7775

Nishijima K, Kashiwa A, Nishikawa T (1994) Preferential stimulation of extracellular release of dopamine in rat frontal cortex to striatum following competitive inhibition of the N-methyl-D-aspartate receptor. J Neurochem 63:375–378

Oades RD, Halliday GM (1987) Ventral tegmental (A10) system: neurobiology. 1. Anatomy and connectivity. Brain Res 434:117–165

Okubo Y, Suhara T, Suzuki K, Kobayashi K, Inoue O, Terasaki O, Someya Y, Sassa T, Sudo Y, Matsushima E, Iyo M, Tateno Y, Toru M (1997) Decreased prefrontal dopamine D1 receptors in schizophrenia revealed by PET. Nature 385(6617):634–636

Olney JW, Farber NB (1995) Glutamate receptor dysfunction and schizophrenia. Arch Gen Psychiatry 52(12):998–1007

Oye I, Paulsen O, Maurset A (1992) Effects of ketamine on sensory perception: evidence for a role of N-methyl-D-aspartate receptors. J Pharmacol Exp Ther 260:1209–1213

Passani MB, Bacciottini L, Mannaioni PF, Blandina P (2000) Central histaminergic system and cognition. Neurosci Biobehav Rev 24(1):107–113

Phillips AG1, Ahn S, Floresco SB (2004) Magnitude of dopamine release in medial prefrontal cortex predicts accuracy of memory on a delayed response task. J Neurosci 24(2):547–553

Porras A, Sanz B, Mora F (1997) Dopamine-glutamate interaction in the prefrontal cortex of the conscious rat: studies on ageing. Mech Ageing Dev 99:9–17

Posner MI, Rothbart MK, Sheese BE, Tang Y (2007) The anterior cingulate gyrus and the mechanism of self-regulation. Cogn Affect Behav Neurosci 7:391–395

Preuss TM (1995) Do rats have prefrontal cortex? The rose-woolsey-akert program reconsidered. J Cogn Neurosci 7(1):1–24

Rao SG, Williams GV, Goldman-Rakic PS (1999) Isodirectional tuning of adjacent interneurons and pyramidal cells during working memory: evidence for microcolumnar organization in PFC. J Neurophysiol 81:1903–1916

Rao SG, Williams GV, Goldman-Rakic PS (2000) Destruction and creation of spatial tuning by disinhibition: GABAA blockade of prefrontal cortical neurons engaged by working memory. J Neurosci 20:485–494

Ridderinkhof KR, Ullsperger M, Crone EA, Nieuwenhuis S (2004) The role of the medial frontal cortex in cognitive control. Science 306:443–447

Riederer P, Lange KW, Kornhuber J, Danielczyk W (1992) Glutamatergic-dopaminergic balance in the brain: its importance in motor disorders and schizophrenia. Drug Res 42:265–268

Robbins TW, Arnsten AF (2009) The neuropsychopharmacology of fronto-executive function: monoaminergic modulation. Annu Rev Neurosci 32:267–287

Robbins TW, Roberts AC (2007) Differential regulation of fronto-executive function by the monoamines and acetylcholine. Cereb Cortex 17(Suppl 1):i151–i160

Romanides AJ, Duffy P, Kalivas PW (1999) Glutamatergic and dopaminergic afferents to the prefrontal cortex regulate spatial working memory in rats. Neuroscience 92:97–106

Rushworth MF, Noonan NP, Boorman ED, Walton ME, Behrens TE (2001) Frontal cortex and reward-guided learning and decision-making. Neuron 70:1054–1069

Santiago M, Machado A, Cano J (1993) Regulation of the prefrontal dopamine release by GABAA and GABAB receptor agonists and antagonists. Brain Res 630:28–31

Sawaguchi T, Goldman-Rakic PS (1991) D1 dopamine receptors in prefrontal cortex: involvement in working memory. Science 251:947–950

Sawaguchi T, Matsumura M (1985) Laminar distributions of neurons sensitive to acetylcholine, noradrenaline and dopamine in the dorsolateral prefrontal cortex of the monkey. Neurosci Res 2:255–273

Sawaguchi T, Matsumura M, Kubota K (1988) Delayed response deficit in monkeys by locally disturbed prefrontal neuronal activity by bicuculline. Behav Brain Res 31:193–198

Sawaguchi T, Matsumura M, Kubota K (1989) Delayed response deficits produces by local injection of bicuculline into the dorsolateral prefrontal cortex in Japanese macaque monkeys. Exp Brain Res 75:457–469

Sawaguchi T, Matsumura M, Kubota K (1990a) Catecholaminergic effects on neuronal activity related to a delayed response task in monkey prefrontal cortex. J Neurophysiol 63:1385–1400

Sawaguchi T, Matsumura M, Kubota K (1990b) Effects of dopamine antagonists on neuronal activity related to a delayed response task in monkey prefrontal cortex. J Neurophysiol 63:1401–1412

Schultz W (1992) Activity of dopamine neurons in the behaving primate. Semin Neurosci 4: 129–139

Smiley JF, Goldman-Rakic PS (1993) Heterogeneous targets of dopamine synapses in monkey prefrontal cortex demonstrated by serial section electron microscopy: a laminar analysis using the silver-enhanced diaminobenzidine sulfide (SEDS) immunolabeling technique. Cereb Cortex 3:223–238

Smiley JF, Williams SM, Szigeti K, Goldman-Rakic PS (1992) Light and electron microscopic characterization of dopamine-immunoreactive axons in human cerebral cortex. J Comp Neurol 321:325–335

Spencer KM, Nestor PG, Perlmutter R, Niznikiewicz MA, Klump MC, Frumin M, Shenton ME, McCarley RW (2004) Neural synchrony indexes disordered perception and cognition in schizophrenia. PNAS 101(49):17288–17293

Takahashi R, Moghaddam B (1998) Glutamatergic regulation of basal and stimulus-activated dopamine release in the prefrontal cortex. J Neurochem 71:1443–1449

Takashima A, Petersson KM, Rutters F, Tendolkar I, Jensen O, Zwarts MJ, McNaughton BL, Fernández G (2006) Declarative memory consolidation in humans: a prospective functional magnetic resonance imaging study Proc. Natl Acad Sci U S A 103:756–761

Thakkar MM (2011) Histamine in the regulation of wakefulness. Sleep Med Rev 15(1):65–74

Verma A, Moghaddam B (1996) NMDA receptor antagonists impair prefrontal cortex function as assessed via spatial delayed alternation performance in rats: modulation by dopamine. J Neurosci 16:373–379

Watanabe M (1996) Reward expectancy in primate prefrontal neurons. Nature 382(6592):629–632

Watanabe M, Hikosaka K, Sakagami M, Shirakawa S (2002) Coding and monitoring of motivational context in the primate prefrontal cortex. J Neurosci 22:2391–2400

Watanabe M, Kodama T, Hikosaka K (1997) Increase of extracellular dopamine in primate prefrontal cortex during a working memory task. J Neurophysiol 78:2795–2798

Wedzony K, Klimek V, Golembiowska K (1993) MK-801 elevates the extracellular concentration of dopamine in the rat prefrontal cortex and increases the density of striatal dopamine D1 receptors. Brain Res 622:325–329

Williams GV, Castner SA (2000) Under the curve: critical issues for elucidating D1 receptor function in working memory. Neuroscience 139:263–276

Williams SM, Goldman-Rakic PS (1993) Characterization of the dopaminergic innervation of the primate frontal cortex using a dopamine-specific antibody. Cereb Cortex 13(3):199–222

Wilson FA, O'Scalaidhe SP, Goldman-Rakic PS (1994) Functional synergism between putative gamma-aminobutyrate-containing neurons and pyramidal neurons in prefrontal cortex. Proc Natl Acad Sci U S A 91:4009–4013

Yonezawa Y, Kuroki T, Kawahara T, Tashiro N, Uchimura H (1998) Involvement of gamma-aminobutyric acid neurotransmission in phencyclidine-induced dopamine release in the medial prefrontal cortex. Eur J Pharmacol 341:45–56

# Chapter 6
# Neuronal Risk Processing in Human and Monkey Prefrontal Cortex

**Wolfram Schultz**

**Abstract** The usual variation of rewards can be conceptualised as risk. Risk evokes specific emotions, as humans hate or love risk in specific situations, and influences risky decisions, as risk avoidance and risk seeking testify. The analysis of decisions under risk suggests that individual decision-makers benefit from perceiving the risk of rewards, weight the risk into their decisions and adapt their behaviour to references and spreads of rewards. Laboratory tests in humans and monkeys use the well-defined form of variance risk and investigate its influence on behaviour. The tests estimate indifference points in choices between risky and safe rewards, establish usually nonlinear utility functions and assess various forms of stochastic dominance that demonstrate meaningful, consistent and rational choices. Based on well-characterised risk functions, human neuroimaging and monkey neurophysiological studies describe several forms of risk processing in the prefrontal cortex. These activities signal reward risk distinct from reward value, integrate risk into subjective reward value and adapt reward processing to the spread of rewards. Some human prefrontal signals reflect the subjective, nonlinear weighting of reward probabilities that conceptualise preference reversals in economic choices. These prefrontal risk processes form important components of the brain's management of risky rewards and provide a biological basis for mathematical and economic concepts of risk and decision-making under uncertainty.

**Keywords** Variance • Mean-preserving spread • Stochastic dominance • Neuron • Orbitofrontal cortex • Utility • Probability distribution • Adaptation • Response slope

---

W. Schultz (✉)
Department of Physiology, Development and Neuroscience, University of Cambridge, Cambridge CB2 3DY, UK
e-mail: ws234@cam.ac.uk



M. Watanabe (ed.), *The Prefrontal Cortex as an Executive, Emotional, and Social Brain*, DOI 10.1007/978-4-431-56508-6_6

## 6.1 Introduction

Rewards are objects or goods that are beneficial or necessary for the life of individuals. However, we often don't know whether we will find these objects or which ones exactly we are going to obtain. The reason is that these objects are often only partly predictable because their occurrence is incompletely known (epistemological uncertainty) or inherently stochastic. Thus, uncertainty impacts on our plans and chances for obtaining rewards. It seems important to better understand the nature of reward uncertainty and how to deal with it. With the brain being the mediator of behaviour, we should understand how the brain manages the uncertainty. Thus, we should include uncertainty in the investigation of neuronal reward processing.

The uncertainty of rewards is associated with probability. If we know the probabilities of reward occurrence completely, we are dealing with risk. If the probabilities are only incompletely known, as it is often the case, we are dealing with ambiguity as a more profound form of uncertainty. To begin investigating neuronal processes for reward uncertainty, we will begin with its simpler and better characterised form and therefore limit the scope of this review to risk.

The investigation of neuronal processing of reward risk should identify a physiological basis for risk and validate theories about the role of risk in reward-directed behaviour and decision-making under uncertainty. Risk is important for reward processing in at least three ways. First, being a fundamental property of reward, risk needs to be detected and estimated, irrespective of making a decision at that moment. Thus, neurons should process risk irrespective of its influence on current choice. Second, risk contributes importantly to economic decisions between outcomes. It affects the way we value rewards, as the terms of risk avoidance and risk seeking indicate. Therefore, risk should affect neuronal reward value signals. Third, risk indicates how much and over which range rewards vary. That information is useful for adjusting the limited information processing capacity of neurons to the currently available rewards and thus makes neuronal reward processing more efficient. This review will discuss these three functions separately and in addition describe the subjective weighting of reward probabilities that are closely related to risk.

The frontal cortex plays a particularly prominent role in risky choices (Lishman 1998). Human patients and animals with lesions in orbitofrontal cortex show altered risk and ambiguity sensitivity (Miller 1985; Gaffan et al. 1993; Bechara et al. 1994; Rahman et al. 1999; Hsu et al. 2005). Therefore, the frontal cortex seems to be a good brain structure to investigate neuronal signals for reward risk. Monkeys are particularly appropriate for investigating the activity of single neurons with neurophysiological methods during specific behavioural acts; they have a superb behavioural repertoire that allows particularly precise measures while controlling for confounds, such as sensory processing and movements. Once a risk signal has been characterised in single neurons of the monkey frontal cortex, it would be important to extend these studies to humans in order to identify, confirm and localise similar signals using functional neuromagnetic imaging (fMRI). More details and relations to reward processing are described in recent reviews (Schultz et al. 2011; Schultz 2015), from which some of the current material, text and figures have been taken.

## 6.2 Concepts on Risk

### *6.2.1 Basic Definitions*

Rewards have specific values that vary usually. Physical reward values are millilitres of juice for animals and money amounts for humans. Then rewards can be characterised as probability distributions of reward values with statistical 'moments'. The first moment is the expected value (EV), which is defined as the summed product of reward amount and its probability over all elements of a reward distribution. The mean of a distribution converges to the EV with increasing sample size. The second (central) moment, variance and its square root, standard deviation (SD), reflect the spread of a distribution and indicate how far values are expected to deviate on average from the EV (Fig. 6.1a). Thus, the variance indicates the risk of rewards (of receiving more or less reward value compared to EV). Note that risk is not a physical event nor are there sensory physiological receptors for risk; risk is a theoretical construct that captures the known deviations from EV, and variance defines these deviations mathematically. EV and SD fully define Gaussian distributions, which are symmetric, and binary equiprobable probability distributions, which concern two outcomes occurring with the same probability. A binary, symmetric gamble constitutes the most simple and pure form of variance risk; the risk can be varied by increasing or shrinking the interval between the two rewards while keeping EV constant ('mean-preserving spread', Rothschild and Stiglitz 1970). At a probability of $p = 0.5$, subjective probability distortions are symmetric and minor, there is no difference in EV, and skewness risk is absent. The coefficient of variation (CV; standard deviation divided by expected value) is also a good measure for symmetric risk in economic choices (Weber et al. 2004). Although variance risk constitutes the most basic and well-conceptualised form of risk, many real-world situations do not conform to symmetric distributions. This form of risk can be conceptually approached by considering skewness, the third (central) moment of probability distributions. The degree of positive or negative skew indicates the asymmetry induced by positive or negative outliers, respectively. Higher central moments, such as kurtosis, constitute also measures of risk but are not commonly investigated in neuroscience. Thus, different from many reduced tests in the laboratory, rewards are not singular events but constitute probability distributions that have a mean value and some variability. Choices between rewards are choices between probability distributions of reward value.

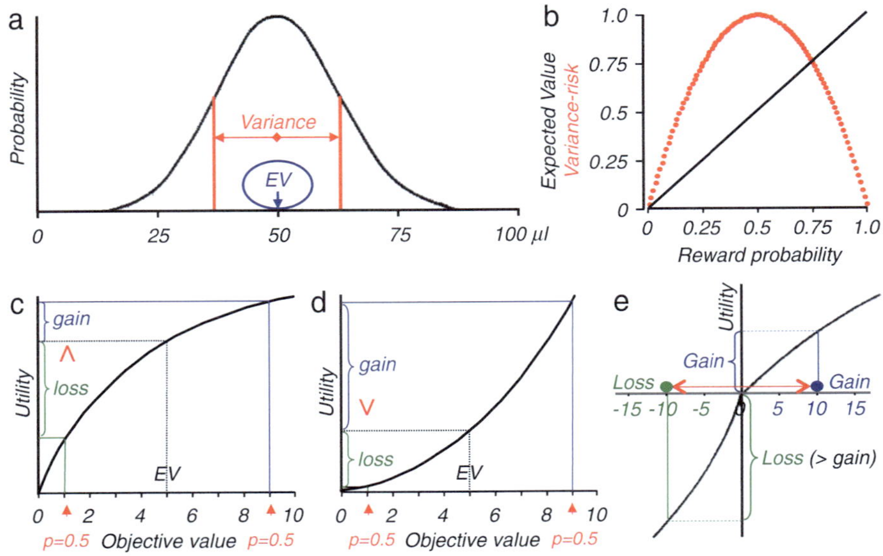


**Fig. 6.1** Risk and utility functions. (**a**) Two principal measures in probability distributions (density function). Expected value (*EV*, first raw moment) denotes value. Variance (second central moment) denotes spread of values and is a good measure for symmetric risk. *Red arrows* show ±1 units of standard deviation (*SD*; square root of variance). Not included are other important risk measures, such as informational entropy and higher statistical moments such as skewness and kurtosis. (**b**) Variance risk (normalised) as a non-monotonic function of probability (*red*). By contrast, normalised expected value increases monotonically with probability (*black line*). (**c**) Concave utility function associated with risk avoidance. The risky gamble (*red*, 1 vs. 9, each occurring with $p = 0.5$) induces stronger utility reduction from losing the gamble (*green*) than the gain from winning the gamble (*blue*) relative to the utility of the expected value of the safe outcome (EV = 5). (**d**) Convex utility function associated with risk seeking. Gain from gamble win is stronger than loss from gamble losing. (**e**) Explaining psychological notions of risk by gain-loss functions. Due to the steeper loss slope ('loss aversion'), the loss from the low outcome of the binary gamble looms larger than the gain from the high outcome. Hence, risk is often associated with the notion of loss (Value function from Kahneman and Tversky 1979)

These definitions are based on physical quantities of reward. During the development of probability theory, Pascal has contended that humans and animals try to maximise EV in their choices between different rewards (Pascal 1658–1662). However, the evolution of economic choice theory revealed that rewards are valued, and maximised, on a subjective basis (Bernoulli 1738). The term subjective refers to the individual decision-maker, not to a conscious or unconscious process. Subjective reward value is an imagined variable that can be estimated by choices that elicit internal preferences; in its specific, formal definition in economics, it is called utility and forms the basis of economic decisions. The utility axioms by von Neumann and Morgenstern (1944) assess whether individual decision-makers behave as if they are maximising utility.

Although probabilistic rewards are intrinsically risky, probability is not a monotonic measure for risk but defines value. In the simplest experiments testing rewards with specific probabilities, binary probability distributions contain a non-zero outcome occurring at a set probability p and a zero outcome occurring with (1-p). In this distribution, reward value (EV) increases monotonically with probability, whereas variance risk follows probability in an inverted-U form (Fig. 6.1b). It peaks at $p = 0.5$ where the chance to gain or miss a reward is equal. Variance risk is lower at all other probabilities, where gains or misses are more certain. Variations in probability affect also skewness risk, as binary probability distributions other than those with $p = 0.5$ are asymmetric. More advanced tests use two or more non-zero outcomes, and EV varies independently from variance and skewness.

### 6.2.2 *Influences on Choices*

The terms of risk avoidance and risk seeking suggest that risk affects economic choices, which can be inferred from the curvature of the utility function. This phenomenon can be best illustrated for the simple case of variance risk. With concave utility (Fig. 6.1c), the initially steeper slope indicates steeper gains in utility (higher marginal utility) in lower ranges. Losses relative to the EV move the outcome into the lower, steeper utility range and thus appear subjectively more important than gains that move the outcome into the upper, flatter utility range. The stronger loss compared to gain results in the avoidance of risk and preference for a safe (riskless) outcome of same value as the gamble's EV. By contrast, with a convex utility function (Fig. 6.1d), the steeper slope in higher ranges makes gains appear subjectively more important than losses, thus favouring risky over safe rewards and encouraging risk seeking. However, these forms are rather schematic and can change in specific situations, often from convex (risk seeking) with low values to concave (risk avoidance) with higher values in monkeys (Stauffer et al. 2014) and humans (Markowitz 1952; Prelec and Loewenstein 1991). Risk seeking at low values may reflect the attraction of gambling that outweighs potential losses, which is called the peanuts effect in human decision-making (Prelec and Loewenstein 1991; Weber and Chapman 2005; Fehr-Duda et al. 2010).

Whereas these considerations concern gains, risk may also involve losses of value. In some cases, the utility function differs between gains and losses in two important ways (Fig. 6.1e) (Kahneman and Tversky 1979). First, gain utility is typically concave, whereas loss utility may be convex, reflecting typical risk avoidance for gains and risk seeking with losses. Second, losses may weigh heavier on utility than gains, which is referred to as loss aversion and shown by an asymmetric gain-loss utility function.

Common perceptions view risk often negatively as the chance of losing, although the opposite view may also prevail. Several subjective factors may explain these risk attitudes. First, the common psychological explanation for risk attitudes is a genuine hate or love of risk. However, it is difficult to find well quantifiable, objective behavioural measures for these emotional reactions, in particular, when working with animals. Second, the curvature of the behaviourally estimated utility function indicates risk attitudes in gain and loss ranges (Fig. 6.1c–e). This dependency includes the convex or concave form of the utility function at the local point at which the gain occurs, as shown by the peanuts effect. A third, related factor is the steepness of the utility function in the loss relative to the gain range, which makes the same physical amounts appear subjectively different between losses and gains. Binary, symmetric gambles across gains and losses appear asymmetric in subjective utility, inducing risk aversion or risk seeking depending on the relative slopes of gains and losses (Fig. 6.1e). Fourth, subjects may distort probabilities of gains and losses subjectively ('today is my bad/lucky day'), which may transform symmetric variance risk into the common asymmetric notion of risk as the danger to lose. A fifth, related factor is the subjective distortion of risk itself, which may be overweighted in an anxious manner or underweighted in an overly confident manner. Besides these factors, risk attitudes are often not stable and vary according to the domain (e.g. money vs. time), perception, situation and familiarity of the risky event (Caraco et al. 1980, 1990; Weber and Milliman 1997). Thus, a number of mostly measurable factors may explain subjective risk attitudes.

## 6.3 Choices Under Risk in Monkeys

Choices under risk can be conveniently conceptualised by different forms of stochastic dominance, which are properties of gambles that relate to the statistical moments of probability distributions and characterise the influence of value and risk on choices (Mas-Colell et al. 1995). The animal chooses between a safe (riskless) reward that can be set to specific values and a fixed, binary, equiprobable gamble in which each reward occurs with $p = 0.5$. Bar stimuli are simple and intuitive predictors of the safe rewards and gambles (Fig. 6.2a).

### *6.3.1 First-Order Stochastic Dominance*

First-order stochastic dominance unambiguously defines the better option as long as the subjective value follows the physical reward amounts in a monotonically increasing and non-saturating fashion. It captures the assumption that more is better without requiring assessment of a utility function; decision-makers choose options with higher value more frequently than options with lower value.

Testing first-order stochastic dominance is straightforward. When the reward amount of the safe option equals exactly the low reward of the gamble, the animal should choose the gamble, because it provides at least as much reward as the safe option plus a 50% chance of getting a higher reward (Fig. 6.2b left bottom). Even the most extreme risk-seeker should choose the gamble in this situation, as it provides at least as much reward as the most pessimistic, lower gamble outcome. Thus, the gamble dominates the safe option. This is a case of statewise dominance,

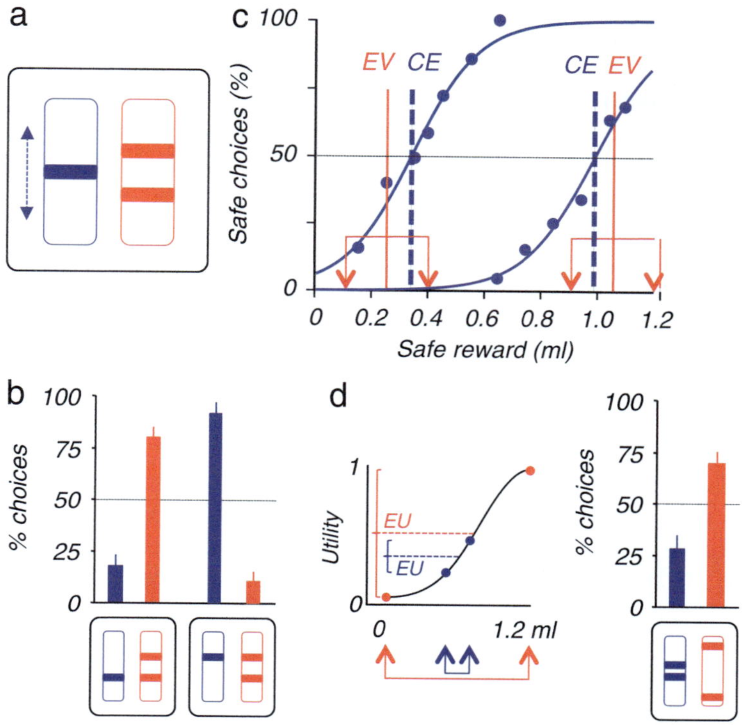


**Fig. 6.2** Choices under risk in monkeys. (**a**) Presentation of a binary choice set. A monkey chooses between the left and the right stimulus. The left stimulus predicts an adjustable safe (riskless) reward (*blue*, diluted fruit juice), and the stimulus to the right predicts a binary gamble (*red*, same fruit juice, two different amounts). Heights of horizontal bars indicate reward amount (higher is more). Each gamble reward occurs with probability $p = 0.5$. (**b**) Choices follow first-order stochastic dominance in a monkey, indicating understanding of stimuli and choice options and demonstrating meaningful and rational choices. In choices against an equiprobable gamble ($p = 0.5$ each reward), the monkey mostly avoids the low safe reward (set at low gamble outcome, *left*) and mostly prefers the high safe reward (set at high gamble outcome, *right*). Insets below bars show the bar stimuli predicting the outcomes to the animal; *left and right insets* indicate separate choice sets. (**c**) Psychophysical assessment of subjective value of binary gambles, eliciting the certainty equivalent (*CE*). The CE is the value of the safe (riskless) reward at which the monkey's choice is indifferent against the risky gamble (oculomotor choices). The CE exceeding expected value (*EV*) indicates risk seeking (gamble at *left*), and the CE<EV indicates risk avoidance (*right*). *Red arrows* indicate the two rewards of each gamble. (**d**) Choices for riskier options follow second-order stochastic dominance for risk seeking in a monkey, indicating meaningful incorporation of risk into choices. *Left*: the riskier gamble with higher mean-preserving spread (*red*) has higher expected utility (*EU*) compared to the less risky gamble (*blue*), thus defining the second-order stochastic dominance of the gambles for risk seeking. *Right*: in direct choices between these gambles, monkeys prefer the riskier to the less risky gamble, thus following the second-order stochastic dominance (**b**–**d** from Stauffer et al. 2014)

which is a simple form of first-order stochastic dominance and requires that all rewards of the choice options are equal except at least one reward that is better. Monkeys' choices follow the dominating option in such situations in at least 80% of their choices (Fig. 6.2b left) (Stauffer et al. 2014). The opposite situation arises when the safe reward equals the high reward of the gamble; the animal should choose the safe option, because the gamble provides on average in half the trials less reward than the safe option. Monkeys follow again the dominating option (Fig. 6.2b right). Thus, monkeys' choices follow largely the first-order stochastic dominance of the gambles and thus can be said to be mostly rational in both situations. Their few choices of the dominated option may reflect exploration, inattention or noise in the decision-making process.

It is important to establish the dominance relationship according to physical values and not simply based on EVs, as nonlinear value functions, nonlinear probability weighting and risk attitude may result in lower subjective values of gambles despite higher EVs. First-order stochastic dominance relationships conveniently allow researchers to assess whether an animal has understood the stimuli and uses them to make meaningful choices. This definition provides the simplest test for rational choices, if we define rationality the way economists do, as the selection of the dominating (best) option in a given choice set. Violations of rationality in the majority of trials of this very simple test usually do not derive from primary irrationality but from incomplete understanding and formalisation of the choice situation.

### 6.3.2 Second-Order Stochastic Dominance

Second-order stochastic dominance defines the meaningful and consistent influence of variance risk on economic choices (Mas-Colell et al. 1995). It relates to the second central moment of probability distributions and applies to binary, equiprobable gambles with different mean-preserving spreads that vary purely in variance risk (Rothschild and Stiglitz 1970). To determine the direction of second-order stochastic dominance of a gamble requires knowing the risk attitude of the decision-maker, which can be estimated from certainty equivalents (CEs) or from utility functions.

Assessment of CEs involves psychophysical choices between an adjustable safe (riskless) reward and a gamble defined solely by variance risk (Fig. 6.2a, c). The safe reward is swept between the lower and higher values of the gamble, and the point of choice indifference (50% choice of safe reward and gamble) is the CE and indicates the subjective value of the gamble (in monkeys typically in millilitres of a specific juice). Typically, the CE is lower than the EV of the gamble in risk avoiders and higher than the EV in risk seekers. Rhesus monkeys show higher CEs than EVs with standard liquid amounts (Fig. 6.2c) (Stauffer et al. 2014) and thus are risk seekers at that reward level. By contrast, CEs exceed EVs at higher liquid amounts, indicating risk seeking (Stauffer et al. 2014). The amount-dependent risk attitude may explain earlier divergent findings concerning risk seeking (McCoy and Platt 2005; O'Neill and Schultz 2010) and risk avoidance in monkeys (Yamada et al. 2013). Thus, controlled experimental tests demonstrate sophisticated variance-risk attitudes in rhesus monkeys that can be expressed by the relationships of CE to EV.

Another way of estimating risk attitudes is based on the concave or convex curvature of utility functions (Fig. 6.1c, d). Utility functions are conveniently established with the 'fractal' procedure that uses binary gambles in a specifically iterative way (Caraco et al. 1980; Machina 1987). In rhesus monkeys, the curvature is convex with low reward amounts, indicating risk seeking; with higher amounts, the curvature becomes linear and then concave, suggesting risk neutrality and then avoidance (Stauffer et al. 2014). This convex-concave curvature of inflected utility is known in humans (Markowitz 1952; Prelec and Loewenstein 1991; Weber and Chapman 2005; Fehr-Duda et al. 2010). These risk attitudes manifested in the utility function correlate well with the CEs established in psychophysically controlled direct choices.

The risk attitudes assessed from CEs or utility functions define the direction of second-order stochastic dominance of the gambles. The utility function allows to calculate the expected utility (EU) from the utilities of the individual gamble outcomes; risk seeking is indicated if the EU of the gamble with the larger mean-preserving spread exceeds the EU of the gamble with the smaller mean-preserving spread (Fig. 6.2d left). Thus, with risk seeking inferred from the comparison of EUs, the riskier of the mean-preserving spread gambles second-order stochastically dominates the less risky gamble (Mas-Colell et al. 1995). Indeed, in direct choices between gambles with different mean-preserving spreads, monkeys prefer the riskier to the less risky gamble, thus following the second-order stochastic dominance of the gambles (Fig. 6.2d right) (Stauffer et al. 2014). In analogy, in choices between safe rewards and risky gambles, monkeys increasingly prefer gambles with wider mean-preserving spreads (O'Neill and Schultz 2010). However, the animals show a minority of dominated choices in both tests. Thus, monkeys make meaningful choices under risk that are consistent with the integration of risk into utility according to expected utility theory.

It is important for the understanding of risk attitude to identify the true value of an outcome. The famous example is the hungry bird with challenged energy balance (Caraco et al. 1980). The bird requires 100 calories to survive the night and has the choice between a safe option of 90 calories and an equiprobable gamble of 0 and 110 calories ($p = 0.5$ each). The bird would hopefully choose the gamble, as this provides it at least with a 50% chance of survival. However, from the perspective of caloric value, the usually risk-avoiding bird seems to be surprisingly risk seeking, as it prefers a risky option with an EV of 55 calories over a safe 90 calories. But the bird's bigger interest is survival, and this is where the value should be primarily placed in challenging times. When considering survival, the bird has a 50% chance when it chooses the calorie-risky option and no chance with the safe option; its behaviour simply follows the first-order stochastic dominance of the gamble valued in terms of survival rather than reflecting irrationally strong risk seeking for calories. Its behaviour is rational.

## 6.4 Neuronal Risk Signals in Monkey Orbitofrontal Cortex

### 6.4.1 Straightforward Risk Signal

A group of neurons in areas 11 and 13 of the monkey orbitofrontal cortex respond to bar stimuli that predict binary, equiprobable gambles (Fig. 6.3a) (O'Neill and Schultz 2010). The responses increase or decrease monotonically with increasing variance risk of reward, as tested by changing the mean-preserving spreads of the gamble. The risk responses occur equally well to fractals that indicate the same gambles with their different risk levels, suggesting risk coding irrespective of visual stimulus properties. Due to the highly constrained definition of the symmetric, mean-preserving variance risk of the gambles (Rothschild and Stiglitz 1970), these responses do not reflect EV, skewness risk or subjective probability distortion. Although groups of orbitofrontal neurons code reward value (Tremblay and Schultz 2000; Wallis and Miller 2003; Padoa-Schioppa and Assad 2006), the orbitofrontal risk responses do not reflect different reward amounts and thus constitute a separate group from the value-coding orbitofrontal neurons. The distinct coding of risk and value suggests also that risk coding does not simply reflect forms of salience that are associated with both risk and value. Although the usually employed stimuli predict higher risk by higher top bars, the control tests with fractals and reward value demonstrate that the pure risk responses derive neither from the higher value bars nor from subjective probability distortions that mistake the upper bar stimuli for almost safe values. Thus, some orbitofrontal neurons show a pure risk signal without coding other decision variables. These neuronal risk responses constitute a biological implementation of the theoretical construct of variance risk.

Indications for orbitofrontal risk coding come also from a categorical discrimination task. A group of rat orbitofrontal neurons shows higher responses to odour mixtures that are closer to discrimination boundaries and thus more difficult and less accurate to distinguish (Kepecs et al. 2008). The lower accuracy makes correct task performance and thus reward less certain. The neuronal responses correlate with decision confidence and may also reflect reward risk.

Apart from pure risk coding, some orbitofrontal neurons code risk in combination with reward value. The risk influences value coding in close correspondence with the risk attitude of the animals (Roitman and Roitman 2010), although the number of such neurons in the orbitofrontal cortex is rather small (O'Neill and Schultz 2010). Risk preference influences also the orbitofrontal coding of specific decision variables, such as object value and chosen value (Raghuraman and Padoa-Schioppa 2014). Thus, rather than coding risk on its own, some orbitofrontal neurons show an integration of risk into value coding.

Taken together, different groups of inhomogeneous orbitofrontal neurons show several sophisticated forms of variance-risk coding. In contrast to reward amounts that impact directly on sensory receptors, the coding of risk is not explained by simple sensory stimulation but requires neuronal mechanisms including memory. The distinction of reward signals according to statistical moments would suggest

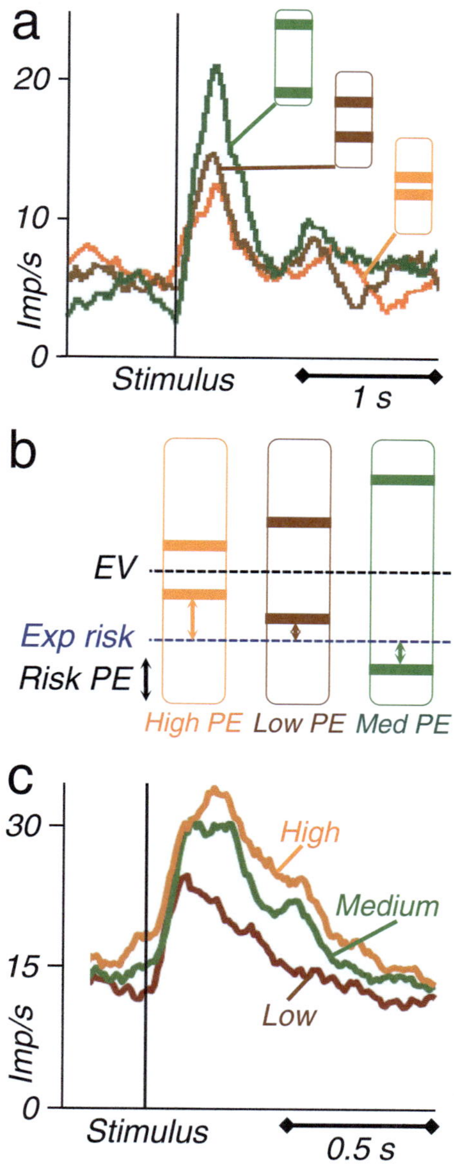


**Fig. 6.3** Neuronal risk responses in monkey orbitofrontal cortex. (**a**) Risk coding in single neuron. Increasing risk, as tested with increasing mean-preserving spread of reward amounts, induces monotonic increases in neuronal responses (*orange* to *brown* to *green*). The *bar heights* in insets indicate liquid reward amounts, the two bars within each *rectangle* indicate an equiprobable gamble ($p = 0.5$ each). The three gambles have same mean but different variance (9, 36 and 144 ml $\times$ $10^{-4}$) (mean-preserving spread). This neuron, as most other orbitofrontal risk neurons, fails to code reward value (not shown) (From O'Neill and Schultz 2010). (**b**) Definition of risk prediction error (*coloured double arrows*) as difference between current risk (vertical distance between bars in each of the three gambles) and predicted (expected) risk (common *dotted blue line*, mean risk from the gambles). Risk is defined as the distance between the two bars in each gamble (representing standard deviation, the square root of variance). The lower reward of the least risky

that neurons process rewards as probability distributions. This provides a physical basis for the mathematical construct of probability distribution and demonstrates the biological plausibility and implementation of the specific terms of EV and variance. The distinct coding of EV and variance in different orbitofrontal neurons makes the mean-variance approach for constructing expected utility a biologically plausible model for economic choices (Markowitz 1952), without refuting other, possibly coexisting expected utility mechanisms. Further experiments may investigate other forms of mathematically defined risk, such as skewness.

### *6.4.2 Risk Prediction Error Signal*

Risk is not constant but varies dynamically in the environment. Therefore, it is important to detect variations in risk, update knowledge to the current risk level and then adapt behavioural reactions. A popular approach to conceptualise and utilise changes in the environment employs the notion of prediction error, which forms the basis of error-driven learning, applies to reward associations and value in Pavlovian and operant conditioning (Rescorla and Wagner 1972; Sutton and Barto 1981) and is implemented in subcortical and cortical neurons (Schultz et al. 1997; Matsumoto and Hikosaka 2007; Seo and Lee 2007; Belova et al. 2007; Apicella et al. 2009; Kim et al. 2009; Ding and Gold 2010; Kennerley et al. 2011; So and Stuphorn 2012). These prediction errors typically reflect changes in reward value. To detect and process changes in risk within the framework of probability distributions would require applying the notion of error to the second central statistical moment, namely, variance risk.

In analogy to value prediction errors, a variance-risk prediction error can be defined as the difference between the current variance and the predicted variance. In a simple experimental implementation of this approach, the current risk is derived from the spread of a binary, equiprobable gamble around the EV, and the predicted risk reflects the average risk from previous trials. Figure 6.3b shows three risk prediction errors that are visualised as differences in deviations from EV (in keeping with the definition of central moments as variations around the EV); with identical EVs of the gambles, the risk prediction error can be directly read

**Fig. 6.3** (continued) gamble (*orange*) deviates most from the average predicted risk (*orange arrow* against *blue line* representing predicted risk), and the lower reward of the intermediately risky gamble deviates least (*centre*, *brown arrow* against *blue line*). *EV* expected value. (**c**) Coding of risk prediction error in orbitofrontal cortex. Each coloured trace shows the averaged population response to one of the gamble stimuli defined by the colours shown in (**b**) (15 neurons). The responses reflect an unsigned (absolute, rectified) risk prediction error derived from the difference in variance risk between one of the gamble stimuli shown in (**b**, **a**) previously presented stimulus predicting the average risk (**b**, **c**, from O'Neill and Schultz 2013)

from the distance between the lower bar of a given gamble and the blue line indicating predicted (average) risk (coloured vertical arrows).

A good structure to search for risk prediction error signals may be the orbitofrontal cortex, because neuronal recording in monkeys reveal reward risk signalling (O'Neill and Schultz 2010) and human neuroimaging suggests prediction error coding of value (O'Doherty et al. 2003; Dreher et al. 2006). Indeed, a group of orbitofrontal neurons in monkeys tracks the discrepancy between the current and predicted variance risk; these neurons respond to visual stimuli that indicate specific levels of variance risk, but the responses reflect not the risk per se but the deviation from the predicted variance risk (Fig. 6.3c) (O'Neill and Schultz 2013). Numerically, the responses increase or decrease with the unsigned difference between the current and predicted variance risk, thus reporting an unsigned, absolute risk prediction error. Most of these neurons code the risk prediction error signal separately from variance risk, although some neurons code risk in both forms.

The function of an orbitofrontal risk prediction error signal may be severalfold. Like any prediction error signal, it could influence other risk neurons to update the level of risk they are coding, in analogy to value prediction error signals updating value signalling. Natural recipients benefitting from risk prediction errors could be the mentioned orbitofrontal neurons coding risk per se. Furthermore, as risk is associated with surprise salience and salience may not be negative, the unsigned nature of the signal may suggest the coding of surprise salience or even a prediction error in surprise salience. According to attentional (associability) learning theories, surprise salience increases the learning constant in the current trial and thus induces faster but more coarse learning of reward values (Pearce and Hall 1980). Thus, besides a role in risk updating, an orbitofrontal risk prediction error signal could be useful for value learning by influencing value neurons in the area or downstream in dorsolateral prefrontal cortex or striatum.

## 6.5 Risk Processing in Human Frontal Cortex

### *6.5.1 Risk Signal*

Given the deficits in decision-making under risk in humans after prefrontal lesions (Bechara et al. 1994; Lishman 1998; Rahman et al. 1999; Hsu et al. 2005), the identification of neuronal risk signals in the monkey frontal cortex should be extended to humans. Initial experiments distinguish risk coding from value coding. Different reward probabilities (probability distributions with one non-zero value and one zero value) are indicated by play cards, slot machines or quantitative stimuli (Preuschoff et al. 2006; Dreher et al. 2006; Tobler et al. 2007). In these designs, reward value increases monotonically with probability, and risk increases in an inverted-U fashion (Fig. 6.1b). Thus, risk can be disambiguated from value, but the risk consists of a mixture of variance risk and skewness risk, with subjective probability distortions likely playing also a role. Subsequent studies use pure variance risk implemented as mean-preserving spreads; EVs are set to specific values rather than changing with probability (Christopoulos et al. 2009).

Human risk preferences are assessed by direct choice preferences between a safe monetary reward and a binary, equiprobable monetary gamble with same EV and non-zero outcomes (Tobler et al. 2007) and by psychophysically estimating the CE and comparing it to EV (Christopoulos et al. 2009). Human risk attitudes with these small gambles vary between risk seeking and risk avoidance, although standard economic theory assumes default overall risk avoidance as expressed in the 'well-behaved' utility function.

Visual stimuli displaying combinations of reward amounts and probabilities predict specific EVs and variance risks. Blood oxygenation level-dependent (BOLD) responses in the orbitofrontal cortex in imperative, no-choice trials follow the inverted-U function of probability and thus code variance risk, without significantly varying with reward value (Tobler et al. 2009). BOLD responses in anterior cingulate cortex follow the mean-preserving spread of binary, equiprobable gambles (Christopoulos et al. 2009). By contrast, BOLD responses in other prefrontal regions and the striatum follow reward probabilities monotonically and thus code EV irrespective of amount-probability combinations. Variance risk, distinct from value, activates also other brain regions, including the ventral striatum, subthalamic nucleus, mediodorsal thalamic nucleus, midbrain and anterior insula (Dreher et al. 2006; Preuschoff et al. 2006). These signals suggest variance-risk coding, although some contribution of skewness risk is possible. Indeed, BOLD responses reflecting skewness risk occur in the human insula (Burke and Tobler 2011), whereas studies have not yet identified skewness-risk signals in the frontal cortex.

The BOLD variance-risk signals, as identified by coding risk according to the inverted-U function of probability, vary with individual risk attitudes in two distinct parts of the frontal cortex (Tobler et al. 2007). In risk avoiders, a risk signal in the lateral orbitofrontal cortex increases monotonically with individual degrees of risk avoidance; it is substantial in risk avoiders but weak in risk seekers (Fig. 6.4a, top). By contrast, in risk seekers, a risk signal in the medial frontal cortex increases with risk seeking; it is strong in risk seekers but weak in risk avoiders (Fig. 6.4a, bottom). In a direct test for risk attitude, BOLD responses in the inferior frontal gyrus increase with stronger risk aversion, thus directly coding risk attitude (Christopoulos et al. 2009). Thus, frontal risk signals reflect individual risk attitudes, suggesting subjective rather than objective coding of risk in these cortical areas. The individual variations of risk signals may explain the different attitudes of individuals towards risk and influence their choices in risky situations. If used by the brain during decision-making, such signals might prevent risk avoiders from choosing risky options and drive risk seekers towards risky options.

Whereas risk is defined as the uncertainty derived from known probability distributions, a more pronounced, genuine form of uncertainty is ambiguity in which probability distributions are only incompletely known. Decision-makers show usually more pronounced risk attitudes towards ambiguity than towards

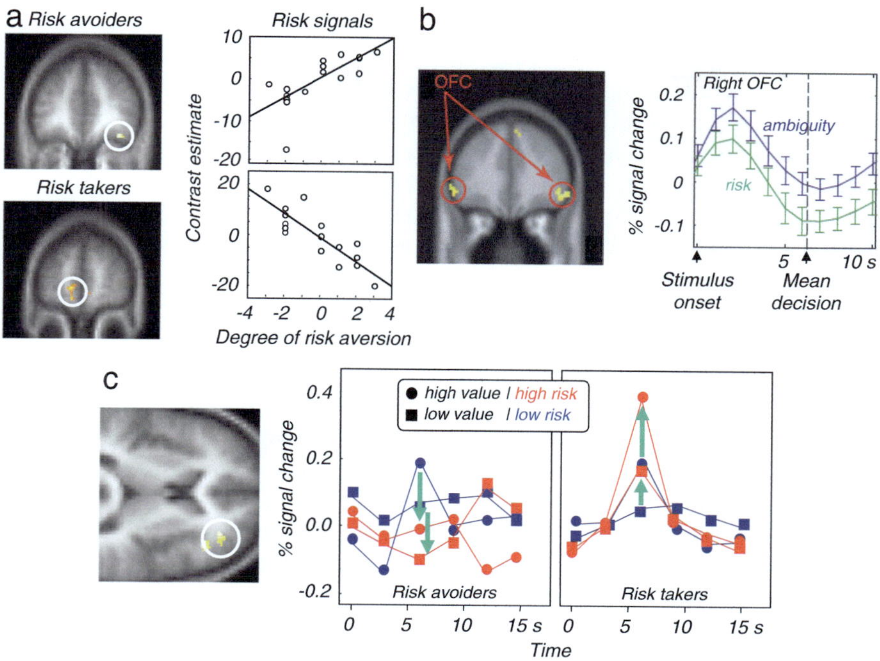


**Fig. 6.4** Reward risk signals in human frontal cortex. (**a**) Human responses to variance risk in the frontal cortex and their relationships to individual risk attitudes. The BOLD responses follow an inverted-U function of reward probability (data not shown). *Top*: location of risk signal in lateral orbitofrontal cortex of risk avoiders (*left*), increasing with increasing risk aversion across participants (*right*). *Bottom*: location of risk signal in medial frontal cortex of risk takers, decreasing with increasing risk aversion (=increasing with risk seeking). Risk aversion (abscissa) ranges from −4 (very risk seeking) to +4 (very risk averse), and 0 indicates risk neutrality (From Tobler et al. 2007). (**b**) Higher responses in human orbitofrontal cortex (*OFC*) to ambiguous than risky gambles (From Hsu et al. 2005). (**c**) Influence of risk on reward value signals in human lateral frontal cortex. *Left*: location of value signal, as identified by regression on safe options with increasing value. *Centre*: time courses of BOLD value signal in risk avoiders. The value signal with low-risk outcomes (*blue curves*) decreases with higher risk (*downward green arrows* to *red curves*), suggesting reduction of neuronal value signal by risk in correspondence with behavioural risk avoidance. *Right*: risk enhances neuronal value signal in risk seekers (*upward green arrows* to *red curves*). The population of 15 participants was median split into seven risk avoiders and seven risk seekers according to their risk preferences in direct choices between safe and risky rewards (From Tobler et al. 2009)

risk. Accordingly, stimuli signalling ambiguity lead to stronger BOLD responses compared to risk in the lateral orbitofrontal and dorsolateral prefrontal cortex (Fig. 6.4b), in correspondence with individual ambiguity attitudes (Hsu et al. 2005; Huettel et al. 2006).

Taken together, BOLD signals in the human frontal cortex reflect the risk of rewards. These neuronal responses suggest that the concept of risk is represented as explicit signals in the human brain. The signals correspond well to the deficits in decision-making under risk observed after lesions in these cortical areas.

### 6.5.2 *Influence of Risk on Subjective Reward Value Signal*

The terms of risk seeking and risk avoidance imply that risk influences the subjective valuation of rewards. This notion is also conceptualised by the curvature of the utility function, in which concavity indicates risk avoidance and convexity indicates risk seeking (Fig. 6.1c, d). Risk reduces subjective reward value in risk avoiders but increases subjective value in risk seekers. The risk attitudes and their influence of risk on subjective value are easily modelled and observed in the experimental laboratory. Risk seekers prefer more risky over less risky outcomes in direct choices and show higher CEs than EVs in psychophysically controlled tests assessing choice indifference. By contrast, risk avoiders show opposite choice preferences and lower CEs than EVs. Thus, risk has major influences on behavioural choices, which can be conceptualised as effects on subjective value.

Given the existence of BOLD signals for risk in the frontal cortex, one wonders whether these signals might be used to influence value coding in these cortical areas, and whether this influence might correlate with the subjective valuation of risky rewards, which is represented in individual risk attitudes and modelled by the curvature of utility functions. Indeed, BOLD signals in the lateral prefrontal cortex that vary with reward value are affected by variance risk (Fig. 6.4c) (Tobler et al. 2009). Specifically, BOLD value signals decrease with increasing risk in risk avoiders (centre) but increase with increased risk in risk seekers (right). As these data derive from no-choice, imperative trials, risk appears to influence the processing of subjective reward value in the prefrontal cortex already at the input stage to decision mechanisms. These data are compatible with the differential influences of risk on behavioural choices in risk avoiders and risk seekers. The integration of risk into subjective value follows the mean-variance approach for constructing expected utility (Markowitz 1952), similar to the activity of individual orbitofrontal neurons with similar integration or combinations of orbitofrontal neurons coding risk and value separately (O'Neill and Schultz 2010; Roitman and Roitman 2010; Raghuraman and Padoa-Schioppa 2014).

Thus, BOLD signals in the prefrontal cortex appear to process risk in distinct forms that correspond to the different functions of risk in behaviour. Risk signals in imperative trials may reflect the perception of risk and would allow individuals to identify and compare risky outcomes before advancing to decisions. Neuronal risk signals occurring together with value coding reflect the influence of risk on subjective value and may be involved in subjective attitudes towards risk.

## 6.6 Adaptive Reward Processing in Monkeys

### *6.6.1 Concepts*

The variance risk of the probability distribution indicates the range over which reward values are spread (Fig. 6.1a) (and skewness risk captures the form of the reward distribution). Different rewards, or the same rewards in different situations or contexts, may show smaller or larger variations in their values, which is captured by different variances of their probability distribution and underlies the notion of risk. Risk implies knowledge of the reward probability distribution and the spread of the reward values, but without knowing exactly which of these rewards will occur at a given instance. With that knowledge, we anticipate also the mean value of the future rewards, called the EV. Knowing risk and EV is very precious, not only for influencing decisions but also for using the limited resources of the brain efficiently.

Brains have a finite number of neurons, which can discharge action potentials only within a limited range defined by their maximal discharge frequency. By contrast, the number of objects that can serve as rewards should be huge in order to assure sufficient supply for the survival of the body (and its genes) in widely differing situations. Thus, the brain should be able to deal with these almost infinite rewards irrespective of how rarely they occur. The job is facilitated by the fact that the number of available rewards at any given moment is limited and that predictions can tell us the current selection of rewards. With this knowledge, neuronal processing could focus onto the available rewards rather than reserving processing power for all possible and impossible rewards.

More formally, the limited frequency of action potentials of each reward neuron would need to cover an almost unlimited number and values of possible rewards. The result would be a rather flat neuronal response slope, defined as the frequency of action potentials per unit of reward value (assuming a rate code). But a flat slope leads to poor discrimination. The slope could steepen if neuronal processing were focused on the currently available rewards. Formally spoken, for best processing efficacy, the probability distribution of neuronal responses should be matched at any instance to the probability distribution of the available rewards.

For simplicity, we consider changes of the first two statistical moments and treat them separately (and disregard higher moments, such as skewness and kurtosis) (Fig. 6.5a), although real-world adaptations would usually concern a mixture of the different moments. In principle, effective adaptations should centre neuronal reward responses on the EV and restrict them according to the variance. A step change of the whole reward probability distribution, which changes the EV without affecting the other moments, moves the processing range without extending it, thus maintaining the response slope (Fig. 6.5b blue vs. green). Thus, the EV serves as reference point for moving processing onto the available rewards and eliminating unnecessary processing capacity for unavailable rewards. Adaptation to the variance would affect the response slope; smaller variance would focus the processing

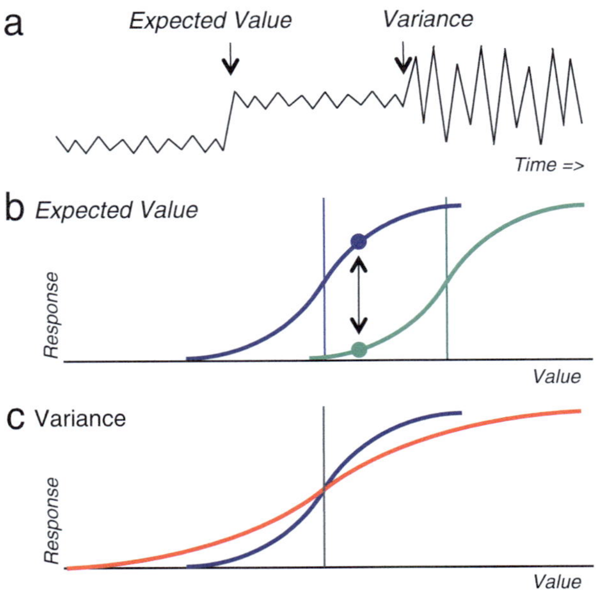


**Fig. 6.5** Schemes of adaptive reward value processing. (**a**) Distinction of step change in expected value and stochasticity change in variance. (**b**) Adaptation to shifts in expected value. The same reward (*blue* and *green dots*) induces strong neuronal responses when presented as part of a low value probability distribution (*blue*) and weak responses when presented as part of a high value probability distribution (*green*). Thus the neuronal responses reflect the relative position of the reward within a given reward distribution (*blue* vs. *green*), irrespective of the actual physical value. The expected value serves as reference point for the adaptation. (**c**) Adaptation to variance changes the value-response slope. Smaller variance of reward probability distribution is associated with steeper slope

range and steepen the slope, whereas larger variance would widen the sensitivity and flatten the slope (Fig. 6.5c blue vs. red). If the reward distribution is skewed, or changes its skew, the adaptation would take local reward densities into account and result in asymmetric slope changes. With these adaptations, the reward neurons could use their full response range for exactly the range of available rewards and thus optimise response slopes and reward discrimination.

Adaptations to reward probability distributions lead to distinct behavioural phenomena. The behavioural consequences of step changes in EV are more frequently documented than changes in the other moments. A downshift of reward leads to slower and lower behavioural reactions and preferences, as compared to the same low reward occurring without preceding higher values, which is called the negative contrast effect (Tinklepaugh 1928; Black 1968). The downshifted reward lies at the lower end of the initial distribution (Fig. 6.5b green) and thus induces a low behavioural or neuronal response (green dot). By contrast, without the preceding higher distribution, the same reward lies at the upper end of the lower distribution (Fig. 6.5b blue) and induces a correspondingly large response (blue dot). Thus, the reward is valued within the distribution in which it occurs; without a preceding distribution and its reference EV, the same reward is valued only relative to the current distribution and its EV. A similar EV adaptation may underlie the frequent observation that cessation of reward is aversive, as conceptualised in the opponent process theory of motivation (Solomon and Corbit 1974). This effect may derive from adaptation to the initially high reward distribution, which positions the nil reward value into its low end where values can be negative, rather than appreciating its nil value within a low reward distribution as valueless but not negative. A similar EV adaptation is captured in the dichotomy between the 'experienced utility' of sequential rewards and their 'decision utility' revealed in overt choices (Kahneman et al. 1997). An underlying mechanism may be the temporal decay of the perceived values of past rewards, akin to leaky integration of evidence (Vestergaard and Schultz 2015). Further, reference-dependent valuation as component of prospect theory captures the change of EV by subtracting a reference value from the reward value when calculating expected utility (Kahneman and Tversky 1979). In contrast to these EV adaptations, adaptations to higher statistical moments are far less documented. In one of the rare examples, reward prediction errors are scaled to the variance of reward distributions (Nassar et al. 2010), and these adaptations improve learning (Diederen and Schultz 2015).

Taken together, the key characteristics of probability distributions are essential for efficient reward processing. Although behavioural adaptations to EV are more frequently documented, variance risk provides useful information for adjusting the limited processing range to the current distribution of available rewards and thus optimises reward discrimination.

### 6.6.2 Neuronal Adaptations in Orbitofrontal Cortex

Some orbitofrontal neurons adapt to reward references and thus seem to undergo adaptations to the EV. Such adaptations occur when different reward distributions are tested in separate blocks of trials. In an experiment on monkey orbitofrontal neurons (Tremblay and Schultz 1999), one trial block presents morsels of apple and cereal in pseudorandom alternation, of which the animal prefers apple (Fig. 6.6a, Block 1, blue). The value of this distribution is determined by the apple and cereal morsels. By contrast, Block 2 tests raisin and apple morsels (green), of which the animal prefers raisin and disprefers apple. The raisin and apple morsels determine the obviously higher value of this distribution. In each block, the same orbitofrontal neuron responds more to whatever is the more preferred reward, which is apple in Block 1 and raisin in Block 2. Thus, the neuronal response to apple shifts from high to low when the value of the distribution increases, conforming to the scheme of Fig. 6.5b (blue to green dot). A second experiment uses the same design but presents an electrical shock instead of the lowest valued reward (Hosokawa et al. 2007). Orbitofrontal neurons in monkeys show a very similar shift of responses depending on the combination of outcomes in different trial blocks. A third

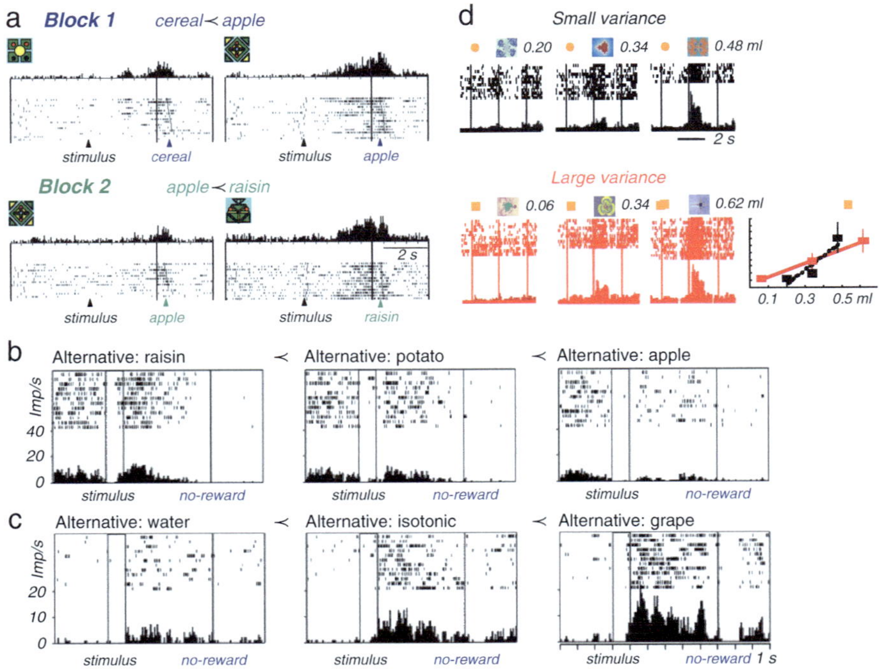


**Fig. 6.6** Adaptive neuronal reward processing in single neurons of monkey frontal cortex. (**a**) Reference-dependent response adaptation to approximate expected value (*EV*) in orbitofrontal neuron. Trial Block 1 offers cereal or apple morsels, and separate Block 2 offers apple or raisin morsels. Behavioural choices reveal preferences raisin \prec apple ≻ cereal. Visual stimuli predict the type of reward. The same physical reward (*apple*) induces lower neuronal responses when being part of a higher valued reward distribution (*green*), thus following the scheme of Fig. 6.5b (*dots*) (From Tremblay and Schultz 1999). (**b**) Reference-dependent EV adaptation of no-reward response in dorsolateral cortex neuron. No-reward and one specific reward alternate pseudorandomly in each of three different trial blocks. The subjective value of the alternative reward is inferred from choice preferences (apple \prec potato ≻ raisin). The neuronal no-reward response decreases with increasing value of the alternative reward, thus following the scheme of Fig. 6.5b (*dots*). (**c**) Reference-dependent EV adaptation of no-reward response in inversely reward value-coding neuron of dorsolateral prefrontal cortex. Note that the neuron shows higher responses to no-reward compared to any of the three rewards (inverse coding, not shown). In the same experimental design as shown in (**b**), the no-reward response increases with higher subjective reward value, as inferred from choice preferences (grape juice \prec isotonic fluid ≻ water), thus reversing the response scheme shown in (**b**) (*dots*) due to inverse coding (**b** and **c**) (From Watanabe et al. 2002). (**d**) Response adaptation to variance risk of distribution of reward amounts (ml) in orbitofrontal neuron. In each of two distributions (*top* and *bottom*), three different reward amounts varied pseudorandomly between trials, and the two distributions varied also pseudorandomly. The distribution shown at the *bottom* is a mean-preserving spread of the distribution shown at the *top* (same EV, different variances). *Inset* shows steeper reward-response slope with narrower distribution (*black*) evidenced by linear regression in same neuron (From Kobayashi et al. 2010)

experiment concerns the dorsolateral prefrontal cortex of monkeys, where neurons show similar reference-dependent coding with different rewards occurring in distinct trial blocks (Watanabe et al. 2002). In each trial block, one specific reward alternates pseudorandomly with no reward. Some dorsolateral prefrontal reward neurons show smaller activities with no-reward than with reward, and the no-reward activities decrease with increasing preference for the specific reward of that trial block (Fig. 6.6b). Thus, the responses to the low outcome (no-reward) adapt to the distribution of (no-reward, reward) in a similar way as schematised (Fig. 6.5b), being low within an overall higher valued distribution (green dot) and higher within a lower valued distribution (blue dot). A fourth experiment on monkeys involves adaptive prefrontal neurons with inverse coding that show stronger responses for no-reward than for reward (Watanabe et al. 2002). Their responses to identical no-reward increase across trial blocks with more valuable rewards (Fig. 6.6c). Because of the inverse coding, this increase appears opposite to the mentioned changes of orbitofrontal and dorsolateral responses but otherwise reflects the same adaptation to higher valued reward distributions. In a fifth experiment, BOLD signals in human orbitofrontal cortex show comparable reward adaptations (Nieuwenhuis et al. 2005) in a similar experimental design with pairs of rewards presented in different trial blocks as in monkeys (Tremblay and Schultz 1999). Taken together, some reward responses in the orbitofrontal and dorsolateral prefrontal cortex of humans and monkeys adapt to changes in EV of reward probability distributions.

In addition to EV-dependent reward coding, some orbitofrontal reward responses adapt to the variance (Kobayashi et al. 2010; Pastor-Bernier and Cisek 2011). The neuron in Fig. 6.6d shows increasing responses with larger reward amounts (top). Importantly, very similar responses occur also with wider mean-preserving spread of reward values (bottom). In particular, the responses to the smallest and the largest reward are indistinguishable between the two reward distributions. The unchanged minimal and maximal responses across different ranges of reward values result in changes of neuronal response slope, as schematised (Fig. 6.5c), which adapts dynamically to the prevailing reward distribution (Fig. 6.6d inset). In providing more information about a reward distribution than EV alone, the result involving changes in variance comes closer to the initially stated matching of probability distributions between neuronal responses and reward values. As the separate adaptations to EV and variance suggest, orbitofrontal neurons show also adaptations with combined changes in EV and variance (Padoa-Schioppa 2009; Cai and Padoa-Schioppa 2012), suggesting that the separate testing of the two statistical moments conveys adequate, and well-controlled, characterisation of the variables underlying the adaptive process.

The neuronal adaptations take time. When varying pseudorandomly on every trial between the two probability distributions shown in Fig. 6.6d, only 13% of orbitofrontal neurons show the described adaptations. With mini-blocks of 4–16 trials with the same distribution, 16% of neurons adapt, and with blocks of mean 28 trials, 39% of neurons adapt (Kobayashi et al. 2010). Apparently, slower changes of reward distributions result in more adaptive coding. Although longer test periods might have increased the number of adapting neurons further, it seems that only a fraction of orbitofrontal reward neurons show adaptation, whereas the remainder codes reward amount irrespective of other rewards (Padoa-Schioppa and Assad 2008; Kobayashi et al. 2010).

Taken together, reward neurons in the frontal cortex adapt their responses to the prevailing probability distributions. These reward adaptations resemble sensory adaptations to ambient visual intensities (Laughlin 1981; Fairhall et al. 2001; Hosoya et al. 2005) and thus seem to be components of a general mechanism for efficient neuronal processing. But this advantage may come at a cost. Although the loss of actual reward values would be immaterial in choices comparing options relative to each other, a reward system relying on adaptive processing alone would not be able to maintain rank-ordering of rewards, insensitivity from menu variance and transitivity of choices, which are requirements for rational choices. However, the mentioned non-adaptive orbitofrontal reward neurons could assure correct reward value processing that is necessary for rational choices. Thus, a combination of adaptive and non-adaptive neurons could provide efficacy from adaptation while maintaining transitivity from non-adaptive coding. Further, information about reward distributions made available to neurons could help to reconstruct distributions even after adaptations. Specifically, EV and risk information, as distinctly coded in orbitofrontal neurons (O'Neill and Schultz 2010), may inform adapted orbitofrontal value neurons and allow correct decision mechanisms to occur. In this way, reward neurons can make efficient use of processing resources for optimal reward discrimination and choice.

## 6.7 Human Neuroimaging of Subjective Probability Distortion

### 6.7.1 *Concepts*

Risk is derived from probability, yet the two concepts differ in important points. Reward probability reflects the predicted frequency of reward and thus is closely related to reward rate, which is a measure of reward value, the first statistical moment. By contrast, variance risk indicates the (symmetric) deviation from the EV and constitutes the second central moment. As Fig. 6.1b shows, reward value increases monotonically with probability, whereas risk follows probability as an inverted-U function. Nevertheless, the formal treatment of utility maximisation involves risk derived from probability (von Neumann and Morgenstern 1944). Therefore, the investigation of decision-making under risk requires consideration of reward probability.

Classic EU theory may show violations when rewards occur with specific, usually low probabilities (Allais 1953). The violations can be often be explained by a nonlinear weighting of probabilities in their influence on EU, called probability distortion. In particular, humans often overweight low probability rewards and underweight high probability rewards (Kahneman and Tversky 1979; Prelec 1998; Gonzalez and Wu 1999). Small probabilities of obtaining a reward may be amplified by factors of 10–10,000, which disproportionally increases EU at low probabilities. Probability distortions are probably not due to misperceptions of the probabilities themselves (d'Acremont et al. 2013), but reflect the misweighting of the influence of probabilities on the utility of rewards. Such misweighting seems intuitively plausible when considering that probabilities at the lower and upper boundaries have references only towards the inside of the probability range, whereas probabilities in the centre of the range can be compared to both lower and higher probabilities (Tversky and Kahneman 1992; Gonzalez and Wu 1999). Probability weighting is incorporated into the calculation of EU by multiplying the objective probability with a weighting factor (Kahneman and Tversky 1979). The probability weighting function is most nonlinear at $0.0<p<\sim0.2$ and $\sim0.5<p<1.0$ and can be modelled with one- or two-parameter functions (Prelec 1998; Gonzalez and Wu 1999) that have often an inverted-S shape and an inflection point at about $p = 0.37$.

Monkeys show nonlinear probability weighting. In choices between probabilistic constant-amount gambles and an adjustable safe reward, the CE (amount of safe reward at choice indifference) exceeds the gambles' EV with lower probabilities and lies below the EV with higher probabilities, even if nonlinear, convex utility is taken into account (Stauffer et al. 2015). Analysis with one- or two-parameter probability weighting functions and incorporating nonlinear utility reveals overweighting of small probability rewards and underweighting of large probability rewards. This distortion is captured by inverted-S probability weighting functions with crossover points below $p = 0.5$ and is best fit by the one-parameter Prelec function (Prelec 1998) with an inflection point around $p = 0.37$ (Stauffer et al. 2015) (Fig. 6.7a).

Humans show similar nonlinear, inverted-S shape probability weighting as monkeys, with similar inflections points around $p = 0.37$ (Fig. 6.7b blue) (Tobler et al. 2008; Hsu et al. 2009). However, individual behaviour is heterogeneous and shows even inverse distortions following a regular-S shape (Fig. 6.7b orange), which are commonly known from verbal descriptions of probabilities without actual experience of outcomes (Hertwig et al. 2004). These curves reflect fits of a two-parameter linear-in-log-odds probability weighting function (Gonzalez and Wu 1999) to pleasantness ratings of stimuli predicting probabilistic rewards in the absence of choice (Tobler et al. 2008). Similar inverted-S- and regular-S-shaped distortions are obtained from fits of the one-parameter Prelec probability weighting function (Prelec 1998) to behavioural choices between probabilistic rewards while accounting for concave utility (Hsu et al. 2009). These experiments replicate the probability distortions known from experimental economics. The divergence between inverted-S and regular-S shapes provides distinct parameters of individual behaviour that can be correlated with BOLD signals across individual participants to identify the involvement of specific brain structures in probability weighting.

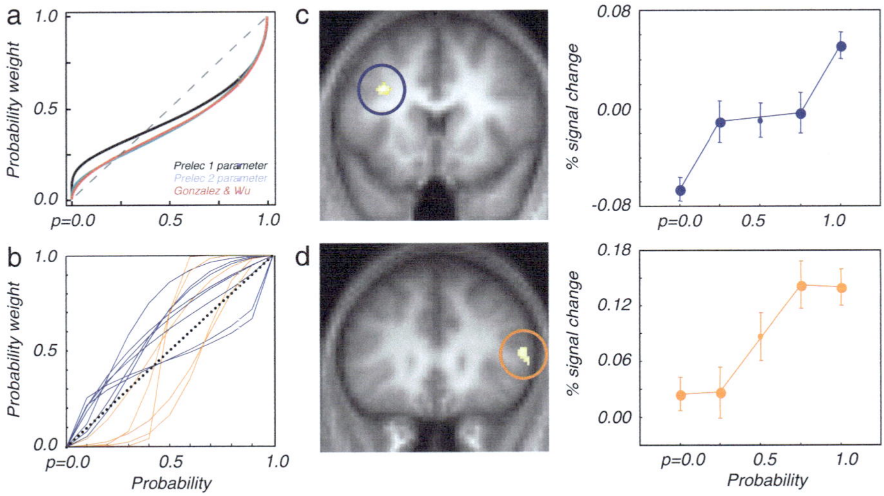


**Fig. 6.7** Subjective probability weighting. (**a**) Behavioural estimation of probability distortion in monkey. The different best-fitting probability weighting functions (Prelect 1998; Gonzalez and Wu 1999) show independently and reproducibly the traditional overweighting of small probability rewards and underweighting of large probability rewards. Each analysis takes the best-fitting convex utility function into account (From Stauffer et al. 2015). (**b**) Behavioural probability weighting functions in individual humans, using a two-parameter linear-in-log-odds weighting function (Gonzalez and Wu 1999). *Blue*: traditional overweighting of small and underweighting of large probabilities in most participants. *Orange*: occasional inverse, regular-S-shaped probability weighting in some participants. (**c**) Activation in dorsolateral prefrontal cortex covarying with a traditional, inverted-S-shaped probability weighting function in participants shows corresponding traditional probability weighting. (**d**) Activation in ventrolateral prefrontal cortex covarying with an inverse, regular-S-shaped probability weighting function in participants shows corresponding inverse probability weighting (**b**–**d**, from Tobler et al. 2008)

### *6.7.2 Neuroimaging*

BOLD responses in the dorsolateral prefrontal cortex show inverted-S shape distortions in humans who overweight small and underweight large probabilities in non-choice pleasantness ratings (Fig. 6.7c). By contrast, BOLD responses in the ventrolateral prefrontal cortex show regular-S probability distortions in individuals who underweight small and overweight large probabilities (Fig. 6.7d). Thus, the neuronal probability distortions correlate with the particular distortion shapes across individuals (Tobler et al. 2008). These probability-distorted BOLD responses contrast with veridical, linear probability coding in the striatum. When assessing probability distortions in choices and including nonlinear utility in the analysis, BOLD responses overweight small and underweight large probabilities in the striatum but not in the prefrontal cortex (Hsu et al. 2009). The striatal responses reflect individual degrees of inverted-S shape distortions across participants.

Taken together, human BOLD responses in the prefrontal cortex show specific forms of probability weighting that correlate with individual behavioural distortions. These neuronal signals occur when probability weighting is assessed with pleasantness ratings without choice, which constitutes a very basic situation in which probabilities are conceivably weighted at perceptual levels without necessarily leading to behavioural choices. When probability weighting is tested during choices, and nonlinear utility is taken into account, probability weighting affects BOLD responses in the striatum.

## 6.8 Conclusions

Rewards are inherently uncertain, except in reduced laboratory situations in which exactly the same sound predicting exactly the same piece of sausage makes a dog salivate. To manage the uncertainty, humans generate cognitive supports such as traditions, myths and religions. And they develop some of the strongest nonsocial emotions in which they hate risk or love risk, depending on the situation, their upbringing, their hormones and many other short-term and long-term factors. However, how the brain as originator of our actions deals with risk is largely unknown.

The emotional attitudes towards risk are difficult to grasp in an objective manner. Risk is very complex and not a physical event but a construct that tries to explain the inherent unpredictability of most events. There are no sensory receptors for risk, and the brain needs to extract risk information from the frequency and unpredictability of past events, whether they were individually experienced, inferred from other information or communicated by other individuals. In an attempt to make these complexities manageable, mathematics and economics have developed a theoretical framework that distinguishes risk from uncertainty on the basis of known probabilities and defines different forms of risk. We are facing risk if we know the probability distribution, irrespective of defining a binary, equiprobable gamble with each outcome being predicted with an equal chance or constituting a fully fledged distribution with many values occurring with lower probability and poor predictability. The different statistical moments of the distribution define different forms of risk, such as variance risk and skewness risk. Otherwise, we are facing ambiguity, the true uncertainty.

Once we define risk in a well-constrained, concise manner, we can use specific tests of experimental economics for assessing meaningful, consistent and rational (utility-maximising) choices under risk, including the estimation of choice indifference points, utility functions and various forms of stochastic dominance. These behavioural tests form the basis for investigating the neuronal processing of risky rewards in the prefrontal cortex in three ways. Individual neurons in the monkey orbitofrontal cortex signal the level of variance risk distinct from reward value and, separately, signal the difference between the current and the predicted risk (risk prediction error). Conceptually, such signals provide a biological basis for the mathematical definitions separating EV from variance. Such neuronal signals may underlie the perception of risk and would be helpful for comparing the riskiness of different rewards before making decisions and for updating these risk signals. Second, in the human and monkey frontal cortex, neuronal signals integrate risk into value in correspondence with subjective risk attitudes that reflect how risk affects choices between different rewards. These neuronal processes follow basic assumptions of the mean-variance approach of financial decision theory. Third, risk information allows the brain with its limited capacity to flexibly acquire the large range of possible rewards. Neurons in the monkey orbitofrontal cortex adapt their response slopes to the spread of the actual reward probability distribution at any given moment. The spread is captured as variance risk. The adaptation to variance risk is part of the matching of neuronal probability distributions to reward probability distributions, which explains such phenomena as contrast effect and reference-dependent processing and constitutes an important component of neuronal processing efficiency. In addition to these specific forms of risk processing, neuronal signals in the human prefrontal cortex reflect the subjective weighting of probabilities that may explain why infrequent rewards may be overrated, a phenomenon shared with the behaviour of monkeys. Such distorted probability weighting may explain some behavioural preference reversals and thus have conceptual consequences for economic choice theory.

Taken together, the well-constrained definition of risk reveals distinct neuronal signals in the prefrontal cortex that correlate with behaviour and inform economic choice theory. Further work will hopefully address how other well-defined forms of risk, such as skewness, affect behavioural choices and influence neuronal processing in the prefrontal cortex and other brain structures.

**Acknowledgements** The author's work mentioned in this article received support from the Wellcome Trust, the European Research Council (ERC) and the Human Frontiers Science Program.

## References

Allais M (1953) Le comportement de l'homme rationnel devant le risque. Critique des postulats de l'ecole américaine. Econometrica 21:503–546

Apicella P, Deffains M, Ravel S, Legallet E (2009) Tonically active neurons in the striatum differentiate between delivery and omission of expected reward in a probabilistic task context. Eur J Neurosci 30:515–526

Bechara A, Damasio AR, Damasio H, Anderson SW (1994) Insensitivity to future consequences following damage to human prefrontal cortex. Cognition 50:7–15

Belova MA, Paton JJ, Morrison SE, Salzman CD (2007) Expectation modulates neural responses to pleasant and aversive stimuli in primate amygdala. Neuron 55:970–984

Bernoulli D (1738) Specimen theoriae novae de mensura sortis. Comentarii Academiae Scientiarum Imperialis Petropolitanae (Papers Imp Acad Sci St Petersburg) 5:175–192 (Translated as: Exposition of a new theory on the measurement of risk. Econometrica 22:23–36, 1954)

Black RW (1968) Shifts in magnitude of reward and contrast effects in instrumental and selective learning: a reinterpretation. Psychol Rev 75:114–126

Burke CJ, Tobler PN (2011) Reward skewness coding in the insula independent of probability and loss. J Neurophysiol 106:2415–2422
Cai X, Padoa-Schioppa C (2012) Neuronal encoding of subjective value in dorsal and ventral anterior cingulate cortex. J Neurosci 32:3791–3808
Caraco T, Martindale S, Whitham TS (1980) An empirical demonstration of risk-sensitive foraging preferences. Anim Behav 28:820–830
Caraco T, Blankenhorn WU, Gregory GM, Newman JA, Recer GM, Zwicker SM (1990) Risk-sensitivity: ambient temperature affects foraging choice. Anim Behav 39:338–345
Christopoulos GI, Tobler PN, Bossaerts P, Dolan RJ, Schultz W (2009) Neural correlates of value, risk, and risk aversion contributing to decision making under risk. J Neurosci 29:12574–12583
d'Acremont M, Fornari E, Bossaerts P (2013) Activity in inferior parietal and medial prefrontal cortex signals the accumulation of evidence in a probability learning task. PLoS Comput Biol 9:e1002895
Diederen KMJ, Schultz W (2015) Scaling prediction errors to reward variability benefits error-driven learning in humans. J Neurophysiol 114:1628–1640
Ding L, Gold JI (2010) Caudate encodes multiple computations for perceptual decisions. J Neurosci 30:15747–15759
Dreher JC, Kohn P, Berman KF (2006) Neural coding of distinct statistical properties of reward information in humans. Cereb Cortex 16:561–573
Fairhall AL, Lewen GD, Bialek W, de Ruyter van Steveninck RR (2001) Efficiency and ambiguity in an adaptive neural code. Nature 412:787–792
Fehr-Duda H, Bruhin A, Epper T, Schubert R (2010) Rationality on the rise: why relative risk aversion increases with stake size. J Risk Uncertain 40:147–180
Gaffan D, Murray EA, Fabre-Thorpe M (1993) Interaction of the amygdala with the frontal lobe in reward memory. Eur J Neurosci 5:968–975
Gonzalez R, Wu G (1999) On the shape of the probability weighting function. Cogn Psychol 38:129–166
Hertwig R, Barron G, Weber EU, Erev I (2004) Decisions from experience and the effect of rare events in risky choice. Psychol Sci 15:534–539
Hosokawa T, Kato K, Inoue M, Mikami A (2007) Neurons in the macaque orbitofrontal cortex code relative preference of both rewarding and aversive outcomes. Neurosci Res 57:434–445
Hosoya T, Baccus SA, Meister M (2005) Dynamic predictive coding by the retina. Nature 436:71–77
Hsu M, Bhatt M, Adolphs R, Tranel D, Camerer CF (2005) Neural systems responding to degrees of uncertainty in human decision-making. Science 310:1680–1683
Hsu M, Krajbich I, Zhao C, Camerer CF (2009) Neural response to reward anticipation under risk is nonlinear in probabilities. J Neurosci 29:2237–2231
Huettel SA, Stowe CJ, Gordon EM, Warner BT, Platt ML (2006) Neural signatures of economic preferences for risk and ambiguity. Neuron 49:765–775
Kahneman D, Tversky A (1979) Prospect theory: an analysis of decision under risk. Econometrica 47:263–291
Kahneman D, Wakker PP, Sarin R (1997) Back to Bentham? Explorations of experienced utility. Q J Econ 112:375–405
Kennerley SW, Behrens TEJ, Wallis JD (2011) Double dissociation of value computations in orbitofrontal and anterior cingulate neurons. Nat Neurosci 14:1581–1589
Kepecs A, Uchida N, Zariwala H, Mainen ZF (2008) Neural correlates, computation and behavioural impact of decision confidence. Nature 455:227–231
Kim S, Hwang J, Seo H, Lee D (2009) Valuation of uncertain and delayed rewards in primate prefrontal cortex. Neural Netw 22:294–304
Kobayashi S, Pinto de Carvalho O, Schultz W (2010) Adaptation of reward sensitivity in orbitofrontal neurons. J Neurosci 30:534–544
Laughlin S (1981) A simple coding procedure enhances a neuron's information capacity. Z Naturforsch 36c:910–912

Lishman WA (1998) Organic psychiatry. Blackwell, Oxford
Machina MJ (1987) Choice under uncertainty: problems solved and unsolved. J Econ Perspect 1:121–154
Markowitz H (1952) Portfolio selection. J Financ 7:77–91
Mas-Colell A, Whinston M, Green J (1995) Microeconomic theory. Oxford Univ Press, New York
Matsumoto M, Hikosaka O (2007) Lateral habenula as a source of negative reward signals in dopamine neurons. Nature 447:1111–1115
McCoy AN, Platt ML (2005) Risk-sensitive neurons in macaque posterior cingulate cortex. Nat Neurosci 8:1220–1227
Miller LA (1985) Cognitive risk-taking after frontal or temporal lobectomy. I: synthesis of fragmented visual information. Neuropsychologia 23:359–369
Nassar MR, Wilson RC, Heasly B, Gold JI (2010) An approximately Bayesian delta-rule model explains the dynamics of belief updating in a changing environment. J Neurosci 30:12366–12378
Nieuwenhuis S, Heslenfeld DJ, Alting van Geusau N, Mars RB, Holroyd CB, Yeung N (2005) Activity in human reward-sensitive brain areas is strongly context dependent. NeuroImage 25:1302–1309
O'Doherty J, Critchley H, Deichmann R, Dolan RJ (2003) Dissociating valence of outcome from behavioral control in human orbital and ventral prefrontal cortices. J Neurosci 23:7931–7939
O'Neill M, Schultz W (2010) Coding of reward risk distinct from reward value by orbitofrontal neurons. Neuron 68:789–800
O'Neill M, Schultz W (2013) Risk prediction error coding in orbitofrontal neurons. J Neurosci 33:15810–15814
Padoa-Schioppa C (2009) Range-adapting representation of economic value in the orbitofrontal cortex. J Neurosci 29:14004–14014
Padoa-Schioppa C, Assad JA (2006) Neurons in the orbitofrontal cortex encode economic value. Nature 441:223–226
Padoa-Schioppa C, Assad JA (2008) The representation of economic value in the orbitofrontal cortex is invariant for changes of menu. Nat Neurosci 11:95–102
Pascal B (1658–1662) Pensées (Translated by R. Ariew. Indianapolis: Hackett Publishing Co. Inc. 2004)
Pastor-Bernier A, Cisek P (2011) Neural correlates of biased competition in premotor cortex. J Neurosci 31:7083–7088
Pearce JM, Hall G (1980) A model for Pavlovian conditioning: variations in the effectiveness of conditioned but not of unconditioned stimuli. Psychol Rev 87:532–552
Prelec D (1998) The probability weighting function. Econometrica 66:497–527
Prelec D, Loewenstein G (1991) Decision making over time and under uncertainty: a common approach. Manag Sci 37:770–786
Preuschoff K, Bossaerts P, Quartz SR (2006) Neural differentiation of expected reward and risk in human subcortical structures. Neuron 51:381–390
Raghuraman AP, Padoa-Schioppa C (2014) Integration of multiple determinants in the neuronal computation of economic values. J Neurosci 34:11583–11603
Rahman S, Sahakian BJ, Hodges JR, Rogers RD, Robbins TW (1999) Specific cognitive deficits in mild frontal variant frontotemporal dementia. Brain 122:1469–1493
Rescorla RA, Wagner AR (1972) A theory of Pavlovian conditioning: variations in the effectiveness of reinforcement and nonreinforcement. In: Black AH, Prokasy WF (eds) Classical conditioning II: current research and theory. Appleton Century Crofts, New York, pp 64–99
Roitman JD, Roitman MF (2010) Risk-preference differentiates orbitofrontal cortex responses to freely chosen reward outcomes. Eur J Neurosci 31:1492–1500
Rothschild M, Stiglitz JE (1970) Increasing risk: I. A definition. J Econ Theory 2:225–243
Schultz W (2015) Neuronal reward and decision signals: from theories to data. Physiol Rev 95:853–951

Schultz W, Dayan P, Montague RR (1997) A neural substrate of prediction and reward. Science 275:1593–1599
Schultz W, O'Neill M, Tobler PN, Kobayashi S (2011) Neuronal signals for reward risk in frontal cortex. Ann N Y Acad Sci 1239:109–117
Seo H, Lee D (2007) Temporal filtering of reward signals in the dorsal anterior cingulate cortex during a mixed-strategy game. J Neurosci 27:8366–8377
So N-Y, Stuphorn V (2012) Supplementary eye field encodes reward prediction error. J Neurosci 32:2950–2963
Solomon RL, Corbit JD (1974) An opponent-process theory of motivation. Psychol Rev 81:119–145
Stauffer WR, Lak A, Schultz W (2014) Dopamine reward prediction error responses reflect marginal utility. Curr Biol 24:2491–2500
Stauffer WR, Lak A, Bossaerts P, Schultz W (2015) Economic choices reveal probability distortion in monkeys. J Neurosci 35:3146–3154
Sutton RS, Barto AG (1981) Toward a modern theory of adaptive networks: expectation and prediction. Psychol Rev 88:135–170
Tinklepaugh OL (1928) An experimental study of representation factors in monkeys. J Comp Psychol 8:197–236
Tobler PN, O'Doherty JP, Dolan R, Schultz W (2007) Reward value coding distinct from risk attitude-related uncertainty coding in human reward systems. J Neurophysiol 97:1621–1632
Tobler PN, Christopoulos GI, O'Doherty JO, Dolan RJ, Schultz W (2008) Neuronal distortions of reward probability without choice. J Neurosci 28:11703–11711
Tobler PN, Christopoulos GI, O'Doherty JP, Dolan RJ, Schultz W (2009) Risk-dependent reward value signal in human prefrontal cortex. Proc Natl Acad Sci U S A 106:7185–7190
Tremblay L, Schultz W (1999) Relative reward preference in primate orbitofrontal cortex. Nature 398:704–708
Tremblay L, Schultz W (2000) Reward-related neuronal activity during go-nogo task performance in primate orbitofrontal cortex. J Neurophysiol 83:1864–1876
Tversky A, Kahneman D (1992) Cumulative prospect theory: an analysis of decision under uncertainty. J Risk Uncertain 5:297–323
Vestergaard MD, Schultz W (2015) Choice mechanisms for past, temporally extended outcomes. Proc R Soc B 282:20141766 1810 (10 pages)
von Neumann J, Morgenstern O (1944) The theory of games and economic behavior. Princeton University Press, Princeton
Wallis JD, Miller EK (2003) Neuronal activity in primate dorsolateral and orbital prefrontal cortex during performance of a reward preference task. Eur J Neurosci 18:2069–2081
Watanabe M, Hikosaka K, Sakagami M, Shirakawa SI (2002) Coding and monitoring of behavioral context in the primate prefrontal cortex. J Neurosci 22:2391–2400
Weber BJ, Chapman GB (2005) Playing for peanuts: why is risk seeking more common for low-stakes gambles? Organ Behav Hum Decis Process 97:31–46
Weber EU, Milliman RA (1997) Perceived risk attitudes: relating risk perception to risky choice. Manag Sci 43:123–144
Weber EU, Shafir S, Blais A-R (2004) Predicting risk sensitivity in humans and lower animals: risk as variance or coefficient of variation. Psychol Rev 111:430–445
Yamada H, Tymula A, Louie K, Glimcher PW (2013) Thirst-dependent risk preferences in monkeys identify a primitive form of wealth. Proc Natl Acad Sci U S A 110:15788–15793

# Chapter 7
# Hierarchical Organization of Frontoparietal Control Networks Underlying Goal-Directed Behavior

**Mathew L. Dixon, Manesh Girn, and Kalina Christoff**

**Abstract** Goal-directed behavior involves a variety of processes that operate over different temporal scales, from the generation and maintenance of distal (long-term) goals to the identification of proximal (immediate) subgoals to the execution of actions in service of those goals. There is also evidence suggesting that the neural underpinnings of goal-directed behavior may be organized along a hierarchical anterior-to-posterior gradient, at least within the frontal cortex. In this review, we examine recently identified large-scale functional networks that are composed of regions spanning the frontal, parietal, and lateral temporal cortices and determine whether there is evidence of a hierarchical organization based on the representation of goals at different temporal scales. Findings from recent functional neuroimaging studies suggest that: (1) the anterior frontoparietal network is involved in generating and planning for distal goals, (2) the posterior frontoparietal is involved in realizing proximal goals by representing desired outcomes and currently relevant rules for action, and (3) the sensorimotor network translates such rules into the execution of motor output. These findings are consistent with the idea that goal-directed behavior can be deconstructed into a temporal hierarchy of goals and corresponding brain networks.

**Keywords** Goals • Prefrontal • Control • Frontoparietal network • fmri • Rewards • Rules

---

M.L. Dixon (✉) • M. Girn
Department of Psychology, University of British Columbia, Vancouver, BC, Canada
e-mail: mattdixon@psych.ubc.ca

K. Christoff (✉)
Department of Psychology, University of British Columbia, Vancouver, BC, Canada

Centre for Brain Health, University of British Columbia, Vancouver, BC, Canada
e-mail: kchristoff@psych.ubc.ca



M. Watanabe (ed.), *The Prefrontal Cortex as an Executive, Emotional, and Social Brain*, DOI 10.1007/978-4-431-56508-6_7

## 7.1 Introduction

Goal-directed behavior involves a variety of processes that operate over different temporal scales. Much of human behavior is guided by *distal* (long-term) goals, including those related to career and family aspirations. For example, based on one's personal preferences and skills, a young adult may decide to become a doctor. To realize such distal goals, it is often necessary to identify and carry out numerous *proximal* (immediate) goals. For example, to become a doctor, the individual will need to study for classes, complete clinical internships, take steps to become licensed, etc. These proximal goals guide much of day-to-day behavior and often require the use of instructions or rules to direct action in a goal-congruent manner. Finally, at the shortest temporal scale, proximal goals need to be translated into the execution of specific actions (e.g., opening a text book and taking notes). This requires the translation of proximal goals and rules for action into corresponding motor commands. As this example illustrates, many instances of goal-directed behavior may emerge as the result of a temporal hierarchy of goals.

Considerable research using humans, nonhuman primates, and rodents has sought to characterize the critical neural substrates of goal-directed behavior. Regions of the lateral prefrontal, posterior parietal, and lateral temporal cortices have been linked to various aspects of goal-directed behavior including rule use, working memory, reasoning, planning, abstract thinking, and response inhibition (Aron et al. 2004; Badre and D'Esposito 2007; Bunge 2004; Bunge et al. 2003; Christoff and Gabrieli 2000; Christof et al. 2009a; Cole and Schneider 2007; Dixon and Christoff 2012; Dosenbach et al. 2006; Duncan 2001, 2010; Koechlin et al. 2003; Kouneiher et al. 2009; Miller and Cohen 2001; Petrides 2005; Rushworth et al. 2007). Furthermore, recent studies suggest that these regions also integrate information about reward and punishment (Beck et al. 2010; Dixon 2015; Dixon and Christoff 2012, 2014; Dixon et al. 2014a; Histed et al. 2009; Jimura et al. 2010; Kobayashi et al. 2006; Padmala and Pessoa 2011; Pochon et al. 2002; Watanabe 1996).

Accumulating evidence suggests that the goal-directed action is instantiated along a hierarchal anterior-to-posterior gradient, at least in the frontal cortex. Generally speaking, more anterior regions have been linked to temporally extended and abstract action goals, while more posterior regions have been linked to concrete information that governs immediate motor output (Badre and D'Esposito 2009a; Bunge and Zelazo 2006; Christoff and Gabrieli 2000; Christoff et al. 2009a; Dixon 2015; Dixon et al. 2014a; Koechlin et al. 2003; Petrides 2005). However, while most of this research has focused on establishing associations between the activity of localized brain regions and specific cognitive functions, ultimately, regions of the frontal cortex operate in concert with parietal and temporal cortices as coordinated large-scale functional networks (Fox and Raichle 2007; Power et al. 2011; Smith et al. 2013; van den Heuvel and Hulshoff Pol 2010; Yeo et al. 2011). This leads to the possibility that there may be a hierarchical organization of functional brain networks that underlies goal-directed action. The literature provides some evidence consistent with this possibility; however, it has yet to be integrated into a coherent framework.

In this review, we examine three functional networks consisting of frontal, parietal, and temporal cortices and identify their roles in goal-directed behavior. We begin by reviewing the basic elements of functional connectivity analyses. Next, we examine the function of each network, in turn, by summarizing findings from recent functional magnetic resonance imaging (fMRI) studies. This review suggests that the functions of the three networks correspond remarkably well with the idea of a temporal hierarchy of goals.

## 7.2 Functional Connectivity Networks

The last decade has seen tremendous progress in understanding the way in which brain regions group together as meaningful systems or networks. Much of this work has emerged from functional connectivity (FC) analyses, which offer a powerful, non-invasive tool for delineating the functional architecture of the human brain. FC analyses typically involve computing the correlation between temporal fluctuations in blood oxygenation level-dependent (BOLD) signal across distributed brain regions (Biswal et al. 1995; Fox et al. 2005; Greicius et al. 2003; Power et al. 2011; Smith et al. 2009; Yeo et al. 2011). These analyses have repeatedly shown that a given brain region will exhibit correlated activation with a very specific collection of other regions, consistent with the idea that they work together as a functional system. For the most part, FC analyses have been performed on data acquired during the "resting state," with the idea that spontaneous fluctuations in activation provide a window into the *intrinsic* neurocognitive architecture of the brain (Biswal et al. 1995; Fox and Raichle 2007; Greicius et al. 2003). A multitude of distinct functional networks have been identified, with each being associated with a specific cognitive function (e.g., visual processing, language, somatosensory-motor control, attention) (Biswal et al. 1995; Buckner et al. 2009; Dosenbach et al. 2007; Fair et al. 2008; Fox et al. 2006; Greicius et al. 2003; Smith et al. 2009; Vincent et al. 2007, 2008). The validity of the networks identified by FC is supported by the fact that they exhibit relative stability across time and context (Damoiseaux et al. 2006; Smith et al. 2009; Zuo et al. 2010) and bear a close resemblance to known anatomical connectivity patterns (Honey et al. 2009; Margulies et al. 2009; Raichle 2009; van den Heuvel and Hulshoff Pol 2010; Van Dijk et al. 2010).

A recent FC-based parcellation using the data of 1000 participants revealed that regions of the frontal, parietal, and temporal cortices cluster into multiple distinct networks (Yeo et al. 2011). This functional organization is important to consider when trying to make links between neural substrates and the different levels of cognitive processing involved in goal-directed behavior. The present discussion will focus on three of the networks identified by Yeo et al. (2011), which we refer to as: (1) the anterior frontoparietal network (aFPN), (2) the posterior frontoparietal

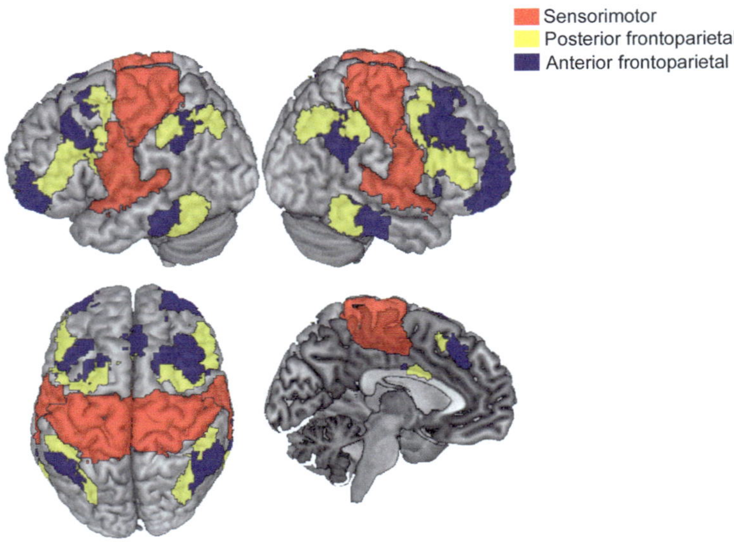


**Fig. 7.1** Frontoparietal networks defined based on patterns of functional connectivity (Adapted from Yeo et al. 2011). Sensorimotor network (*red*), posterior frontoparietal network (*yellow*), anterior frontoparietal network (*blue*)

network (pFPN), and (3) the sensorimotor network (SMN) (Fig. 7.1). These networks are organized along a roughly anterior-to-posterior gradient, particularly in the frontal cortex. Below, we review findings pertaining to the function of each network and demonstrate that they map closely to the different elements of goal-directed behavior described above. This suggests that these networks may reflect a hierarchical organization of cognitive processes that operate over different time scales.

## 7.3 The Anterior Frontoparietal Network and Distal Goals

The aFPN is exclusively comprised of higher-order multimodal association areas in the frontal, parietal, and lateral temporal cortex (Spreng et al. 2010; Vincent et al. 2008; Yeo et al. 2011). Theories have linked regions of the aFPN to complex, internally oriented cognitive processes (Burgess et al. 2007; Christoff and Gabrieli 2000; Ramnani and Owen 2004a) based on its activation across diverse tasks, including complex reasoning (Bunge et al. 2005; Christoff et al. 2001; Kroger et al. 2002; Monti et al. 2007), memory retrieval (Rugg and Wilding 2000; Velanova et al. 2003), multitasking (Braver and Bongiolatti 2002; Koechlin et al. 1999 ), moral decision-making (Greene et al. 2004), abstract thought (Badre and D'Esposito 2009b; Christoff et al. 2009b), spontaneous thought (Christoff et al. 2009a; Christoff et al. 2004b), and future-oriented reward processing (Diekhof and

Gruber 2010; Jimura et al. 2013; McClure et al. 2004). Specifically, the aFPN appears to be involved in a reflective or monitoring function that coordinates, integrates, and evaluates the outputs of prior stages of cognitive processing (Christoff and Gabrieli 2000; Fletcher and Henson 2001; Petrides 2005; Ramnani and Owen 2004b; Tsujimoto et al. 2010). Consistent with this, recent work employing subjective reports and well-controlled paradigms has shown that this network, especially the rostrolateral prefrontal cortex (RLPFC), is critical for *meta-cognitive awareness*—the ability to reflect on and accurately report one's mental contents (Baird et al. 2013; De Martino et al. 2013; Fleming et al. 2010; McCaig et al. 2011; McCurdy et al. 2013). For example, we have shown that participants can learn to modulate RLPFC activation by monitoring real-time feedback and directing attention to their thoughts (McCaig et al. 2011).

Consistent with a role in reflecting on thoughts and feelings, the aFPN is often coactivated with the default mode network (DMN), which has a well-established role in internally directed self-referential processes (Andrews-Hanna et al. 2010; Buckner et al. 2008; Christoff et al. 2009b; Fox et al. 2005; Gusnard and Raichle 2001; Mazoyer et al. 2001; Shulman et al. 1997). Furthermore, resting-state functional connectivity analysis has revealed extensive functional interactions between regions of the aFPN and DMN (Spreng et al. 2013). These networks are activated during periods of waking rest and boring tasks, when individuals tend to reflect on personal concerns and future plans, especially those involving social relationships (Andrews-Hanna 2012; Andrews-Hanna et al. 2010; Fox et al. 2013; Klinger 2008; Mar et al. 2012; Christoff et al. 2004a, 2009b; Fox et al. 2015).

Thus, the aFPN and its interaction with the DMN may enable individuals to transcend the allure of immediate sensory input and, instead, allow attention to be directed to an internal train of thought (Dixon et al. 2014b). One function of internal thought may be to discern long-term goals and plans to achieve them. In one study, Spreng et al. (2010) had participants think about personal future goals (e.g., academic success) and three to five steps necessary to achieve those goals and obstacles that might interfere with goal attainment. They monitored fMRI signal while participants engaged in this autobiographical planning task, and compared it to signal during a control task that did not require keeping in mind a distal goal or means to achieve it. The results demonstrated robust activation of the aFPN and DMN as participants engaged in future goal planning (Spreng et al. 2010). These findings are consistent with the idea that interactions between the aFPN and DMN enable individuals to internally simulate future scenarios and intermediary steps necessary to turn those scenarios into reality. In another study, we had participants perform a task requiring a series of goal-directed actions and informed them that they could earn up to $60 at the end of the experiment based on their performance (Dixon et al. 2014a). Participants had to use rules to respond to visual images on each trial, and following their response, they were presented with a feedback screen that revealed their total cumulative winnings up to that point. Thus, the feedback screen allowed participants to monitor their progress toward the distal goal of earning $60. The results demonstrated robust activation of the aFPN specifically

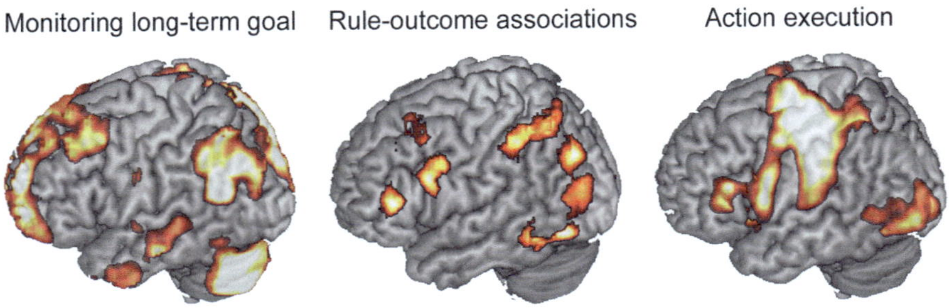


**Fig. 7.2** Hierarchical organization of functional networks (Results adapted from Dixon and Christoff (2012) and Dixon et al. (2014a)). The aFPN is specifically activated when individuals generate and monitor progress toward distal goals. The pFPN is involved in guiding behavior toward proximal goals by specifying currently relevant rule-outcome associations. The SMN is selectively activated during the execution of overt actions

during this task phase, when information about progress toward the distal goal was available (Fig. 7.2) (Dixon et al. 2014a).

Corroborating these findings, regions within the aFPN are recruited when individuals plan steps to attain a future goal (Gerlach et al. 2014), choose to avoid situations that may interfere with the attainment of future rewards (Crockett et al. 2013), and select actions directed toward future rather than immediate rewards (Diekhof and Gruber 2010; Jimura et al. 2013; McClure et al. 2004). Furthermore, several studies have found that the aFPN is critical for the coordination of multiple subgoals (Braver and Bongiolatti 2002; Dreher et al. 2008; Koechlin et al. 1999). Finally, the aFPN plays an important role when individuals decide to shift from current actions to explore new options that may yield better long-term outcomes (Badre et al. 2012; Boorman et al. 2009) suggesting that it may be involved in prioritizing motivational goals (Dixon and Christoff 2014). The involvement of the aFPN in generating distal goals and steps to achieve them is consistent with its activation in reasoning tasks that require the capacity to see the relationships between multiple pieces of information (Bunge et al. 2005; Christoff et al. 2001; Kroger et al. 2002; Monti et al. 2007) and the suggestion that regions of the aFPN supports the serial organization of several subgoals (Koechlin et al. 1999) and the manipulation and evaluation of internally generated information (Christoff and Gabrieli 2000).

To summarize, the findings reviewed above are consistent with the idea that the aFPN is preferentially involved in processes related to distal rather than proximal goals. The aFPN has a close relationship with the DMN and is consistently recruited when individuals direct attention internally to thoughts about personal concerns and steps to achieve desired future outcomes. Thus, the aFPN may sit at the top of the hierarchy supporting goal-directed behavior.

## 7.4 The Posterior Frontoparietal Network and Proximal Goals

Whereas the aFPN plays a role in discerning overarching priorities, the pFPN operates on a shorter time scale, contributing to the attainment of proximal goals. The pFPN is not required for all proximal goals; rather, it is primarily recruited when desired outcomes can only be obtained through the intentional regulation of behavior, often referred to as exerting *cognitive control*. A large literature has shown that the pFPN is invariably engaged by experimental tasks that involve responding to externally presented stimuli (Duncan 2010). The regions comprising this network include frontal, parietal, and lateral temporal regions that are each located posterior to the regions comprising the aFPN. This network facilitates goal-directed behavior by encoding and maintaining task rules within working memory (Bunge et al. 2003; De Baene et al. 2012). Rules often take the form of "if-then" mappings that specify a set of stimulus-response contingencies.

Recent work has shown that individuals are reluctant to employ cognitive control due to an inherent *effort cost* (Botvinick and Braver 2015; Dixon and Christoff 2012; McGuire and Botvinick 2010). For example, when given the choice between two tasks, individuals will reliably choose the easier task (McGuire and Botvinick 2010). It requires considerable effort to intentionally direct action, and this may often be experienced as aversive. Accordingly, individuals only engage of cognitive control if they think that it will produce an emotionally valuable outcome that outweighs the effort cost (Dixon and Christoff 2012). We have reported that when given the choice between selecting a well-practiced habitual action or a rule-based action that requires cognitive control, individuals are only likely to select the latter option if it is expected to result in a larger monetary reward than the habitual action (Dixon and Christoff 2012). This implies that many moment-to-moment decisions about goal-directed behavior require that individuals perceive the relationship between cognitive demands (e.g., rule use) and desired outcomes.

We have found evidence that the pFPN encodes such rule-outcome associations (Dixon and Christoff 2012). Participants performed a task in which they employed simple rules to obtain monetary rewards. There were two different rules (judge whether a face is male/female or judge whether a word has a concrete/abstract meaning) and two different monetary outcomes (25¢ or $0). Importantly, each trial started with an instruction cue that signaled the currently relevant rule and which outcome to expect. On some trials, a second instruction cue appeared prior to the stimulus and signaled either the same (repeated) rules or novel rules and either the same (repeated) outcome or a novel outcome. This 2 × 2 factorial design allowed us to look for fMRI adaptation (i.e., a change in neural activation) when there was repetition of the rules, repetition of the reward outcome, or repetition of a specific rule-outcome pairing. Several areas within the pFPN exhibited an interaction effect, demonstrating fMRI adaptation specifically when there was repetition of a specific rule-outcome pairing, but not when there was repetition of just the rules or just the outcome (Fig. 7.2). Hence, this network represents the *association* between a specific rule and an expected motivational outcome (Dixon and Christoff 2012). Given that rule-outcome combinations changed from trial to trial, this suggests that the pFPN has the capacity to rapidly represent rule-outcome associations based on symbolic instructed information and thereby contribute to the acquisition of proximal goals.

Consistent with our findings, recent evidence indicates that activation of the inferior frontal sulcus—a key node of the pFPN—reflects an interaction between the complexity of rules that are required to respond to stimuli and the size of an expected reward outcome (Bahlmann et al. 2015). Moreover, a study employing multivariate pattern analysis found that several regions of the pFPN exhibited stronger encoding of task rules during a monetary incentive condition relative to the no incentive condition (Etzel et al. 2015). Similarly, Jimura et al. (2010) found that monetary incentives shifted pFPN working memory-related activation toward a sustained pattern indicative of enhanced proactive control (i.e., anticipatory maintenance of task information). Together, these studies indicate that the pFPN integrates cognitive demands and reward outcomes and thereby represents information that is crucial in guiding action toward proximal goals.

The majority of studies to date have focused on the engagement of cognitive control in service of obtaining an immediately available incentive. However, in many real-life situations, cognitive control may be required to complete numerous tasks in service of a distal goal, without immediate reinforcement. In fact, to attain distal goals, it is often necessary to resist the temptation of immediate rewards that may interfere with proximal goals that support the attainment of distal goals (e.g., choosing to stay in and study for an exam instead of going out to the movies). In these cases, input from the aFPN signaling the value of the distal goal may engage the cognitive control processes of the pFPN such that relevant proximal goals are completed, and the aFPN may simultaneously send inhibitory signals to regions such as the ventral striatum that are sensitive to immediate rewards. Findings from delay of gratification paradigms appear to support this idea (Diekhof and Gruber 2010; Jimura et al. 2013; McClure et al. 2004; van den Bos et al. 2014); however, more research is needed that directly examines situations in which the aFPN and pFPN may be in a hierarchical relationship, contributing to the same goal at different levels (temporal scales).

To summarize, the pFPN is involved in specifying proximal goals and how to achieve them. This network is invariably activated in studies that require focused attention on externally presented stimuli, and it represents rule-outcome relationships that guide action selection in a goal-congruent manner. In doing so, the pFPN may exert top-down control over sensorimotor regions that execute actions. In many cases, the pFPN may serve as the second tier of the goal hierarchy, taking information from the aFPN about distal goals and translating this information into corresponding proximal goals.

## 7.5 The Sensorimotor Network and the Execution of Goal-Directed Actions

Once the pFPN has established a proximal goal and the rules specifying goal-appropriate behavior, this information needs to be translated into the execution of specific voluntary actions. The SMN supports this process (Bunge 2004; Koechlin et al. 2003; Petrides 2005). The SMN is a set of interconnected brain regions that serve to integrate concrete sensory and motor information for the initiation of voluntary movement. The main constituents of this network have had a long history of study in neuroscience (Ferrier 1873; Penfield and Boldrey 1937). Resting-state functional connectivity studies have revealed that the SMN consists of primary motor and somatosensory cortices, as well as lateral and medial premotor regions (Biswal et al. 1995; De Luca et al. 2005; Power et al. 2011; Xiong et al. 1999; Yeo et al. 2011).

The primary and premotor cortices, as their names suggest, mediate motor sequences. Historically, research has viewed these regions as in a hierarchical relationship with each other, with the premotor cortex concerned with complex higher-order movements and primary motor cortex concerned with breaking these movements down into their basic constituents to allow for execution (Picard and Strick 1996; Rizzolatti and Luppino 2001). Specifically, the premotor cortex consists of a number of specialized subregions which correspond to different high-level functions such as the planning, preparing, or imagining of movement (Muakkassa and Strick 1979; Rizzolatti et al. 1988, 2002). Petrides (2005) found that lesions to the macaque premotor cortex caused selective impairments in performance on visuomotor conditional learning tasks, which involve learning associations between visual stimuli and movement sequences (Petrides 2005). Thus, the premotor cortex is vital for the ability to translate visual information into correct motor responses. It seems likely that the premotor cortex uses information about task rules represented by the pFPN to discern goal-appropriate actions (Bunge 2004; Koechlin et al. 2003).

In comparison, the primary motor cortex contains a body map, with different sectors linked to the control of specific groups of muscles or body parts (Grafton et al. 1991; Penfield and Boldrey 1937). As such, it has typically been viewed as a slave to the commands of the premotor cortex, responsible for translating action commands into sequences of muscle recruitment. Recent work, however, has suggested that this may be an overly simplistic view. Originating from the finding that complex behavioral sequences could be evoked through stimulation of primary motor subregions (Graziano et al. 2002), multiple lines of research have suggested that the primary motor cortex additionally contains a map of ethologically meaningful action sequences (Graziano 2006, 2016). For example, electrical stimulation of certain regions of primary motor cortex in primates has been demonstrated to reliably evoke sequences such as eating-related hand to mouth movements or tree-climbing movements (Graziano 2016). While the exact functional relationship between the primary and premotor cortices is still under debate, both are central for the execution of voluntary movement.

The primary somatosensory cortex is associated with the conscious sense of touch, including feelings of pressure, temperature, vibration, position, and movement that arise from the skin, joints, and muscles. It contains a somatotopic map of the body, with greater cortical surface area being devoted to areas possessing greater tactile sensitivity, and this has been pictorially rendered in the so-called sensory homunculus (Penfield and Boldrey 1937). The coordinated interaction between primary and premotor cortices on the one hand, and somatosensory cortex on the other hand, may contribute to an embodied sense of self and agency (Christoff et al. 2011). When performing an action, a copy of the motor commands can be compared with the resulting changes in sensory input by a comparator mechanism (likely instantiated by the SMN), to implicitly signal that one has acted upon the world (Christoff et al. 2011). "Reafferent" sensory signals that match initiated motor commands are self-specific, as they are intrinsically related to agent's executed action, whereas all other sensory changes are nonself-specific (Christoff et al. 2011). Thus, by tracking the relationship between efferent motor signals and afferent sensory consequences, the SMN may contribute to the bodily experience of agency associated with the expression of goal-directed behavior.

The aforementioned findings suggest that the SMN is involved in processes that are closely related to the actual initiation of goal-directed actions and not other high-level processes. We recently found evidence consistent with this idea. We analyzed neural activation during several distinct phases of a cognitive task: (1) instruction cue period, during which the relevant rules and outcome for the current trial were indicated; (2) delay period, which required working memory; (3) stimulus-response period during which participants executed an action; and (4) feedback period. We found that medial and lateral premotor regions were preferentially activated during the stimulus-response period, when participants executed an overt goal-directed action (Fig. 7.2) (Dixon et al. 2014a). This suggests that the SMN mainly contributes to goal-directed action at the finest temporal scale, when goals are realized through embodied action output.

To summarize, the SMN has been associated with specifying movement plans and monitoring reafferent feedback that can be used to fine-tune the execution of goal-directed actions. This network facilitates the execution of context appropriate actions over competing actions by representing sensorimotor associations, likely based on rules specified by the pFPN (Bunge 2004; Koechlin et al. 2003; Petrides 2005). Thus, the SMN is involved in realizing proximal goals by specifying action patterns and their sensory consequences and contributes to an embodied sense of being an agent acting on the world.

## 7.6 Conclusions

In this chapter, we have outlined a simple deconstruction of goal-directed behavior based on the components operating over different time scales, from distal goals that span considerable periods of time to proximal goals that govern moment-to-moment interactions with the environment to the overt execution of specific actions. These components map onto three distinct brain networks that are organized along a roughly anterior-to-posterior axis in functional networks composed of frontal, parietal, and lateral temporal cortices. This cognitive and neural architecture underlying goal-directed behavior is consistent with the notion of a hierarchical organization, with neurocognitive components pertaining to longer time scales governing those pertaining to shorter time scales. The findings reviewed here offer a systems-level account of how we achieve personal goals and offer a window into the constituent elements making up complex human behavior. While we have sketched a view of the spatial topography and functional relevance of three brain networks relevant to goal-directed behavior, it is important to note that network properties change across time and context (Dixon et al. 2017). Thus, our review provides a "bird's-eye view" of brain networks and goal-directed behavior; however, future work may reveal more precise cognitive and neural components.

## References


Andrews-Hanna JR, Reidler JS, Sepulcre J, Poulin R, Buckner RL (2010) Functional-anatomic fractionation of the brain's default network. Neuron 65(4):550–562

Andrews-Hanna JR (2012) The brain's default network and its adaptive role in internal mentation. Neuroscientist 18(3):251–270

Aron AR, Robbins TW, Poldrack RA (2004) Inhibition and the right inferior frontal cortex. Trends Cogn Sci 8(4):170–177

Badre D, D'Esposito M (2007) Functional magnetic resonance imaging evidence for a hierarchical organization of the prefrontal cortex. J Cogn Neurosci 19(12):2082–2099

Badre D, D'Esposito M (2009a) Is the rostro-caudal axis of the frontal lobe hierarchical? Nat Rev Neurosci 10(9):659–669

Badre D, D'Esposito M (2009b) Is the rostro-caudal axis of the frontal lobe hierarchical? Nat Rev Neurosci 10(9):659–669

Badre D, Doll BB, Long NM, Frank MJ (2012) Rostrolateral prefrontal cortex and individual differences in uncertainty-driven exploration. Neuron 73(3):595–607

Bahlmann J, Aarts E, D'Esposito M (2015) Influence of motivation on control hierarchy in the human frontal cortex. J Neurosci 35(7):3207–3217

Baird B, Smallwood J, Gorgolewski KJ, Margulies DS (2013) Medial and lateral networks in anterior prefrontal cortex support metacognitive ability for memory and perception. J Neurosci 33(42):16657–16665

Beck SM, Locke HS, Savine AC, Jimura K, Braver TS (2010) Primary and secondary rewards differentially modulate neural activity dynamics during working memory. PLoS One 5(2): e9251

Biswal B, Yetkin FZ, Haughton VM, Hyde JS (1995) Functional connectivity in the motor cortex of resting human brain using echo-planar MRI. Magn Reson Med 34(4):537–541

Boorman ED, Behrens TE, Woolrich MW, Rushworth MF (2009) How green is the grass on the other side? Frontopolar cortex and the evidence in favor of alternative courses of action. Neuron 62(5):733–743

Botvinick M, Braver T (2015) Motivation and cognitive control: from behavior to neural mechanism. Annu Rev Psychol 66:83–113

Braver TS, Bongiolatti SR (2002) The role of frontopolar cortex in subgoal processing during working memory. NeuroImage 15(3):523–536

Buckner RL, Andrews-Hanna JR, Schacter DL (2008) The brain's default network: anatomy, function, and relevance to disease. Ann N Y Acad Sci 1124:1–38

Buckner RL, Sepulcre J, Talukdar T, Krienen FM, Liu H, Hedden T et al (2009) Cortical hubs revealed by intrinsic functional connectivity: mapping, assessment of stability, and relation to Alzheimer's disease. J Neurosci 29(6):1860–1873

Bunge SA (2004) How we use rules to select actions: a review of evidence from cognitive neuroscience. Cogn Affect Behav Neurosci 4(4):564–579

Bunge SA, Zelazo PD (2006) A brain-based account of the development of rule use in childhood. Curr Dir Psychol Sci 15(3):118–121

Bunge SA, Kahn I, Wallis JD, Miller EK, Wagner AD (2003) Neural circuits subserving the retrieval and maintenance of abstract rules. J Neurophysiol 90(5):3419–3428

Bunge SA, Wendelken C, Badre D, Wagner AD (2005) Analogical reasoning and prefrontal cortex: evidence for separable retrieval and integration mechanisms. Cereb Cortex 15 (3):239–249

Burgess PW, Dumontheil I, Gilbert SJ (2007) The gateway hypothesis of rostral prefrontal cortex (area 10) function. Trends Cogn Sci 11(7):290–298

Christoff K, Gabrieli JDE (2000) The frontopolar cortex and human cognition: evidence for a rostrocaudal hierarchical organization within the human prefrontal cortex. Psychobiology 28 (2):168–186

Christoff K, Prabhakaran V, Dorfman J, Zhao Z, Kroger JK, Holyoak KJ et al (2001) Rostrolateral prefrontal cortex involvement in relational integration during reasoning. NeuroImage 14 (5):1136–1149

Christoff K, Ream JM, Gabrieli JD (2004a) Neural basis of spontaneous thought processes. Cortex; J Devoted Study Nerv Syst Behav 40(4–5):623–630

Christoff K, Ream JM, Gabrieli JD (2004b) Neural basis of spontaneous thought processes. Cortex 40(4–5):623–630

Christoff K, Gordon AM, Smallwood J, Smith R, Schooler JW (2009a) Experience sampling during fMRI reveals default network and executive system contributions to mind wandering. Proc Natl Acad Sci U S A 106(21):8719–8724

Christoff K, Gordon AM, Smallwood J, Smith R, Schooler JW (2009b) Experience sampling during fMRI reveals default network and executive system contributions to mind wandering. Proc Natl Acad Sci U S A 106(21):8719–8724

Christoff K, Keramatian K, Gordon AM, Smith R, Madler B (2009c) Prefrontal organization of cognitive control according to levels of abstraction. Brain Res 1286:94–105

Christoff K, Keramatian K, Gordon AM, Smith R, Madler B (2009d) Prefrontal organization of cognitive control according to levels of abstraction. Brain Res 1286:94–105

Christoff K, Cosmelli D, Legrand D, Thompson E (2011) Specifying the self for cognitive neuroscience. Trends Cogn Sci 15(3):104–112

Cole MW, Schneider W (2007) The cognitive control network: integrated cortical regions with dissociable functions. NeuroImage 37(1):343–360

Crockett MJ, Braams BR, Clark L, Tobler PN, Robbins TW, Kalenscher T (2013) Restricting temptations: neural mechanisms of precommitment. Neuron 79(2):391–401

Damoiseaux J, Rombouts S, Barkhof F, Scheltens P, Stam C, Smith SM et al (2006) Consistent resting-state networks across healthy subjects. Proc Natl Acad Sci 103(37):13848–13853

De Baene W, Kuhn S, Brass M (2012) Challenging a decade of brain research on task switching: brain activation in the task-switching paradigm reflects adaptation rather than reconfiguration of task sets. Hum Brain Mapp 33(3):639–651

De Luca M, Smith S, De Stefano N, Federico A, Matthews PM (2005) Blood oxygenation level dependent contrast resting state networks are relevant to functional activity in the neocortical sensorimotor system. Exp Brain Res 167(4):587–594
De Martino B, Fleming SM, Garrett N, Dolan RJ (2013) Confidence in value-based choice. Nat Neurosci 16(1):105–110
Diekhof EK, Gruber O (2010) When desire collides with reason: functional interactions between anteroventral prefrontal cortex and nucleus accumbens underlie the human ability to resist impulsive desires. J Neurosci 30(4):1488–1493
Dixon ML (2015) Cognitive control, emotional value, and the lateral prefrontal cortex. Front Psychol 6:758
Dixon ML, Christoff K (2012) The decision to engage cognitive control is driven by expected reward-value: neural and behavioral evidence. PLoS One 7(12):1–12
Dixon ML, Christoff K (2014) The lateral prefrontal cortex and complex value-based learning and decision making. Neurosci Biobehav Rev 45:9
Dixon ML, Fox KCR, Christoff K (2014a) Evidence for rostro-caudal functional organization in multiple brain areas related to goal-directed behavior. Brain Research 1572:26
Dixon ML, Fox KCR, Christoff K (2014b) A framework for understanding the relationship between externally and internally directed cognition. Neuropsychologia 62:321–330
Dixon ML, Andrews-Hanna JR, Spreng RN, Irving Z, Mills C, Girn M, Christoff K (2017) Interactions between the default network and dorsal attention network vary across default subsystems, time, and cognitive states. Neuroimage 147:632–649
Dosenbach NU, Visscher KM, Palmer ED, Miezin FM, Wenger KK, Kang HC et al (2006) A core system for the implementation of task sets. Neuron 50(5):799–812
Dosenbach NU, Fair DA, Miezin FM, Cohen AL, Wenger KK, Dosenbach RA et al (2007) Distinct brain networks for adaptive and stable task control in humans. Proc Natl Acad Sci U S A 104(26):11073–11078
Dreher JC, Koechlin E, Tierney M, Grafman J (2008) Damage to the fronto-polar cortex is associated with impaired multitasking. PLoS One 3(9):e3227
Duncan J (2001) An adaptive coding model of neural function in prefrontal cortex. Nat Rev Neurosci 2(11):820–829
Duncan J (2010) The multiple-demand (MD) system of the primate brain: mental programs for intelligent behaviour. Trends Cogn Sci 14(4):172–179
Etzel JA, Cole MW, Zacks JM, Kay KN, Braver TS (2015) Reward motivation enhances task coding in frontoparietal cortex. Cereb Cortex 26:1647
Fair DA, Cohen AL, Dosenbach NU, Church JA, Miezin FM, Barch DM et al (2008) The maturing architecture of the brain's default network. Proc Natl Acad Sci U S A 105(10):4028–4032
Ferrier D (1873) The localization of function in the brain. Proc R Soc Lond 22(148–155):228–232
Fleming SM, Weil RS, Nagy Z, Dolan RJ, Rees G (2010) Relating introspective accuracy to individual differences in brain structure. Science 329(5998):1541–1543
Fletcher PC, Henson RN (2001) Frontal lobes and human memory: insights from functional neuroimaging. Brain 124(Pt 5):849–881
Fox M, Raichle ME (2007) Spontaneous fluctuations in brain activity observed with functional magnetic resonance imaging. Nat Rev Neurosci 8(9):700–711
Fox M, Snyder AZ, Vincent JL, Corbetta M, Van Essen DC, Raichle ME (2005) The human brain is intrinsically organized into dynamic, anticorrelated functional networks. Proc Natl Acad Sci U S A 102(27):9673–9678
Fox M, Corbetta M, Snyder AZ, Vincent JL, Raichle ME (2006) Spontaneous neuronal activity distinguishes human dorsal and ventral attention systems. Proc Natl Acad Sci U S A 103 (26):10046–10051
Fox KCR, Nijeboer S, Solomonova E, Domhoff GW, Christoff K (2013) Dreaming as mind wandering: evidence from functional neuroimaging and first-person content reports. Front Hum Neurosci 7(412):1–18

Fox KC, Spreng RN, Ellamil M, Andrews-Hanna JR, Christoff K (2015) The wandering brain: meta-analysis of functional neuroimaging studies of mind-wandering and related spontaneous thought processes. Neuroimage 111:611

Gao W, Lin W (2012) Frontal parietal control network regulates the anti-correlated default and dorsal attention networks. Hum Brain Mapp 33(1):192–202

Gerlach KD, Spreng RN, Madore KP, Schacter DL (2014) Future planning: default network activity couples with frontoparietal control network and reward-processing regions during process and outcome simulations. Soc Cogn Affect Neurosci 9:1942

Grafton ST, Woods RP, Mazziotta JC, Phelps ME (1991) Somatotopic mapping of the primary motor cortex in humans: activation studies with cerebral blood flow and positron emission tomography. J Neurophysiol 66(3):735–743

Graziano M (2006) The organization of behavioral repertoire in motor cortex. Annu Rev Neurosci 29:105–134

Graziano M (2016) Ethological action maps: a paradigm shift for the motor cortex. Trends Cogn Sci 20(2):121–132

Graziano M, Taylor CS, Moore T (2002) Complex movements evoked by microstimulation of precentral cortex. Neuron 34(5):841–851

Greene JD, Nystrom LE, Engell AD, Darley JM, Cohen JD (2004) The neural bases of cognitive conflict and control in moral judgment. Neuron 44(2):389–400

Greicius M, Krasnow B, Reiss AL, Menon V (2003) Functional connectivity in the resting brain: a network analysis of the default mode hypothesis. Proc Natl Acad Sci U S A 100(1):253–258

Gusnard DA, Raichle ME (2001) Searching for a baseline: functional imaging and the resting human brain. Nat Rev Neurosci 2(10):685–694

Histed MH, Pasupathy A, Miller EK (2009) Learning substrates in the primate prefrontal cortex and striatum: sustained activity related to successful actions. Neuron 63(2):244–253

Honey CJ, Sporns O, Cammoun L, Gigandet X, Thiran JP, Meuli R et al (2009) Predicting human resting-state functional connectivity from structural connectivity. Proc Natl Acad Sci U S A 106(6):2035–2040

Jimura K, Locke HS, Braver TS (2010) Prefrontal cortex mediation of cognitive enhancement in rewarding motivational contexts. Proc Natl Acad Sci U S A 107(19):8871–8876

Jimura K, Chushak MS, Braver TS (2013) Impulsivity and self-control during intertemporal decision making linked to the neural dynamics of reward value representation. J Neurosci 33 (1):344–357

Klinger E (2008) 15 Daydreaming and fantasizing: thought flow and motivation In Markman KD, Klein WMP, Suh JA (eds) Handbook of imagination and mental simulation

Kobayashi S, Nomoto K, Watanabe M, Hikosaka O, Schultz W, Sakagami M (2006) Influences of rewarding and aversive outcomes on activity in macaque lateral prefrontal cortex. Neuron 51 (6):861–870

Koechlin E, Basso G, Pietrini P, Panzer S, Grafman J (1999) The role of the anterior prefrontal cortex in human cognition. Nature 399(6732):148–151

Koechlin E, Ody C, Kouneiher F (2003) The architecture of cognitive control in the human prefrontal cortex. Science 302(5648):1181–1185

Kouneiher F, Charron S, Koechlin E (2009) Motivation and cognitive control in the human prefrontal cortex. Nat Neurosci 12(7):939–945

Kroger JK, Sabb FW, Fales CL, Bookheimer SY, Cohen MS, Holyoak KJ (2002) Recruitment of anterior dorsolateral prefrontal cortex in human reasoning: a parametric study of relational complexity. Cereb Cortex 12(5):477–485

Mar RA, Mason MF, Litvack AD (2012) How daydreaming relates to life satisfaction, loneliness, and social support: the importance of gender and daydream content. Conscious Cogn 21:401–407

Margulies DS, Vincent JL, Kelly C, Lohmann G, Uddin LQ, Biswal BB et al (2009) Precuneus shares intrinsic functional architecture in humans and monkeys. Proc Natl Acad Sci U S A 106 (47):20069–20074

Mazoyer B, Zago L, Mellet E, Bricogne S, Etard O, Houde O et al (2001) Cortical networks for working memory and executive functions sustain the conscious resting state in man. Brain Res Bull 54(3):287–298

McCaig RG, Dixon M, Keramatian K, Liu I, Christoff K (2011) Improved modulation of rostrolateral prefrontal cortex using real-time fMRI training and meta-cognitive awareness. NeuroImage 55(3):1298–1305

McClure SM, Laibson DI, Loewenstein G, Cohen JD (2004) Separate neural systems value immediate and delayed monetary rewards. Science 306(5695):503–507

McCurdy LY, Maniscalco B, Metcalfe J, Liu KY, de Lange FP, Lau H (2013) Anatomical coupling between distinct metacognitive systems for memory and visual perception. J Neurosci 33(5):1897–1906

McGuire JT, Botvinick MM (2010) Prefrontal cortex, cognitive control, and the registration of decision costs. Proc Natl Acad Sci U S A 107(17):7922–7926

Miller EK, Cohen JD (2001) An integrative theory of prefrontal cortex function. Annu Rev Neurosci 24:167–202

Monti MM, Osherson DN, Martinez MJ, Parsons LM (2007) Functional neuroanatomy of deductive inference: a language-independent distributed network. NeuroImage 37(3):1005–1016

Muakkassa KF, Strick PL (1979) Frontal lobe inputs to primate motor cortex: evidence for four somatotopically organized 'premotor' areas. Brain Res 177(1):176–182

Padmala S, Pessoa L (2011) Reward reduces conflict by enhancing attentional control and biasing visual cortical processing. J Cogn Neurosci 23(11):3419–3432

Penfield W, Boldrey E (1937) Somatic motor and sensory representation in the cerebral cortex of man as studied by electrical stimulation. Brain 60(4):389–443

Petrides M (2005) Lateral prefrontal cortex: architectonic and functional organization. Philos Trans R Soc Lond Ser B Biol Sci 360(1456):781–795

Picard N, Strick PL (1996) Motor areas of the medial wall: a review of their location and functional activation. Cereb Cortex 6(3):342–353

Pochon JB, Levy R, Fossati P, Lehericy S, Poline JB, Pillon B et al (2002) The neural system that bridges reward and cognition in humans: an fMRI study. Proc Natl Acad Sci U S A 99 (8):5669–5674

Power JD, Cohen AL, Nelson SM, Wig GS, Barnes KA, Church JA et al (2011) Functional network organization of the human brain. Neuron 72(4):665–678

Raichle ME (2009) A paradigm shift in functional brain imaging. J Neurosci 29(41):12729–12734

Ramnani N, Owen AM (2004a) Anterior prefrontal cortex: insights into function from anatomy and neuroimaging. Nat Rev Neurosci 5(3):184–194

Ramnani N, Owen AM (2004b) Anterior prefrontal cortex: insights into function from anatomy and neuroimaging. Nat Rev Neurosci 5(3):184–194

Rizzolatti G, Luppino G (2001) The cortical motor system. Neuron 31(6):889–901

Rizzolatti G, Camarda R, Fogassi L, Gentilucci M, Luppino G, Matelli M (1988) Functional organization of inferior area 6 in the macaque monkey. Exp Brain Res 71(3):491–507

Rizzolatti G, Fogassi L, Gallese V (2002) Motor and cognitive functions of the ventral premotor cortex. Curr Opin Neurobiol 12(2):149–154

Rugg MD, Wilding EL (2000) Retrieval processing and episodic memory. Trends Cogn Sci 4 (3):108–115

Rushworth MF, Behrens TE, Rudebeck PH, Walton ME (2007) Contrasting roles for cingulate and orbitofrontal cortex in decisions and social behaviour. Trends Cogn Sci 11(4):168–176

Shulman GL, Fiez JA, Corbetta M, Buckner RL, Miezin FM, Raichle ME et al (1997) Common blood flow changes across visual tasks: II. Decreases in cerebral cortex. J Cogn Neurosci 9 (5):648–663

Smith SM, Fox PT, Miller KL, Glahn DC, Fox PM, Mackay CE et al (2009) Correspondence of the brain's functional architecture during activation and rest. Proc Natl Acad Sci U S A 106 (31):13040–13045

Smith SM, Vidaurre D, Beckmann CF, Glasser MF, Jenkinson M, Miller KL et al (2013) Functional connectomics from resting-state fMRI. Trends Cogn Sci 17(12):666–682
Spreng RN, Stevens WD, Chamberlain JP, Gilmore AW, Schacter DL (2010) Default network activity, coupled with the frontoparietal control network, supports goal-directed cognition. NeuroImage 53(1):303–317
Spreng RN, Sepulcre J, Turner GR, Stevens WD, Schacter DL (2013) Intrinsic architecture underlying the relations among the default, dorsal attention, and frontoparietal control networks of the human brain. J Cogn Neurosci 25(1):74–86
Tsujimoto S, Genovesio A, Wise SP (2010) Evaluating self-generated decisions in frontal pole cortex of monkeys. Nat Neurosci 13(1):120–126
van den Bos W, Rodriguez CA, Schweitzer JB, McClure SM (2014) Connectivity strength of dissociable striatal tracts predict individual differences in temporal discounting. J Neurosci 34 (31):10298–10310
van den Heuvel MP, Hulshoff Pol HE (2010) Exploring the brain network: a review on resting-state fMRI functional connectivity. Eur Neuropsychopharmacol 20(8):519–534
Van Dijk KR, Hedden T, Venkataraman A, Evans KC, Lazar SW, Buckner RL (2010) Intrinsic functional connectivity as a tool for human connectomics: theory, properties, and optimization. J Neurophysiol 103(1):297–321
Velanova K, Jacoby LL, Wheeler ME, McAvoy MP, Petersen SE, Buckner RL (2003) Functional-anatomic correlates of sustained and transient processing components engaged during controlled retrieval. J Neurosci 23(24):8460–8470
Vincent JL, Patel GH, Fox MD, Snyder AZ, Baker JT, Van Essen DC et al (2007) Intrinsic functional architecture in the anaesthetized monkey brain. Nature 447(7140):83–86
Vincent JL, Kahn I, Snyder AZ, Raichle ME, Buckner RL (2008) Evidence for a frontoparietal control system revealed by intrinsic functional connectivity. J Neurophysiol 100(6):3328–3342
Watanabe M (1996) Reward expectancy in primate prefrontal neurons. Nature 382(6592):629–632
Xiong J, Parsons LM, Gao JH, Fox PT (1999) Interregional connectivity to primary motor cortex revealed using MRI resting state images. Hum Brain Mapp 8(2–3):151–156
Yeo BT, Krienen FM, Sepulcre J, Sabuncu MR, Lashkari D, Hollinshead M et al (2011) The organization of the human cerebral cortex estimated by intrinsic functional connectivity. J Neurophysiol 106(3):1125–1165
Zuo X-N, Kelly C, Adelstein JS, Klein DF, Castellanos FX, Milham MP (2010) Reliable intrinsic connectivity networks: test–retest evaluation using ICA and dual regression approach. NeuroImage 49(3):2163–2177

# Part III
# The Prefrontal Cortex as an Integrator of Executive and Social Function

# Chapter 8
# Self–Other Differentiation and Monitoring Others' Actions in the Medial Prefrontal Cortex of the Monkey

**Masaki Isoda**

**Abstract** Actions of other individuals are outward manifestations of their unobservable mental states, such as beliefs, desires, and intentions. The correct identification and monitoring of others' actions are of vital importance for determining adaptive behavior in social environments. The neuronal architecture and mechanisms underlying others-action monitoring have been poorly understood. Using two monkeys monitoring each other's actions for their own action planning, we found that one's own actions and the other's actions are coded by largely distinct populations of neurons in the medial prefrontal cortex (MPFC), a key structure in the mentalizing network. Notably, neurons coding the other's actions were significantly more prevalent in a dorsal convexity region (MPFC-convexity), whereas neurons coding one's own actions were significantly more frequent in a ventral, cingulate sulcus region (MPFC-sulcus). Furthermore, a sizable number of MPFC neurons exhibited a phasic increase in activity when another animal made erroneous actions. Such "partner-error" neurons were also distributed in the two segregated regions: the MPFC-convexity, where the other's error was detected, and the MPFC-sulcus, where one's correct action following the other's error was encoded. These findings suggest that the MPFC plays pivotal roles in differentiating between one's own actions and others' actions and in monitoring the correctness of others' actions for adaptive social decisions. Continuing efforts in this research direction will clarify the neuronal basis whereby primates have become such successful social beings in the animal kingdom.

**Keywords** Self–other distinction • Action monitoring • Self-actions • Others' actions • Medial prefrontal cortex • Monkeys • Single-unit recording • Mentalizing system

---

M. Isoda (✉)
Department of System Neuroscience, National Institute for Physiological Sciences, 38 Nishigonaka Myodaiji, Okazaki, Aichi 444-8585, Japan
e-mail: isodam@nips.ac.jp



M. Watanabe (ed.), *The Prefrontal Cortex as an Executive, Emotional, and Social Brain*, DOI 10.1007/978-4-431-56508-6_8

## 8.1 Monitoring of Others' Actions

The action of others provides an important window into their unobservable states of mind, such as their beliefs, desires, and intentions. This process, related to the theory of mind, allows us to explain others' behavior in mentalistic terms. In addition, the action of others can tell us about what action to take or avoid without trial and error. This process, related to observational learning, is particularly helpful when individuals encounter a dangerous or novel situation in which just one slip of attention to others could result in fatal consequences. Thus, others' actions contain a wealth of information that is useful for our effective and sensible social functioning.

The primate brain is equipped with two separate systems that play a central role in monitoring others' actions (Fig. 8.1) (Wheatley et al. 2007; Van Overwalle and Baetens 2009). One system, called the mirror system, consists primarily of the ventral premotor cortex (PMv) and the inferior parietal lobule. The other system, called the mentalizing system, consists mainly of the superior temporal sulcus (STS) and the medial prefrontal cortex (MPFC). Several hypothetical roles played by the mirror system have been documented since the discovery of "mirror neurons" in the macaque brain (Rizzolatti et al. 2001; Rizzolatti and Fabbri-Destro 2010; Rizzolatti and Sinigaglia 2010). Neurons of this class, originally found in the

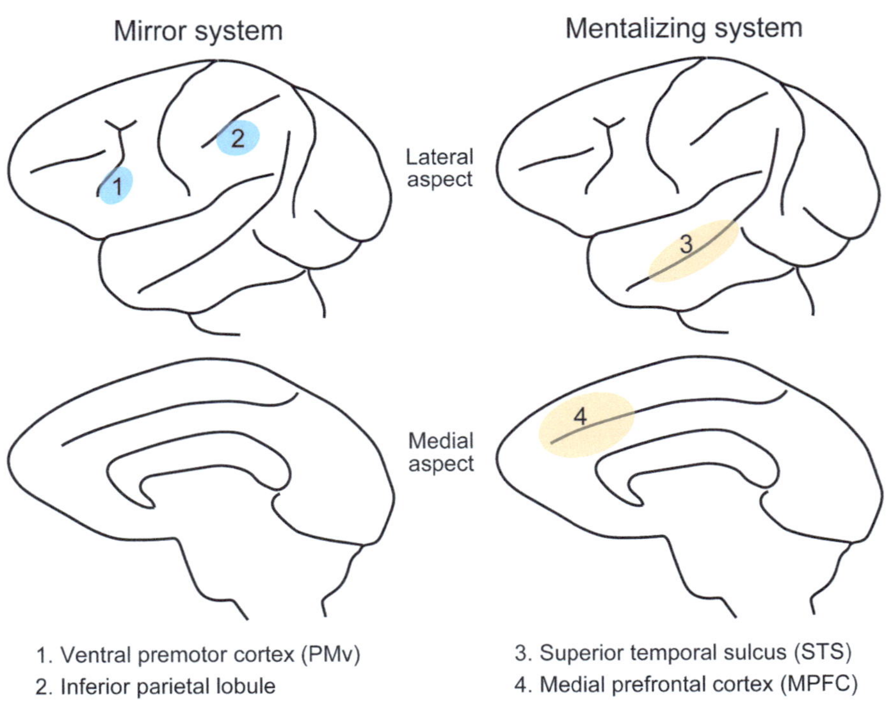


**Fig. 8.1** Illustration of major cortical areas involved in monitoring of others' actions. Shown are the lateral (*top*) and medial (*bottom*) surfaces of the macaque brain

PMv (di Pellegrino et al. 1992), fire not only when a monkey performs a certain action but also when the monkey sees another agent performing the same or a similar action. Such shared coding of one's own actions and others' actions suggests that visual aspects of a given action are identified with motor aspects of that action at the single-cell level, leading to the hypothesis that mirror neurons are involved in monitoring and understanding others' actions via embodied simulation (Gallese and Goldman 1998). In the mirror system, perceiving others' actions automatically activates the equivalent motor representation of one's own. The existence of such a mirror-matching mechanism is supported by other studies in different disciplines (Luria 1980; Brass et al. 2009; Liepelt et al. 2009).

Then what might be the neuronal basis for the monitoring of others' actions in the mentalizing system? How do individual neurons in the mentalizing system respond to others' actions?

## 8.2 Distinction Between One's Own Actions and Others' Actions

Our successful social interaction depends on not only the ability to identify with others but also the ability to distinguish between aspects of the self and others (Isoda and Noritake 2013). In social life, the actions of others are ubiquitous, and yet people do not normally confuse others' actions with their own. Mirror neurons respond similarly to actions of the self and others; however, the classical mirror-matching theory is silent on how the brain distinguishes between self-actions and others' actions, thereby preventing self–other confusion and chaotic social interactions. Despite ample evidence in favor of the overlapping neuronal representation of the two types of action, their distinct representation has been comparatively much less studied.

Studies in developmental psychology have shown that the ability to distinguish between the self and others emerges during infancy (Sebastian et al. 2008; Burnett and Husain 2011). Newborn babies orient their face toward the source of tactile stimulation more frequently to external touch than to spontaneous self-touch to the cheek (Hespos and Rochat 1997). By 5–6 months of age, infants preferentially view a video of another infant compared with a video of themselves (Bahrick et al. 1996). In the second and third years, infants start to understand that others are similarly self-aware and differentiate between themselves and another in speech (Bates 1990). These empirical observations suggest the existence of neuronal mechanisms that enable the distinction between the self and others.

The self–other distinction in the domain of motor actions is of crucial importance for adaptive social exchanges. In social life, people must accurately assign a shared outcome to one's own actions or those of others in order to decide the optimal level of cooperation and reciprocity. In this light, the monitoring of others' actions and the distinction between self-actions and others' actions constitute a core component of social intelligence. As mirror neurons, by definition, do not specify which social agent is acting, the self–other distinction requires separate neuronal mechanisms.

## 8.3 Search for Distinct Neuronal Substrates of One's Own Actions and Others' Actions

Research conducted on human subjects suggests that aspects of the self and others are differentially represented within the MPFC, a key structure in the mentalizing system (Fig. 8.1). A meta-analysis of neuroimaging studies has shown that self-related judgments are associated with the ventral MPFC, whereas others-related judgments are related to the dorsal MPFC (Van Overwalle and Baetens 2009). Importantly, the z-coordinate of activation foci in individual studies can predict whether each study involves self- or others-related judgments (Denny et al. 2012). Such areal segregation may depend, at least partly, on the perceived overlap between the self and others: mentalizing about a *similar* other (in terms of socio-political views) engages a region of the ventral MPFC that is linked to self-referential thoughts, whereas mentalizing about a *dissimilar* other engages a more dorsal region in the MPFC (Mitchell et al. 2006).

The focus of the studies mentioned above has been on mentalizing-related processing, rather than action-related processing. However, the ability to mentalize can be considered to have evolved from a system for representing higher-order aspects of motor actions (Frith and Frith 1999), as actions are one of the main channels used for interpersonal communication and are virtually the only window into an actor's hidden mental states. It seems reasonable, therefore, to investigate the activity of MPFC neurons to identify distinct neuronal substrates of self-actions and others' actions. The identification of such neuronal codes would, in turn, illuminate the neural architecture and mechanisms underlying the self–other distinction in other behavioral domains.

## 8.4 Design of a Role-Reversal Choice Task

To address the issue, Yoshida et al. (2011) designed a task, called the role-reversal choice task, for monkeys. In this task, two monkeys sat facing each other across a table and took turns making a choice for a juice reward (Fig. 8.2). In each trial, one monkey was assigned the role of an actor, which was signaled by a red illuminated start button, and the other monkey the role of an observer. The actor made a choice between a green or yellow illuminated target button, the position of which changed randomly across trials, whereas the observer pressed its start button without making a choice. The two roles were switched every two trials. Both monkeys could see the other's actions. If the actor made the "correct" target choice, both monkeys were rewarded. However, incorrect choices resulted in no reward to either monkey.

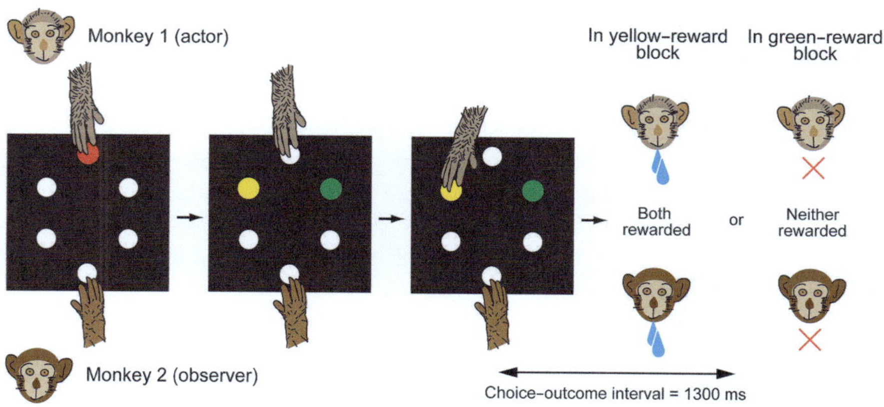


**Fig. 8.2** Temporal sequence of events in role-reversal choice task. Shown is an example of a single trial in which monkey 1 was the actor and monkey 2 was the observer. On the table, two sets of three *white buttons* were placed, each set composed of a start button and two target buttons. In each trial, the actor was required to press the *red* illuminated start button. After 1–1.5 s, the start button turned *white*, and simultaneously the two target buttons on the actor's side turned on: one turned *green* and the other turned *yellow*. The actor then had to select one of the target buttons within 3 s. The reward outcome was determined on the basis of the actor's choice and the color–reward contingency. The temporal interval between the choice and reward feedback was 1.3 s

The correct target color remained constant for 5–17 consecutive trials and then was switched without external signals. Thus, the monkeys were unable to know precisely when the color–reward contingency would switch.

Unlike tasks for studying mirror neurons, the animal's reward outcomes in the role-reversal choice task depended on the partner's choices in *observer* trials. Moreover, the monitoring and use of partner's choice information could increase the likelihood of getting a reward in *actor* trials. For example, if the actor chose the target that had been reward-contingent in preceding trials and yet the current choice ended in no reward, it would mean that the reward contingency was switched and the next turn-taker should change their upcoming choice from previous ones (case 1). Alternatively, if the actor chose the target that had been no-reward-contingent in preceding trials and the current choice ended in no reward, it would mean that the reward contingency remains unchanged and the next turn-taker should keep choosing the previously rewarded target (case 2). Thus, the monitoring of which button the actor chose and whether the choice was rewarded was crucial to determine the continuation or switching of the color–reward contingency. Behavioral analyses revealed that the monkeys performed correctly in both cases with a probability of more than 90% (Yoshida et al. 2011), supporting the interpretation that the monkeys were actively monitoring and utilizing the other's action information for guiding one's own actions.

## 8.5 Distinction Between One's Own Actions and Others' Actions by MPFC Neurons

Recordings were made from the MPFC during both actor and observer trials (Yoshida et al. 2011). On the basis of activity profiles quantified during a 200-ms period before the actor's target pressing, three types of actor-related neuron were identified. Specifically, consistent with a human electrophysiological study (Mukamel et al. 2010), there was a group of neurons that fired similarly in both self-actions and others' actions (Fig. 8.3a, *right*). However, this neuronal population, referred to as the "mirror type," was in the minority (Table 8.1). What was

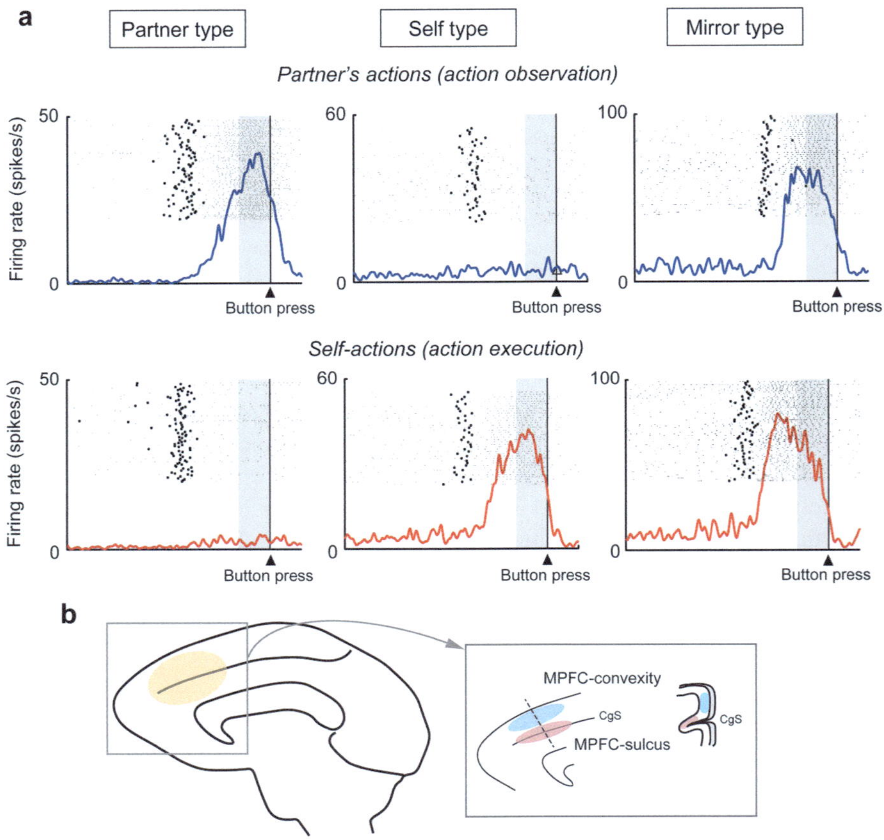


**Fig. 8.3** Actor-related neurons in MPFC. (**a**) Examples of partner-type neuron (*left*), self-type neuron (*middle*), and mirror-type neuron (*right*). Shown are raster displays and spike density functions aligned on the time of the actor's target button pressing (*vertical lines* and *black triangles*). Larger *black dots* denote the time of target button onset, and smaller *black dots* indicate the time of individual action potentials. *Gray rectangles* represent a 200-ms period during which firing rates were quantified for statistical analyses. (**b**) Locations of MPFC-convexity (*blue*) and MPFC-sulcus (*pink*). CgS, cingulate sulcus. The *broken line* indicates the rostrocaudal level through which a coronal section at the rightmost is made

**Table 8.1** Number of actor-related MPFC neurons (Data from two monkeys combined)

| | Partner type | Self type | Mirror type |
|---|---|---|---|
| MPFC-convexity (460) | 100 | 44 | 40 |
| MPFC-sulcus (402) | 38 | 56 | 20 |
| Total (862) | 138 | 100 | 60 |

Values in parentheses denote the total numbers of neurons sampled

most striking was the existence of neurons that selectively encoded the other's actions ("partner type"; Fig. 8.3a, *left*) in addition to those selectively encoding one's own actions ("self type"; Fig. 8.3a, *middle*). In particular, partner-type neurons accounted for as many as 46% of all actor-related neurons. These findings suggest that self-actions and others' actions are represented by largely distinct populations of neurons in the MPFC.

Among the partner-type neurons recorded ($n = 138$), the activity of 45 neurons (33%) was selective for, or affected by, the position of the correct target button, i.e., whether the partner selected the right or left button. In contrast, only five neurons (4%) showed activity changes depending on the target color, i.e., whether the partner selected the yellow or green button. It can be hypothesized that non-position-selective partner-type neurons may play a role in monitoring which agent is acting, whereas position-selective partner-type neurons may play a role in the monitoring of which target the partner is choosing. Although there were only a minority of partner-type neurons selective for target color, which was critical information for determining upcoming choices, it does not compromise the significance of partner-type neurons in monitoring the other's actions. In fact, position-selective partner-type neurons could facilitate the identification of the physical properties (e.g., color) of the target chosen by the other. Recent evidence shows that spatial selection precedes object identification during visual search. Specifically, in a task in which monkeys are required to report object identity, but not object position, neuronal activity specifying the object position nonetheless increases in the frontal cortex *before* the emergence of neuronal activity specifying the object identity in the temporal cortex (Monosov et al. 2010).

The recording was made in two adjacent regions in the MPFC (Fig. 8.3b), i.e., a dorsomedial convexity region ("MPFC-convexity") and a more ventral, cingulate sulcus region ("MPFC-sulcus"). The MPFC-convexity consisted of the pre-supplementary motor area (pre-SMA) and its rostrally adjoining site (area 8/9), whereas the MPFC-sulcus consisted of the rostral cingulate motor area (CMAr) and its rostrally adjoining site (area 24c). The three types of actor-related neuron could be recorded in both the MPFC-convexity and MPFC-sulcus (Table 8.1) and were spatially intermingled. Indeed different types of neuron – e.g., a self-type neuron and a partner-type neuron – were often recorded simultaneously from the tip of the same electrode (Table 8.2). However, it was also found

**Table 8.2** Number of recording sessions in which different types of actor-related neuron were simultaneously recorded from the same electrode tip

| Partner type | Self type | Mirror type | Number of sessions |
|---|---|---|---|
| √ | √ | | 10 |
| √ | | √ | 12 |
| | √ | √ | 5 |
| √ | √ | √ | 2 |

that partner-type neurons were encountered significantly more frequently in the MPFC-convexity, whereas self-type neurons were significantly more prevalent in the MPFC-sulcus (Table 8.1). This regional difference was consistent between the two monkeys tested (Yoshida et al. 2011).

## 8.6 Monitoring of Others' Errors

The ability to detect errors is important for living organisms and can be vital for their survival. In social life, errors can arise from not only one's own actions but also the actions of other individuals. Although much learning occurs via direct experience of errors, many social animals can learn through observation of others' errors. The saying that "The wise learn by the mistakes of others, fools by their own" points to the importance of others' error monitoring for the control of one's own actions in socially adaptive manners.

Previous studies have consistently implicated the MPFC in error monitoring, and yet their attention has been directed mostly to self-generated errors (Niki and Watanabe 1979; Falkenstein et al. 1991; Gehring et al. 1993). It is only recently that researchers have begun to study the neural basis for observed error monitoring using event-related potentials (van Schie et al. 2004; Miltner et al. 2004; Bates et al. 2005) and functional neuroimaging (Shane et al. 2008). These studies have shown that the MPFC is concerned with the monitoring and detection of others' errors as well as self-generated errors. However, several fundamental issues remain unexplored. For example, it is unclarified whether the monitoring of self-errors and that of others' errors share exactly the same neuronal substrates. Moreover, it is unknown how the neural system involved in others' error processing contributes to the adjustment and control of one's own actions. The role-reversal choice task offers a unique opportunity to address these issues at the cellular level.

During the performance of the role-reversal choice task, the monkeys properly detected others' choice errors. In the task, the temporal interval between the actors pressing the target and the receipt of the reward feedback was 1300 ms (Fig. 8.2). During this period, the observer started licking movements in anticipation of a reward if the actor pressed the target button that was most likely correct (Yoshida et al. 2012). Such anticipatory licking movements were not observed, however, if the actor selected the opposite target. The occurrence of the differential licking behavior indicates that the animals judged the correctness of others' actions as soon as the actor made a choice.

## 8.7 MPFC Neurons Specifically Responding to Others' Errors

When the actor committed a choice error, a group of MPFC neurons were phasically activated in the observer's brain (Yoshida et al. 2012). These neurons, which will be called "partner-error neurons," accounted for 17% ($n = 97$) of all the tested neurons ($n = 552$). An example of a partner-error neuron is shown in Fig. 8.4a. The neuron was almost silent when the partner pressed the correct target button (Fig. 8.4a, *left*). However, when the partner pressed the wrong target button, the neuron immediately increased its activity (Fig. 8.4a, *middle*). A subset of partner-error neurons exhibited differential activity depending on the target position (18%) or color (8%) chosen by the partner erroneously. The partner-error neurons were

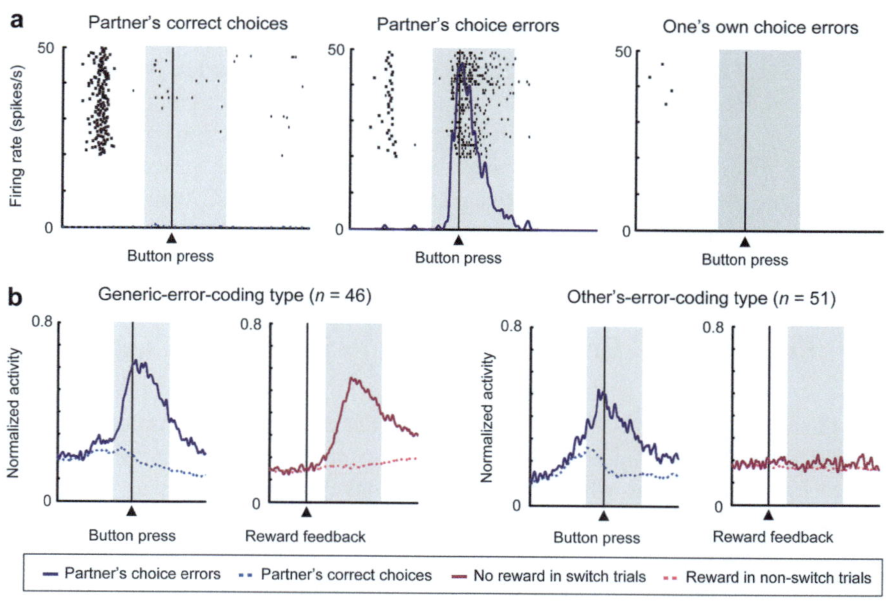


**Fig. 8.4** Response profile of partner-error neurons. (**a**) Activity of partner-error neuron in relation to partner's correct choices (*left*), partner's choice errors (*middle*), and one's own choice errors (*right*). Same conventions as in Fig. 8.3a. Here, gray rectangles represent a 600-ms period (200 ms before and 400 ms after the actor's target button pressing) in which firing rates were quantified for statistical analyses. (**b**) Two types of partner-error neuron. Shown are ensemble-averaged activities of the populations of generic-error-coding-type neurons (*left*) and others-error-coding-type neurons (*right*). For quantifying the reward-feedback-related activity, a 600-ms analysis window was set starting at 200 ms following the feedback onset (*gray rectangles*). Switch trials refer to the first trials in individual trial blocks in which the color–reward contingency remained unchanged

identified in both regions of the MPFC, with their proportions being comparable (MPFC-convexity, 16%; MPFC-sulcus, 19%; $P = 0.23$).

To further characterize the response properties of partner-error neurons, additional analyses were carried out. First, about 90% of the partner-error neurons were nonresponsive to one's own error commission (Fig. 8.4a, *right*). Second, although the frequency of occurrence of the partner's errors was low (~6%), the activity of partner-error neurons was unlikely to reflect a nonselective response to infrequent events. This consideration is important, as a human neuroimaging study shows that the activation of the MPFC is associated with the occurrence of unexpected outcomes rather than errors per se (Jessup et al. 2010). Using a control task in which a novel target color (blue) was introduced in lieu of a green or yellow button only in a fraction of trials (~6%), it was confirmed that none of the partner-error neurons was activated by the appearance of the rare, blue target. Finally, it has been proposed that the MPFC constitutes a cortical system for generic-error processing that can respond to *both* the commission of erroneous actions and the external signal indicating errors (Holroyd and Coles 2002; Gentsch et al. 2009). However, it remains unknown whether the two types of error-related information activate the same set of neurons. During performance of the role-reversal choice task, about half (47%) of the partner-error neurons responded to both the partner's erroneous action and unexpected no-reward feedback due to reward-contingency switches (Fig. 8.4b, *left*), a finding consistent with the generic-error processing model. Notably, however, the remaining neurons (53%) responded only to the partner's erroneous actions (Fig. 8.4b, *right*). Thus, the monitoring of erroneous actions and the monitoring of error-indicating signals can be distinct neuronal processes. Taken together, the results mentioned above indicate that the MPFC contains neurons that selectively monitor the errors of others' actions.

## 8.8 From Others' Error Monitoring to Self-Action Control

Did the response of partner-error neurons in trials when the recorded monkey observed the other's choice errors have any impact on the correctness of performance in the next trials when the recorded monkey made choices as the actor? Although the MPFC has been consistently implicated in error monitoring (Taylor et al. 2007), the mechanisms underlying the implementation of subsequent performance adjustments are less well understood (Ridderinkhof et al. 2004; Gentsch et al. 2009). Moreover, it is unknown which region of the MPFC is involved in such behavioral adjustments. To address these issues, the partner's error trials were classified into two groups on the basis of the success or failure of the subsequent self-choices. Specifically, one group consisted of the partner's error trials that were followed by one's own correct choices (denoted by "Ec" trials; Fig. 8.5a, top). The second group consisted of the partner's error trials that were followed by one's own choice errors ("Ee" trials; Fig. 8.5a, bottom). For comparison, a control group consisting of the partner's correct trials ("C" trials) was also examined. Neuronal

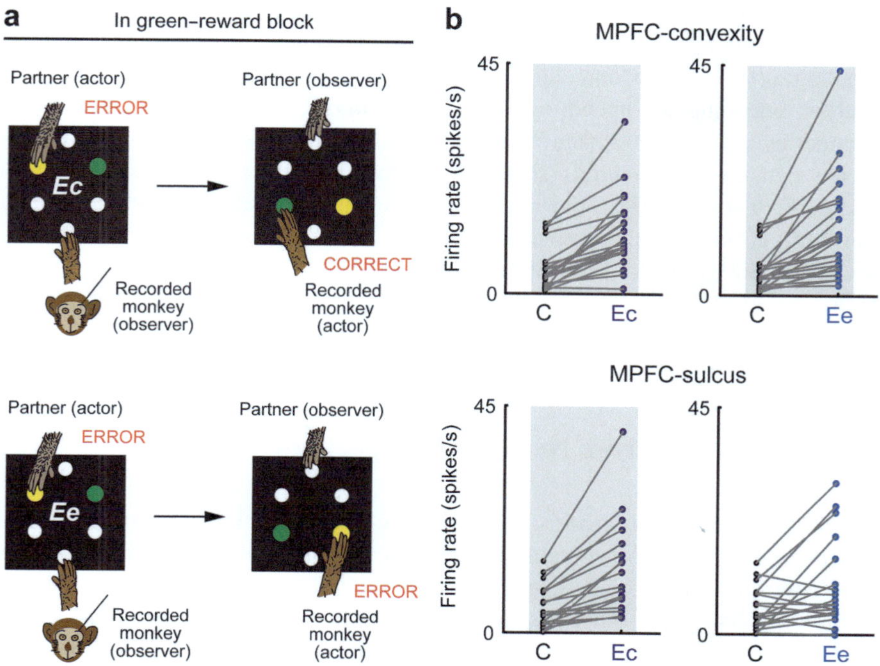


**Fig. 8.5** Relationships between activity of partner-error neurons in observer trials and correctness of choice performance in next actor trials. (**a**) Illustration of two types of partner's error trial, i.e., Ec trials (*top*) and Ee trials (*bottom*). Shown are examples in which the choice of the *green button* is associated with rewards. (**b**) Statistical analyses. Each line represents activity changes of each neuron ($n = 23$, MPFC-convexity; $n = 19$, MPFC-sulcus). *Gray rectangles* indicate comparisons with statistically significant differences ($P < 0.01$)

activity was then compared between Ec and C, and Ee and C, separately for MPFC-convexity neurons and MPFC-sulcus neurons (Fig. 8.5b).

The population-level analysis revealed that neurons in the MPFC-convexity significantly increased their activity in both Ec and Ee trials (Fig. 8.5b, *top*), suggesting that the activity increase was associated with the monitoring or detection of the other's action errors regardless of the correctness of the self-action in the next trial. In contrast, the activity of MPFC-sulcus neurons significantly increased only in Ec trials (Fig. 8.5b, *bottom*), suggesting that the activity increase was associated with the correct selection of one's own actions following the other's action errors. The net error effect (Ec–C and Ee–C) was then directly compared between the two MPFC regions by two-way analysis of variance, revealing a marginally significant interaction ($P = 0.066$). The observed interaction was ascribed to a significant difference in the error effect for MPFC-sulcus neurons. These findings suggest that the MPFC-convexity may be more closely related to the monitoring of others' erroneous actions, whereas the MPFC-sulcus may play a more important role in the determination and regulation of one's own actions based on the information about others' errors.

The MPFC-sulcus that contained neurons associated with self-action adjustments is located 7–16 mm anterior to the genu of the arcuate sulcus. This cortical site largely overlaps a cingulate sulcus region in the MPFC that encompasses neurons encoding an upcoming reward size during a working-memory delay (Kennerley and Wallis 2009). The site also overlaps, albeit partially, an MPFC region that contains neurons signaling a prediction error of self-action values (Matsumoto et al. 2007), those involved in reward-based action switching (Shima and Tanji 1998), and those responding to the proximity to rewards (Shidara and Richmond 2002), fictive rewards (Hayden et al. 2009), commission of one's own errors (Ito et al. 2003; Amiez et al. 2005), and feedback signals during a search (Quilodran et al. 2008).

## 8.9 Summaries from Network Perspectives

Single-neuron recording made during the performance of the role-reversal choice task in monkeys has shown novel findings on the role of the MPFC in self–other distinction and monitoring of others' actions. First, regarding actor selectivity, neurons encoding the other's actions were significantly more prevalent in the MPFC-convexity, whereas neurons encoding the self-actions were significantly more frequent in the MPFC-sulcus. Regarding the other's error processing, the activity increase in MPFC-convexity neurons was associated with a correct detection of the other's erroneous actions, whereas that for MPFC-sulcus neurons was associated with the correct selection of one's own action following the other's erroneous choices. These findings suggest that the MPFC-convexity, a dorsal sector of the MPFC, is more specialized for specifying and monitoring others' actions, whereas the MPFC-sulcus, a ventral sector of the MPFC, is more intimately associated with monitoring and controlling one's own actions (Fig. 8.6). This functional differentiation regarding action processing along the dorsoventral axis accords well with other aspects of self vs. other representations. As mentioned earlier, neuroimaging studies have repeatedly shown that self-related judgments are associated with the ventral MPFC and others-related judgments with the dorsal MPFC (Van Overwalle and Baetens 2009); moreover, mentalizing about a *similar* other engages a region of the ventral MPFC and mentalizing about a *dissimilar* other engages a more dorsal region in the MPFC (Mitchell et al. 2006).

The MPFC-convexity may receive information about others' actions from the dorsal bank and fundus of the anterior portion of the STS (Fig. 8.6), which appear to correspond to the superior temporal polysensory area (Bruce et al. 1981). The basis for this claim is twofold. First, similarly to neurons in the MPFC, STS neurons distinguish actions of others from those of one's own. Specifically, STS neurons, albeit in the minority (~5%), respond vigorously to the sight of others' limb movements; the sight of own limb movements with similar trajectory has negligible effects on the activity of those neurons (Hietanen and Perrett 1993). Second, medial frontal regions including part of the MPFC-convexity receive preferential projections from the STS (Luppino et al. 2001). These findings indicate a close,

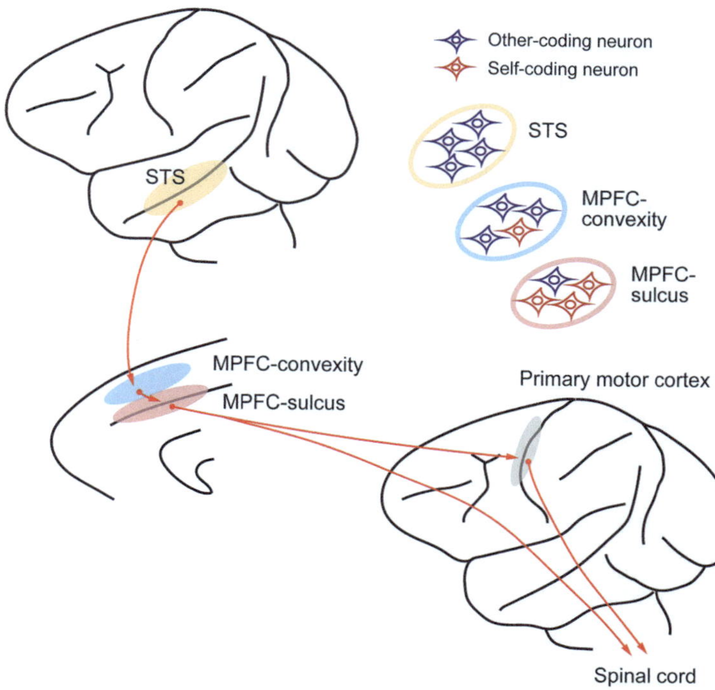


**Fig. 8.6** Hypothetical scheme illustrating information flow for detection of others' actions and control of one's own actions

functional coupling between the STS and the MPFC. However, there are also notable differences in the firing property between the two cortical areas. First, it is generally considered that STS neurons do not fire in relation to one's own actions (Fig. 8.6) (Rizzolatti and Luppino 2001; Rizzolatti and Craighero 2004). Second, STS neurons are activated by observing others' arm movements that are made outside any task context (e.g., without targets or rewards) (Hietanen and Perrett 1993). In contrast, partner-type neurons in the MPFC responded almost exclusively to others' arm movements that were directed toward the target button; the same set of neurons did not fire in response to others' return movements back to the start button after each target choice (Yoshida et al. 2011). It seems likely that the response of partner-type MPFC neurons was more contingent on the goal-directedness of others' actions.

The MPFC-sulcus may receive information about others' action from the MPFC-convexity and regulate one's own actions in a socially adaptive manner (Fig. 8.6). In support of this idea, it has been shown that the MPFC-sulcus, particularly the CMAr, and the MPFC-convexity, particularly the pre-SMA, are richly interconnected (Hatanaka et al. 2003). In addition, the CMAr, but not the pre-SMA, has direct outputs to the primary motor cortex and spinal cord (Picard and Strick 1996), crucial structures for generating one's own actions. The MPFC-sulcus occupies a unique position to control one's own actions in response to, and via the use of, social signals.

Note, however, that the functional differentiation between the MPFC-convexity and the MPFC-sulcus should be considered in a relative context. As mentioned above, both cortical regions contain neurons coding self-actions and others' actions. A simplistic dichotomy is therefore inappropriate. Furthermore, the MPFC is not the only region that plays a role in monitoring others' actions. For example, the PMv in the mirror system is involved in understanding others' actions and intentions (Rizzolatti and Sinigaglia 2007). Neurons in the striatum of the basal ganglia participate in the determination of which social actions – either self-actions or others' actions – yield one's own reward (Baez-Mendoza et al. 2013). An important question for future research is how these neural structures orchestrate adaptive behavioral monitoring and organization in social settings.

## 8.10 Concluding Remarks

People tend to view others as analogous to oneself, but also identify them as unique. Although converging evidence indicates that both the mirror and mentalizing systems function in monitoring and understanding others' actions, the basic principle underlying their function seems to be different from each other – i.e., self–other identification in the mirror system and self–other differentiation in the mentalizing system. The findings presented here demonstrate that neurons in the MPFC, a key structure in the mentalizing system, distinguish between one's own actions and others' actions and monitor the correctness of others' actions for adaptive social decisions. The MPFC may provide a neuronal mechanism whereby social animals prevent self–other confusion and chaotic interactions. Continuing efforts in refining social paradigms for monkeys, combined with technological advancement, will offer groundbreaking opportunities to uncover brain mechanisms underlying various aspects of social cognition and behavior.

**Acknowledgments** This work was supported by JSPS KAKENHI 15H04262.

## References

Amiez C, Joseph JP, Procyk E (2005) Anterior cingulate error-related activity is modulated by predicted reward. Eur J Neurosci 21(12):3447–3452 doi:EJN4170 [pii] 10.1111/j.1460-9568.2005.04170.x

Baez-Mendoza R, Harris CJ, Schultz W (2013) Activity of striatal neurons reflects social action and own reward. Proc Natl Acad Sci U S A 110(41):16634–16639. doi:10.1073/pnas.1211342110

Bahrick LE, Moss L, Fadil C (1996) Development of visual self-recognition in infancy. Ecol Psychol 8(3):189–208

Bates E (1990) Language about me and you: pronominal reference and the emerging concept of self. In: Cicchetti D, Beeghly M (eds) The self in transition: infancy to childhood. University of Chicago Press, Chicago, pp 165–182

Bates AT, Patel TP, Liddle PF (2005) External behavior monitoring mirrors internal behavior monitoring. J Psychophysiol 19(4):281–288

Brass M, Ruby P, Spengler S (2009) Inhibition of imitative behaviour and social cognition. Philos Trans R Soc Lond Ser B Biol Sci 364(1528):2359–2367 doi:364/1528/2359 [pii] 10.1098/rstb.2009.0066

Bruce C, Desimone R, Gross CG (1981) Visual properties of neurons in a polysensory area in superior temporal sulcus of the macaque. J Neurophysiol 46(2):369–384

Burnett S, Husain M (2011) Cognitive neuroscience: distinguishing self from other. Curr Biol 21(5):R189–R190 doi:S0960-9822(11)00112-6 [pii] 10.1016/j.cub.2011.01.056

Denny BT, Kober H, Wager TD, Ochsner KN (2012) A meta-analysis of functional neuroimaging studies of self- and other judgments reveals a spatial gradient for mentalizing in medial prefrontal cortex. J Cogn Neurosci 24(8):1742–1752. doi:10.1162/jocn_a_00233

di Pellegrino G, Fadiga L, Fogassi L, Gallese V, Rizzolatti G (1992) Understanding motor events: a neurophysiological study. Exp Brain Res 91(1):176–180

Falkenstein M, Hohnsbein J, Hoormann J, Blanke L (1991) Effects of crossmodal divided attention on late ERP components. II. Error processing in choice reaction tasks. Electroencephalogr Clin Neurophysiol 78(6):447–455

Frith CD, Frith U (1999) Interacting minds – a biological basis. Science 286(5445):1692–1695 doi:8043 [pii]

Gallese V, Goldman A (1998) Mirror neurons and the simulation theory of mind-reading. Trends Cogn Sci 2(12):493–501

Gehring WJ, Gross B, Coles MGH, Meyer DE, Donchin E (1993) A neural system for error detection and compensation. Psychol Sci 4(6):385–390. doi:10.1111/j.1467-9280.1993.tb00586.x

Gentsch A, Ullsperger P, Ullsperger M (2009) Dissociable medial frontal negativities from a common monitoring system for self- and externally caused failure of goal achievement. NeuroImage 47(4):2023–2030 doi:S1053-8119(09)00589-8 [pii] 10.1016/j.neuroimage.2009.05.064

Hatanaka N, Tokuno H, Hamada I, Inase M, Ito Y, Imanishi M, Hasegawa N, Akazawa T, Nambu A, Takada M (2003) Thalamocortical and intracortical connections of monkey cingulate motor areas. J Comp Neurol 462(1):121–138. doi:10.1002/cne.10720

Hayden BY, Pearson JM, Platt ML (2009) Fictive reward signals in the anterior cingulate cortex. Science 324(5929):948–950 doi:324/5929/948 [pii] 10.1126/science.1168488

Hespos SJ, Rochat P (1997) Dynamic mental representation in infancy. Cognition 64(2):153–188 doi:S0010-0277(97)00029-2 [pii]

Hietanen JK, Perrett DI (1993) Motion sensitive cells in the macaque superior temporal polysensory area. I. Lack of response to the sight of the animal's own limb movement. Exp Brain Res 93(1):117–128

Holroyd CB, Coles MG (2002) The neural basis of human error processing: reinforcement learning, dopamine, and the error-related negativity. Psychol Rev 109(4):679–709

Isoda M, Noritake A (2013) What makes the dorsomedial frontal cortex active during reading the mental states of others? Front Neurosci 7:232. doi:10.3389/fnins.2013.00232

Ito S, Stuphorn V, Brown JW, Schall JD (2003) Performance monitoring by the anterior cingulate cortex during saccade countermanding. Science 302(5642):120–122. doi:10.1126/science.1087847302/5642/120[pii]

Jessup RK, Busemeyer JR, Brown JW (2010) Error effects in anterior cingulate cortex reverse when error likelihood is high. J Neurosci 30(9):3467–3472 doi:30/9/3467 [pii] 10.1523/JNEUROSCI.4130-09.2010

Kennerley SW, Wallis JD (2009) Encoding of reward and space during a working memory task in the orbitofrontal cortex and anterior cingulate sulcus. J Neurophysiol 102(6):3352–3364 doi:00273.2009 [pii] 10.1152/jn.00273.2009
Liepelt R, Ullsperger M, Obst K, Spengler S, von Cramon DY, Brass M (2009) Contextual movement constraints of others modulate motor preparation in the observer. Neuropsychologia 47(1):268–275 doi:S0028-3932(08)00276-5 [pii] 10.1016/j.neuropsychologia.2008.07.008
Luppino G, Calzavara R, Rozzi S, Matelli M (2001) Projections from the superior temporal sulcus to the agranular frontal cortex in the macaque. Eur J Neurosci 14(6):1035–1040 doi:ejn1734 [pii]
Luria AR (1980) Higher cortical functions in man. Consultants Bureau, New York
Matsumoto M, Matsumoto K, Abe H, Tanaka K (2007) Medial prefrontal cell activity signaling prediction errors of action values. Nat Neurosci 10(5):647–656 doi:nn1890 [pii]10.1038/nn1890
Miltner WH, Brauer J, Hecht H, Trippe R, Coles MG (2004) Parallel brain activity for self-generated and observed errors. In: Ullsperger M, Falkenstein M (eds) Errors, conflicts, and the brain. Current opinions on performance monitoring. Max Plank Institute of Cognitive Neuroscience, Leipzig, pp 124–129
Mitchell JP, Macrae CN, Banaji MR (2006) Dissociable medial prefrontal contributions to judgments of similar and dissimilar others. Neuron 50(4):655–663 doi:S0896-6273(06)00266-2 [pii] 10.1016/j.neuron.2006.03.040
Monosov IE, Sheinberg DL, Thompson KG (2010) Paired neuron recordings in the prefrontal and inferotemporal cortices reveal that spatial selection precedes object identification during visual search. Proc Natl Acad Sci U S A 107(29):13105–13110. doi:10.1073/pnas.1002870107
Mukamel R, Ekstrom AD, Kaplan J, Iacoboni M, Fried I (2010) Single-neuron responses in humans during execution and observation of actions. Curr Biol 20(8):750–756 doi:S0960-9822(10)00233-2 [pii] 10.1016/j.cub.2010.02.045
Niki H, Watanabe M (1979) Prefrontal and cingulate unit activity during timing behavior in the monkey. Brain Res 171(2):213–224 doi:0006-8993(79)90328-7 [pii]
Picard N, Strick PL (1996) Motor areas of the medial wall: a review of their location and functional activation. Cereb Cortex 6(3):342–353
Quilodran R, Rothe M, Procyk E (2008) Behavioral shifts and action valuation in the anterior cingulate cortex. Neuron 57(2):314–325 doi:S0896-6273(07)01021-5 [pii] 10.1016/j.neuron.2007.11.031
Ridderinkhof KR, Ullsperger M, Crone EA, Nieuwenhuis S (2004) The role of the medial frontal cortex in cognitive control. Science 306(5695):443–447 doi:306/5695/443 [pii] 10.1126/science.1100301
Rizzolatti G, Craighero L (2004) The mirror-neuron system. Annu Rev Neurosci 27:169–192. doi:10.1146/annurev.neuro.27.070203.144230
Rizzolatti G, Fabbri-Destro M (2010) Mirror neurons: from discovery to autism. Exp Brain Res 200(3–4):223–237. doi:10.1007/s00221-009-2002-3
Rizzolatti G, Luppino G (2001) The cortical motor system. Neuron 31(6):889–901
Rizzolatti G, Sinigaglia C (2007) Mirror neurons and motor intentionality. Funct Neurol 22(4):205–210
Rizzolatti G, Sinigaglia C (2010) The functional role of the parieto-frontal mirror circuit: interpretations and misinterpretations. Nat Rev Neurosci 11(4):264–274 doi:nrn2805 [pii]10.1038/nrn2805
Rizzolatti G, Fogassi L, Gallese V (2001) Neurophysiological mechanisms underlying the understanding and imitation of action. Nat Rev Neurosci 2(9):661–670. doi:10.1038/3509006035090060[pii]
Sebastian C, Burnett S, Blakemore SJ (2008) Development of the self-concept during adolescence. Trends Cogn Sci 12(11):441–446 doi:S1364-6613(08)00216-7 [pii] 10.1016/j.tics.2008.07.008

Shane MS, Stevens M, Harenski CL, Kiehl KA (2008) Neural correlates of the processing of another's mistakes: a possible underpinning for social and observational learning. NeuroImage 42(1):450–459 doi:S1053-8119(08)00007-4 [pii] 10.1016/j.neuroimage.2007.12.067

Shidara M, Richmond BJ (2002) Anterior cingulate: single neuronal signals related to degree of reward expectancy. Science 296(5573):1709–1711. doi:10.1126/science.1069504296/5573/1709[pii]

Shima K, Tanji J (1998) Role for cingulate motor area cells in voluntary movement selection based on reward. Science 282(5392):1335–1338

Taylor SF, Stern ER, Gehring WJ (2007) Neural systems for error monitoring: recent findings and theoretical perspectives. Neuroscientist 13(2):160–172 doi:13/2/160 [pii] 10.1177/1073858406298184

Van Overwalle F, Baetens K (2009) Understanding others' actions and goals by mirror and mentalizing systems: a meta-analysis. NeuroImage 48(3):564–584. doi:10.1016/j.neuroimage.2009.06.009

van Schie HT, Mars RB, Coles MG, Bekkering H (2004) Modulation of activity in medial frontal and motor cortices during error observation. Nat Neurosci 7(5):549–554. doi:10.1038/nn1239nn1239[pii]

Wheatley T, Milleville SC, Martin A (2007) Understanding animate agents: distinct roles for the social network and mirror system. Psychol Sci 18(6):469–474. doi:10.1111/j.1467-9280.2007.01923.x

Yoshida K, Saito N, Iriki A, Isoda M (2011) Representation of others' action by neurons in monkey medial frontal cortex. Curr Biol 21(3):249–253 doi:S0960-9822(11)00027-3 [pii] 10.1016/j.cub.2011.01.004

Yoshida K, Saito N, Iriki A, Isoda M (2012) Social error monitoring in macaque frontal cortex. Nat Neurosci 15(9):1307–1312 doi:nn.3180 [pii]10.1038/nn.3180

# Chapter 9
# Neural Correlates of Competition in the Primate Prefrontal Cortex

**Takayuki Hosokawa**

**Abstract** Humans and animals must struggle in order to survive. Since resources are usually limited, competing successfully is vitally important. However, the neuronal mechanisms underlying competitive behavior have been poorly studied. Hosokawa and Watanabe (J Neurosci 32(22):7662–7671, 2012) examined whether neurons in the prefrontal cortex showed response sensitivity related to competition. Monkeys were trained to play a video shooting game, either competing with another monkey or the computer, or playing alone without a rival. The monkeys' motivation level was elevated in the competitive games. Prefrontal neurons showed higher outcome-related activity in the competitive than in the noncompetitive games. A group of prefrontal neurons showed differential activities depending on whether the rival was another monkey or the computer. Furthermore, Hosokawa and Watanabe (Front Neurosci 9:165, 2015) examined the monkey's behavior and prefrontal neuronal activity in competitive situations with unordinary performance-reward contingencies, where both the winner and loser, or neither of them, got a reward (egalitarian outcome conditions). The monkey's behavioral performance greatly deteriorated in trials with these outcome conditions. Prefrontal neurons showed activities that reflected the performance-reward contingency. Some of them showed reward-related activity in the normal, but not in the egalitarian outcome conditions, even though the same reward was given to the animal. These results indicate that the prefrontal cortex may play an important role in monitoring the current context such as the rival's identity (who is the rival) and performance-reward contingency (whether winning a competition leads to a reward or not) and integrating the performance outcome (win/loss) with the current context for better performance in competition.

**Keywords** Monkey • Prefrontal cortex • Neurons • Competition • Reward • Video shooting game • Win • Loss • Egalitarian reward contingency

T. Hosokawa (✉)
Division of Systems Neuroscience, Tohoku University Graduate School of Life Sciences, Sendai, Japan
e-mail: t-hosokawa@m.tohoku.ac.jp



M. Watanabe (ed.), *The Prefrontal Cortex as an Executive, Emotional, and Social Brain*, DOI 10.1007/978-4-431-56508-6_9

## 9.1 Introduction

Competition is a part of our lives. We are often in competitive situations since childhood. At school, children compete with other students to get better grades. They also compete with others in various activities such as sports and contests. Later, teenagers in some cultures have to compete with their peers in entrance exams to enter university. In adulthood, competition can get even fiercer. There are usually competitions to obtain a good job. Even after we get a job, we may have to compete with our colleagues to seek a promotion or higher position. In this sense, winning a competition may improve, while losing may degrade, one's social status. Thus, winning competitions is crucially important in our lives.

In nature, animals typically compete against each other to obtain limited resources such as food and mates. Actually, many species may show aggressive behavior to obtain what they need (Robbins 1999; Janson 1985). There is evidence that dominant individuals tend to have better nutrition and higher success rates of reproduction (Tutin 1979; Robbins 1999; Pusey et al. 1997; Janson 1985; Leboeuf 1974; Cox and Leboeuf 1977; Rubenstein and Nuñez 2009). For example, male elephant seals fight against each other, and only a few dominant males form harems, in which they monopolize the females (Cox and Leboeuf 1977; Leboeuf 1974). Thus, winning a competition improves, while losing diminishes, chances of survival and reproduction.

Competition can be a driving force. If we have rivals, it can force us to go above and beyond what we otherwise would do. We would try to outdo them. Once we have done that, they usually try to outdo us in turn. This process forces people to do not just what is required, but to do the best they can. Actually, psychological and sociological studies have shown that competition can elevate motivation (Reeve et al. 1985; Washburn et al. 1990; Garcia et al. 2006), although it depends on the competence of the subjects (Tauer and Harackiewicz 2004; Epstein and Harackiewicz 1992; Deci et al. 1981). Many neurophysiological studies have reported on neural activity related to motivation or reward in the prefrontal cortex (Watanabe 1996; Kim et al. 2012; Roesch and Olson 2003; Ichihara-Takeda and Funahashi 2008). However, little is known about the neural mechanisms underlying the behavior in competitive situations.

## 9.2 Competitive and Noncompetitive Shooting Games

How do animals behave in competitive situations? Also, how does the brain respond in competitive situations? To study the behavior and neuronal responses of monkeys in competitive situations, Hosokawa and Watanabe (2012) trained Japanese monkeys to play video shooting games. Previous behavioral studies showed that monkeys are able to skillfully manipulate joysticks to shoot bullets at targets in computer video games (Rumbaugh et al. 1989; Washburn et al. 1990).

In the study of Hosokawa and Watanabe (2012), two types of game were employed: competitive and noncompetitive. In the competitive game, two monkeys competed against each other. They faced a computer monitor, arranged at an angle so that they could see each other (Fig. 9.1a). In front of each monkey were a joystick and a button. They manipulated the joysticks to shoot bullets. A trial started when both monkeys pushed their own button. This ensured that both monkeys were ready to play the game. After that, two colored (white and yellow) triangles appeared, one on the left side and the other on the right side of the monitor (Fig. 9.1b, c). The triangle represented a gun turret, and a bullet was launched from the triangle in the direction that the joystick was tilted. The monkeys learned to tilt the joystick and shoot at the turret on the other side (target). The color of the turret was white for the monkey on the left and yellow for the monkey on the right. The positions of the turrets were randomly selected from the top, middle, or bottom and left or right (Fig. 9.1c).

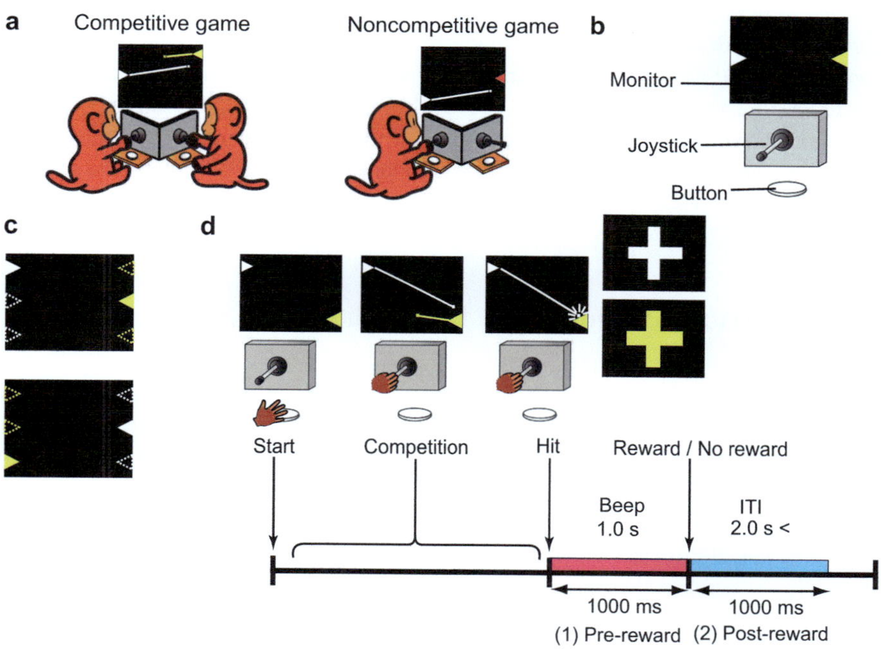


**Fig. 9.1** Description of competitive and noncompetitive conditions in the tasks. (**a**) (*Left*) Schematic diagram of the monkey–monkey competitive condition. Each monkey shot bullets from a turret (*triangle*) of its own color. The *lines* from the turrets represent the trajectories of the bullets and did not appear in the actual game. (*Right*) Schematic diagram of the (one-monkey) noncompetitive condition. One monkey played a noncompetitive shooting game. In this condition, no bullets came from the other side. (**b**) Experimental setup. There was a joystick and a button in front of each monkey. The monkeys shared one PC monitor. (**c**) Position and spatial configuration of the turrets. The turret positions were randomly selected from the top, middle, or bottom, and left or right. These positions changed from trial to trial, but they were fixed within a trial. *Open triangles* represent possible positions at which turrets could appear. (**d**) Time course of the competitive shooting game and the analysis periods. ITI, Intertrial interval (Figure modified with permission from Hosokawa and Watanabe 2012)

These positions changed from trial to trial, but they were fixed within a trial. The monkeys were able to recognize their own turret because they had been trained in the game with the fixed color of turret (white or yellow). When one of the monkeys succeeded in hitting the target, a white or yellow cross appeared on the monitor depending on which monkey won (Fig. 9.1d). The color of the cross corresponded to the turret color of the winning monkey. The monkey that made the first successful hit was the winner and received a drop of grape juice. The loser did not receive any reward. In the noncompetitive game, on the other hand, only one monkey played a shooting game without a rival. As in the competitive game, when the monkey pushed the button, two colored triangles appeared on the left and right sides of the monitor. The color of the target triangle was always red, and the color of the turret triangle was the monkey's own color (white or yellow), so that the monkey could recognize which side of the triangle was the target. The monkey tried to hit the target with no bullets coming from the other side. Hence, the monkey would never lose in this game. When the monkey hit the target, the monkey was randomly rewarded in half of the trials. The reward rate was made comparable between the competitive and noncompetitive games so that any difference in reward rate would not have an effect on behavior or neuronal activity. When the winning rate was unequal between the monkeys in the competition, the experimenter increased the bullet speed of the weaker monkey. The resultant reward rates were almost 50% in all monkeys.

## 9.3 Behavioral Performance in the Shooting Games

The monkeys' behavior significantly differed between the competitive and noncompetitive games. The monkeys played the competitive game more accurately than they did the noncompetitive game. The hit rate of the first shot was significantly higher in the competitive than in the noncompetitive conditions. Also, the monkeys' response speed was faster in the competitive game. The latency of the first shot was significantly shorter in the competitive than in the noncompetitive game. These results suggest that the monkeys were highly motivated and concentrated on playing the competitive game.

In this shooting game study, the monkeys showed faster and more accurate behavior in the competitive than in the noncompetitive game. In usual cases, however, as the response speed increases, its accuracy decreases (i.e., speed-accuracy tradeoff). Actually, in the behavioral studies in which monkeys played competitive games (Rumbaugh et al. 1989; Washburn et al. 1990), the hit rate became worse in the competitive conditions. This discrepancy is probably due to the difference between the tasks employed in the studies. In the tasks of Rumbaugh et al. (1989) and Washburn et al. (1990), the monkeys could reshoot a bullet at any time by pulling back the joystick. In the task of Hosokawa and Watanabe (2012), on the other hand, the monkeys had to wait until one bullet went out of the view before shooting another bullet. Because of this constraint in the task of Hosokawa and

Watanabe (2012), the monkeys probably shot bullets as quickly and accurately as possible.

## 9.4 Prefrontal Neuronal Activity in the Shooting Games

While the monkeys were playing these video shooting games, neuronal activity was recorded from the lateral prefrontal cortex (LPFC) (Fig. 9.2a). LPFC is known to be involved in integration of cognitive and motivational information (Kobayashi et al. 2002; Pan et al. 2008; Watanabe and Sakagami 2007) and in the processing of social information (Fujii et al. 2009; Zink et al. 2008). So it would be interesting to

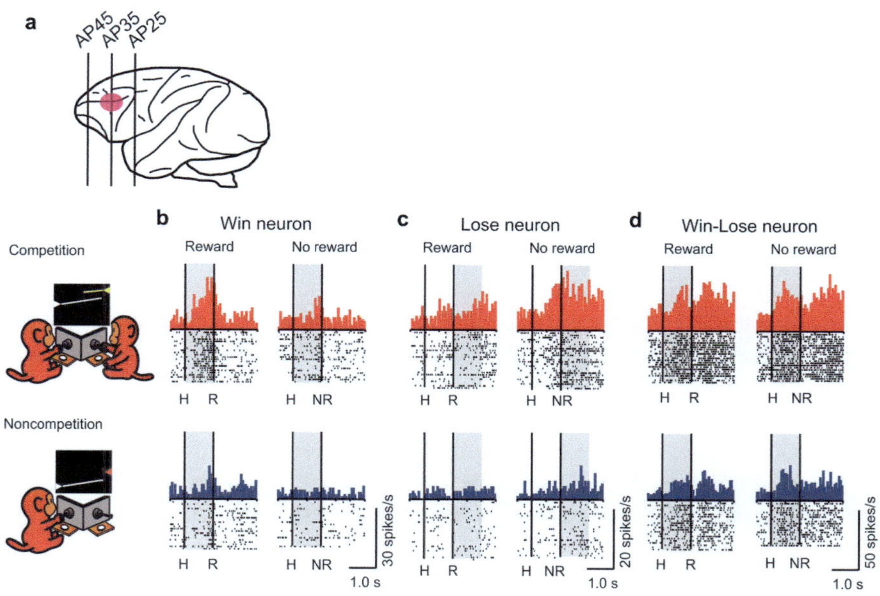


**Fig. 9.2** Examples of LPFC neurons that showed competitive activity. (**a**) Recording areas. Recording areas are highlighted by a magenta ellipse on a lateral view of the monkey brain. Neuronal activity was recorded in the upper and lower banks of the principal sulcus. Most of the recordings were made in the region between AP30 and AP40. AP, Anterior–posterior. (**b**–**d**) Raster and histogram displays for LPFC neurons that showed competitive activities. (**b**) An LPFC neuron that showed win-related activity. (**c**) An LPFC neuron that showed loss-related activity. (**d**) An LPFC neuron that showed higher activity in the competitive than in the noncompetitive conditions irrespective of the presence or absence of a reward. Displays in the *top* and *bottom rows* show activity in the competitive and noncompetitive conditions, respectively. The *left* and *right* columns show activity during the reward and no-reward trials, respectively. The *left vertical line* in each display indicates the timing of a successful hit by either monkey. The *right vertical line* indicates the timing of a reward delivery (reward trials) or 1 s after a successful hit (no-reward trials). Each *shaded area* indicates the period when the typical activity of each type was observed. H, Successful hit; R, reward delivery; NR, 1 s after a successful hit (Figure modified with permission from Hosokawa and Watanabe 2012)

study how LPFC neurons respond in competitive situations. It was hypothesized that the activities of LPFC neurons would differ between the competitive and noncompetitive situations and would also differentiate the competitive situation where the rival was another monkey from the situation where the rival was an inanimate computer. In the data analyses, neuronal activity in the epochs after either monkey succeeded in hitting the target (i.e., the 1,000-ms epoch after a successful hit and the 1000-ms epoch after a reward delivery) was mainly focused. This is because there were many factors that might influence the neuronal activity during the competition epoch, such as arm movements and various visual stimuli (flying bullets and turrets in different positions). In the epochs after a successful hit, on the other hand, the visual stimuli on the monitor was the same (i.e., white or yellow cross in the 1000-ms epoch after a successful hit, and blank screen in the 1000-ms epoch after a reward delivery).

Many LPFC neurons showed differential outcome-related activity between the competitive and noncompetitive games even though the monkey received the same reward. These neurons are called "competitive neurons" and classified into three types: Win, Lose, and Win-Lose neurons. Figure 9.2 shows examples of competitive neurons. The neuron in Fig. 9.2b showed higher activity when the monkey expected to receive a reward after winning a competition (Win neuron). Although the monkey would receive the same reward in the noncompetitive condition, the activity in the reward trials under the noncompetitive condition was much smaller than that in the winning trials under the competitive condition. The neuron in Fig. 9.2c showed higher activity when the monkey lost a competition and could not obtain a reward under the competitive condition than when the monkey hit the target, but there was no reward under the noncompetitive condition (Lose neuron). The neuron in Fig. 9.2d showed higher activity in the competitive than in the noncompetitive condition irrespective of the presence or absence of the reward (Win-Lose neuron). Of 321 LPFC neurons, 181 showed competition-related activity (37 Win neurons, 69 Lose neurons, and 75 Win-Lose neurons). Most of the competitive neurons showed higher activity in the competitive than in the noncompetitive condition.

## 9.5 Effects of Animacy of the Rival

As shown above, LPFC neurons were highly activated in the competitive situation. Next, the effect of a rival's animacy (a monkey or an inanimate computer) on competitive neuronal activities was examined. Competition is a kind of social situation, where one competes with another individual. It has been reported that LPFC neurons of the monkey show differential activity depending on the social rank of the competing subject in a food grab task (Fujii et al. 2009). Neuroimaging studies on humans have found that the brain shows differential activity depending on whether the subject believes that the rival is another human or a computer during competitive games (Fukui et al. 2006; Gallagher et al. 2002; Rilling et al. 2004;

Sanfey et al. 2003). To examine the effect of the rival's animacy on neuronal activity, Hosokawa and Watanabe (2012) introduced a new competitive condition, where the rival was a computer instead of another monkey (monkey–computer competition). This condition was the same as the original condition (monkey–monkey competition), except that the rival was a computer and only one monkey played the game in the experimental booth. When the computer won the game, the solenoid valve at the winner's side operated (emitting an audible click) and the juice was delivered into a reservoir (bucket).

There was not any significant difference in the monkey's behavior in terms of accuracy or speed (i.e., the hit rate and the latency of the first shot) between the monkey–monkey and monkey–computer competitions. However, several LPFC neurons showed modulated activity depending on the animacy of the rival. Figure 9.3 shows competitive neurons that had differential activity between the monkey–monkey and monkey–computer competitions. The neuron in Fig. 9.3a showed higher reward-related activity when the rival was another monkey than when it was a computer program. Likewise, the neuron in Fig. 9.3b showed higher activity when the monkey lost a competition when the rival was another monkey than when it was a computer program. Nineteen (48%) of the 40 competitive neurons that were tested in both the monkey–monkey and monkey–computer competitive conditions showed modulated activity depending on whether the rival was a monkey or a computer. Most of them (17 of 19 neurons) showed higher

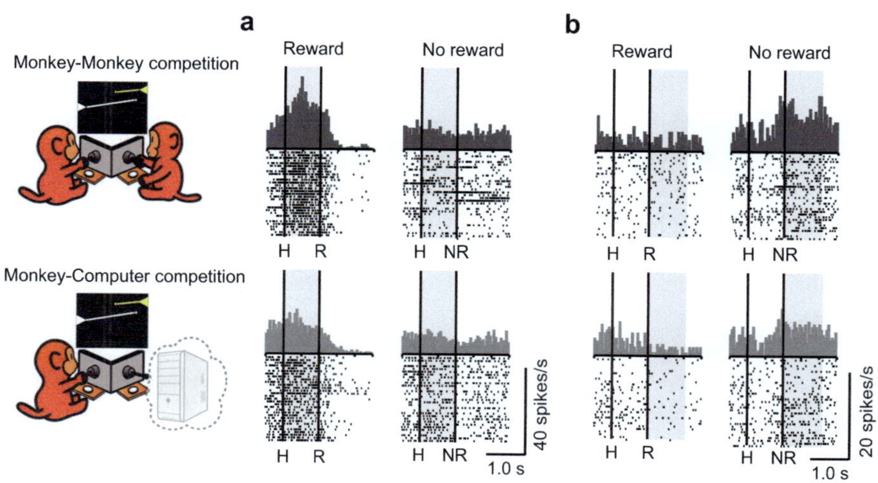


**Fig. 9.3** Effects of rival's animacy on LPFC neuronal activity. (**a**, **b**) Examples of neurons that showed greater activity in win (**a**) and loss (**b**) trials during the monkey–monkey than during the monkey–computer competition. The *top* and *bottom rows* respectively show activity in the monkey–monkey and monkey–computer competitions. The *left* and *right* columns respectively show activity during reward and no-reward trials. Each *shaded area* indicates the period when typical animacy-related activity was observed. The configuration of each raster and histogram display is the same as in Fig. 9.2b–d (Figure modified with permission from Hosokawa and Watanabe 2012)

activity when the monkey competed with another monkey than when it competed with a computer.

A question now arises. Since a higher activity was observed in the competitive condition, especially when the rival was another monkey, there was a possibility that the enhanced activity of competitive neurons might be due to the mere presence of another monkey nearby, rather than the presence of an animate rival. Thus, Hosokawa and Watanabe (2012) examined the effect of the presence of another monkey on LPFC neuronal activity to clarify whether the competitive neuronal activity was caused by the competitive nature of the game or by the presence of another monkey within view. To do so, a modified version of the noncompetitive condition was introduced. In this new noncompetitive condition, the monkey played the game in the presence of another monkey that was sitting nearby, but did not participate in the game (this new condition is called "two-monkey noncompetition," as compared with the original version of "one-monkey noncompetition"). The two-monkey noncompetitive game was the same as the one-monkey noncompetitive game, except that there was another monkey near the game-playing monkey. The other monkey just sat in the same position as in the competitive condition, but did not participate in the game.

Figure 9.4 shows examples of competitive neurons whose activities were not modulated by the mere presence of another monkey. Both neurons showed higher reward-related (Fig. 9.4a) and no-reward-related (Fig. 9.4b) activity in the competitive conditions (monkey–monkey and monkey–computer competitions) than in the noncompetitive conditions (one-monkey and two-monkey noncompetitions). However, they did not show differential activity depending on whether another monkey was sitting nearby or not. Figure 9.4c indicates how much the neuronal activity was modulated by (1) whether the current game was competitive or noncompetitive (competition factor), (2) whether there was another monkey nearby or not (another monkey's presence factor), and (3) their interaction effect. Only the competition factor was statistically significant. Thus, the effect of competition on the competitive neuronal activity was more profound than the effect of the presence of another monkey, indicating that competitive neurons in the LPFC showed differential activities not because there was another monkey present, but because the game was competitive.

These results suggest that social interactions are important for the competitive neuronal activity and are compatible with the results of previous studies that indicate the involvement of the LPFC in processing of social information (Fujii et al. 2009; Zink et al. 2008).

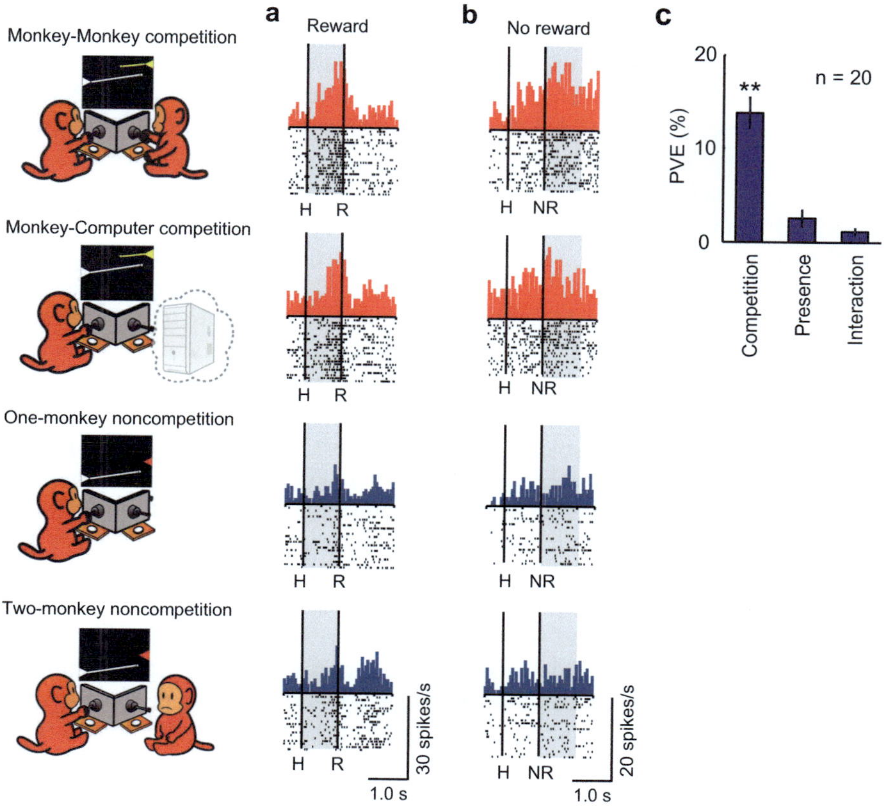


**Fig. 9.4** Effects of competition and presence of another monkey on LPFC neuronal activity. (**a**, **b**) Examples of neurons that showed greater activity in reward trials (**a**) and in no-reward trials (**b**) during the competitive compared with during the noncompetitive conditions. Neuronal activities during reward trials (**a**) and no-reward trials (**b**) under each condition are shown (from *top* to *bottom*, monkey–monkey competition, monkey–computer competition, one-monkey noncompetition, and two-monkey noncompetition). Each *shaded area* indicates the period when typical activity in relation to competition and the presence of another monkey was observed. (**c**) Mean percentage of variance explained (PVE) for the competitive neurons that were recorded in all of the four conditions. The plot shows the mean PVE of each factor in the two-way ANOVA (competition and presence factors, and their interaction) (mean ± SEM). Mann-Whitney U-test was used to determine whether these PVE values were significantly greater than the chance level that was calculated from the randomized data. $^{**}p < 0.01$ (Figure modified with permission from Hosokawa and Watanabe 2012)

## 9.6 Competition with Normal and Egalitarian Reward Contingencies

As described above, LPFC neurons showed higher activity in the competitive than in the noncompetitive condition, especially when the rival was another monkey rather than a computer. However, the performance-reward contingency in the competition was fixed in the study of Hosokawa and Watanabe (2012): winning (losing) a competition always led to the presence (absence) of a reward. Thus, it was not possible to examine the effects of the performance outcome (win/loss) separately from those of the presence or absence of the reward on prefrontal neuronal activity. Prefrontal neurons are known to code the correctness of one's own response independently of the presence or absence of the reward (Watanabe 1989). So, it would be also interesting to examine whether prefrontal neurons are concerned with coding a reward-independent performance outcome, namely, whether they code the win or loss of the game independently of the presence or absence of the reward.

To address this question, Hosokawa and Watanabe (2015) introduced egalitarian outcome conditions, in which both the winner and loser or neither of them received a reward, besides the normal outcome condition in which only the winner received a reward. In the egalitarian outcome conditions, winning and losing experiences were independent of the presence and absence of a reward. It was hypothesized that the egalitarian reward contingency in competition would greatly affect the monkey's behavior and that LPFC neurons would distinguish the context of reward delivery between the normal and egalitarian reward contingencies in competition. It was also predicted that there would be LPFC neurons that code the win/loss of the game independently of the presence/absence of the reward.

The new competitive game was almost the same as the competitive one in the previous study (Hosokawa and Watanabe 2012), except for the reward contingencies. Whether or not the monkeys would obtain a reward was determined not only by which monkey won the competition but also by the trial condition indicated by the background color of the monitor. When it was black, the winner got a reward and the loser did not get any reward (Win+Lose− trials, *normal competitive reward condition*). When it was green, both the winner and the loser got a reward (Win+Lose+ trials, *egalitarian reward condition*). When it was blue, neither the winner nor the loser got any reward (Win−Lose− trials, *egalitarian no-reward condition*). In Win−Lose− trials, to advance to the next trial, the current trial had to be terminated by either monkey's winning response even though neither of them would receive a reward. These three types of trial were intermingled and randomly presented in the same session.

## 9.7 Behavioral Performance in Trials with Normal and Egalitarian Outcome Conditions

The monkey's behavioral performance deteriorated in trials with the egalitarian outcome conditions. There were significant behavioral differences in accuracy and speed of the response between the trial types: the monkeys performed more accurately and quickly in the Win+Lose− trials than in the Win+Lose+ and Win−Lose− trials. The hit rate for the first shot was much better in the

Win+Lose− trials than in the other trials. There was no significant difference in the hit rates between the Win+Lose+ and Win−Lose− trials in either monkey. The latency of the first shot was shortest in Win+Lose− and longest in Win−Lose− trials in all monkeys. These findings suggest that the monkey's motivation was higher in trials with the normal competitive reward condition (i.e., when they had to win to obtain a reward) than in trials with the egalitarian outcome conditions (i.e., when the presence/absence of reward was independent of winning or losing).

Since winning the game had nothing to do with obtaining a reward in the Win−Lose− trials, it is reasonable that the monkeys lost their motivation. Interestingly, they also lost their motivation in the Win+Lose+ trials even though they could obtain a reward by winning a competition. The situation in the Win+Lose+ trials, where there was no need for the monkeys to win a competition to obtain a reward, may have been like social loafing, in which people exert less effort to achieve the goal when they work in groups than when they work alone (Karau and Williams 1993). These results indicate that an egalitarian outcome (irrespective of a reward or not) in competition reduces the monkeys' motivation.

## 9.8 Prefrontal Neuronal Activity in Trials with Normal and Egalitarian Outcome Conditions

Of 257 LPFC neurons recorded while the monkeys played the competitive game with normal and egalitarian outcome contingencies, many reflected the normal/egalitarian reward contingency indicated by the background color irrespective of whether the monkey won or lost a competition: Black (Win+Lose−, 31/257, 12%), Green (Win+Lose+, 33/257, 13%), or Blue (Win−Lose−, 68/257, 26%) type. Figure 9.5 shows examples of LPFC neurons that showed higher or lower activity in trials with a specific reward contingency. The neuron in Fig. 9.5a showed higher activity during the pre-reward period in the Win+Lose− (Black) trials than in the Win+Lose+ (Green) and Win−Lose− (Blue) trials. The neuron in Fig. 9.5b showed lower activity in the Win+Lose+ (Green) trials than in the Win+Lose− (Black) and Win−Lose− (Blue) trials. The neuron in Fig. 9.5c showed higher activity in the Win−Lose− (Blue) trials than in the Win+Lose− (Black) and Win+Lose+ (Green) trials.

Also, a substantial number of LPFC neurons (48/257, 19%) showed significantly higher activity in the normal competitive than in the egalitarian reward condition in relation to the presence or absence of a reward. This type of neuron is called Win/Lose x Black. The neuron in Fig. 9.5d showed a reward-related activity in the post-reward period. The reward-related activity was much higher in the Win+Lose− trials than in the Win+Lose+ trials, even though the monkey received the same reward in these trials, suggesting that the response of this neuron did not simply reflect the presence of the reward. Likewise, the activity of the neuron in Fig. 9.5e was much higher in the no-reward trials of Win+Lose− than in the no-reward trials of the Win−Lose−.

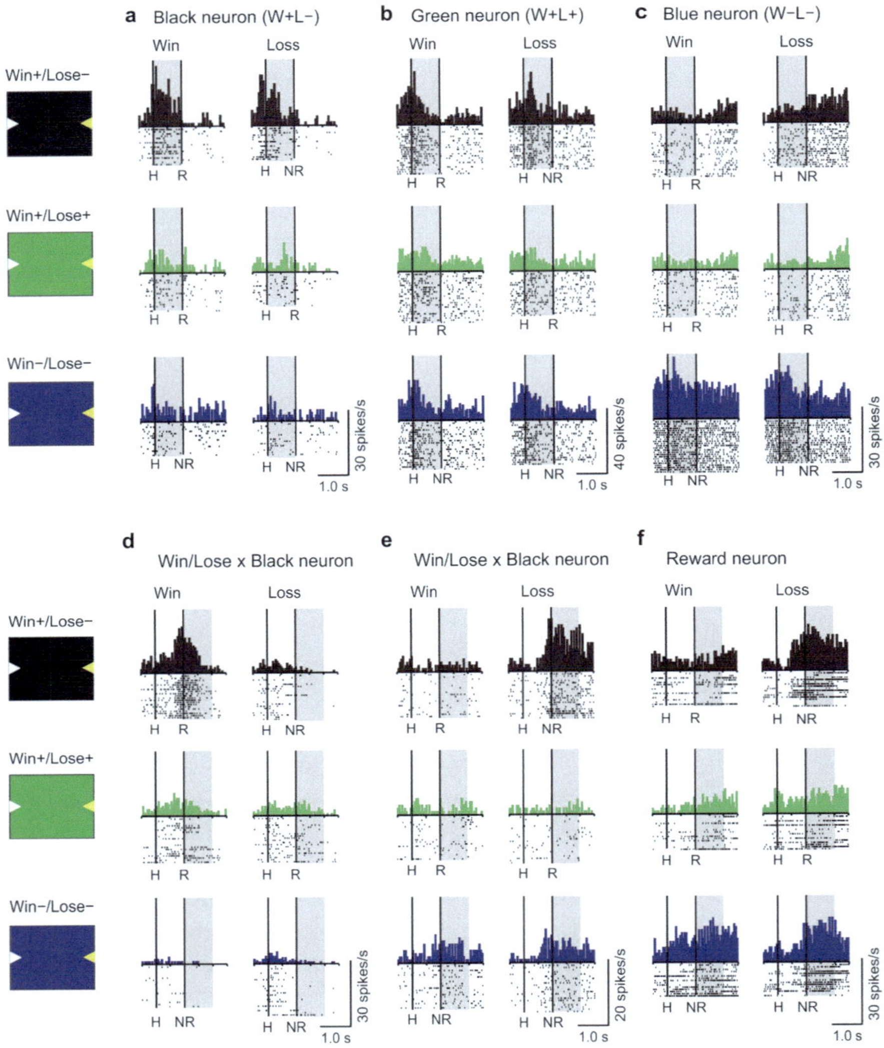


**Fig. 9.5** Examples of LPFC neurons that showed differential activity depending on the types of performance-reward contingency. (**a**) An LPFC neuron that showed higher activity in the normal (*Black*) than in the egalitarian conditions. (**b**) An LPFC neuron that showed lower activity in the egalitarian reward condition (*Green*) than in the other conditions. (**c**) An LPFC neuron that showed higher activity in the egalitarian no-reward condition (*Blue*) than in the other conditions. (**d**) An LPFC neuron that showed differential reward-related activity between the normal and egalitarian conditions (*Win/Loss* x *Black*). (**e**) An LPFC neuron that showed differential no-reward-related activity between the normal and egalitarian conditions (*Win/Loss* x *Black*). (**f**) An LPFC neuron that showed activation in no-reward trials irrespective of whether the reward contingency was normal or egalitarian (Reward). The configuration of each raster and histogram display is the same as in Fig. 9.2b–d (Figure modified with permission from Hosokawa and Watanabe 2015)

In addition, many LPFC neurons (91/257, 35%) showed activity depending on the presence or absence of the reward. This type of neuron is called Reward type. The neuron in Fig. 9.5f showed higher activity in the no-reward trials during the post-reward period.

Since there are prefrontal neurons that code the correctness of one's own response (Watanabe1989), it was expected that there would be neurons exclusively coding the performance outcome (win/loss). Contrary to the expectation, the percentage of such kind of neurons was not significantly higher than that from the chance level. Only a small number of LPFC neurons (16/257, 6%) that showed win/loss-related activity independently of either the presence/absence of the reward or the performance-reward contingency, suggesting that LPFC neurons do not code the winning or losing of the game independently of the presence or absence of a reward.

In this study, most of the LPFC neurons were concerned with the current context about whether the current reward contingency was normal or egalitarian (Win/Lose x Black, Black, Green, and Blue neurons) or whether the reward was given or not (Win/Lose x Black and Reward neurons). Thus, LPFC neurons appear to code the performance (win/loss) outcome in conjunction with the current context of the reward contingency. This result is consistent with previous studies indicating that the LPFC is concerned with both cognitive (such as rule) and motivational context (White and Wise 1999; Wallis et al. 2001; Kennerley and Wallis 2009; Watanabe et al. 2002; Amemori and Sawaguchi 2006a, b; Ichihara-Takeda and Funahashi 2008).

## 9.9 Summary and Future Prospects

In a series of experiments, the following results were found: (1) Monkeys were more motivated to play a competitive than a noncompetitive game. (2) LPFC neurons showed higher outcome-related (reward/no-reward) activity in the competitive than in the noncompetitive games. (3) LPFC neurons showed higher activity when the rival was another monkey than when it was a computer program. (4) Even in the competitive situations, the monkey's motivation greatly decreased when the performance-reward contingency was egalitarian. (5) LPFC neurons showed context-dependent activity reflecting the performance-reward contingency (Black, Green, and Blue neurons) or differential outcome-related (reward/no reward) activity depending on the performance-reward contingency (Win/Lose x Black neurons). These results indicate that the main role of LPFC may be monitoring the current context such as the rival's identity (i.e., who is the rival) and performance-reward contingency (i.e., whether winning a competition led to a reward or not) and integrating the performance outcome (win/loss) with the current context. This kind of integrated information may be important for the animal's survival in nature. Also, these results are compatible with those of previous studies reporting the involvement of LPFC in coding the rule of the behavioral task (Wallis et al. 2001) and in processing social information (Zink et al. 2008; Barbey et al. 2009).

Several studies have focused on the neuronal activity of the monkey brain in social situations. It was reported that neurons in the orbitofrontal cortex (OFC) and in the anterior cingulate cortex (ACC) showed differential activity depending on whether the reward would be given to oneself only or to both oneself and another monkey (partner) (Azzi et al. 2012; Chang et al. 2013). OFC neurons also coded the partner's identity and social rank (Azzi et al. 2012). Neurons in the medial prefrontal cortex (MPFC) code the action of the partner and the error that the partner made (Yoshida et al. 2011, 2012). In those studies, however, there were no interactions between the monkeys: only one of the monkeys made a choice (actor) and the other monkey passively observed the actor or received a reward depending on the actor's choice. In the studies conducted by Hosokawa and Watanabe (2012, 2015), on the other hand, two monkeys competed against each other in the shooting game. Prefrontal neurons showed differential activity only when the other monkey actively participated in the game as a rival, and most of the prefrontal neurons did not show differential activity to the mere presence of another monkey. These results suggest that a group of prefrontal neurons are sensitive to social interactions between oneself and other individuals.

As a future direction of research, it would be interesting to further study the effect of the identity or social rank of the rival on neuronal activity by using different ranks of subject during the same competitive shooting games. As for the monkeys used in the study of Hosokawa and Watanabe (2012), there was no clear dominant-submissive relation between the monkeys. However, a previous study reported that prefrontal neuronal activity was suppressed when one monkey faced with a dominant monkey (Fujii et al. 2009). Also, it would be interesting to study other brain regions using the same shooting games and compare the responses of neurons in different regions. Although many neuroimaging studies suggest that the MPFC especially is involved in processing social information (Behrens et al. 2008; Decety et al. 2004; Zink et al. 2008; Marsh et al. 2009; Zahn et al. 2009; Tricomi et al. 2010; McCabe et al. 2001), there are discrepancies between the results of neuroimaging studies and those of neurophysiological studies (Maier et al. 2008), and the neuronal signals of individual neurons cannot be detected by magnetic resonance imaging. As reward-related neurons are found everywhere in the brain (Arsenault et al. 2013; Haenny and Schiller 1988; Maier et al. 2008; Mogami and Tanaka 2006; Persichetti et al. 2015; Pooresmaeili et al. 2014; Serences 2008; Shmuel et al. 2006), social information may be encoded by neurons in many brain regions. This possibility is plausible given that social information is so important for social animals including humans.

## References


Amemori K, Sawaguchi T (2006a) Contrasting effects of reward expectation on sensory and motor memories in primate prefrontal neurons. Cereb Cortex 16(7):1002–1015. doi:10.1093/cercor/bhj042

Amemori K, Sawaguchi T (2006b) Rule-dependent shifting of sensorimotor representation in the primate prefrontal cortex. Eur J Neurosci 23(7):1895–1909. doi:10.1111/j.1460-9568.2006.04702.x

Arsenault JT, Nelissen K, Jarraya B, Vanduffel W (2013) Dopaminergic reward signals selectively decrease fMRI activity in primate visual cortex. Neuron 77(6):1174–1186. doi:10.1016/j.neuron.2013.01.008

Azzi JC, Sirigu A, Duhamel JR (2012) Modulation of value representation by social context in the primate orbitofrontal cortex. Proc Natl Acad Sci U S A 109(6):2126–2131. doi:10.1073/pnas.1111715109

Barbey AK, Krueger F, Grafman J (2009) An evolutionarily adaptive neural architecture for social reasoning. Trends Neurosci 32(12):603–610. doi:10.1016/j.tins.2009.09.001

Behrens TE, Hunt LT, Woolrich MW, Rushworth MF (2008) Associative learning of social value. Nature 456(7219):245–249. doi:10.1038/nature07538

Chang SW, Gariepy JF, Platt ML (2013) Neuronal reference frames for social decisions in primate frontal cortex. Nat Neurosci 16(2):243–250. doi:10.1038/nn.3287

Cox CR, Leboeuf BJ (1977) Female incitation of male competition – mechanism in sexual selection. Am Nat 111(978):317–335. doi:10.1086/283163

Decety J, Jackson PL, Sommerville JA, Chaminade T, Meltzoff AN (2004) The neural bases of cooperation and competition: an fMRI investigation. NeuroImage 23(2):744–751. doi:10.1016/j.neuroimage.2004.05.025

Deci EL, Betley G, Kahle J, Abrams L, Porac J (1981) When trying to win – competition and intrinsic motivation. Personal Soc Psychol Bull 7(1):79–83. doi:10.1177/014616728171012

Epstein JA, Harackiewicz JM (1992) Winning is not enough: the effects of competition and achievement orientation on intrinsic interest. Personal Soc Psychol Bull 18(2):128–138. doi:10.1177/0146167292182003

Fujii N, Hihara S, Nagasaka Y, Iriki A (2009) Social state representation in prefrontal cortex. Soc Neurosci 4(1):73–84. doi:10.1080/17470910802046230

Fukui H, Murai T, Shinozaki J, Aso T, Fukuyama H, Hayashi T, Hanakawa T (2006) The neural basis of social tactics: an fMRI study. NeuroImage 32(2):913–920. doi:10.1016/j.neuroimage.2006.03.039

Gallagher HL, Jack AI, Roepstorff A, Frith CD (2002) Imaging the intentional stance in a competitive game. NeuroImage 16(3 Pt 1):814–821

Garcia SM, Tor A, Gonzalez R (2006) Ranks and rivals: a theory of competition. Personal Soc Psychol Bull 32(7):970–982. doi:10.1177/0146167206287640

Haenny PE, Schiller PH (1988) State dependent activity in monkey visual cortex. I. Single cell activity in V1 and V4 on visual tasks. Exp Brain Res 69(2):225–244

Hosokawa T, Watanabe M (2012) Prefrontal neurons represent winning and losing during competitive video shooting games between monkeys. J Neurosci 32(22):7662–7671. doi:10.1523/JNEUROSCI.6479-11.2012

Hosokawa T, Watanabe M (2015) Egalitarian reward contingency in competitive games and primate prefrontal neuronal activity. Front Neurosci 9:165. doi:10.3389/fnins.2015.00165

Ichihara-Takeda S, Funahashi S (2008) Activity of primate orbitofrontal and dorsolateral prefrontal neurons: effect of reward schedule on task-related activity. J Cogn Neurosci 20(4):563–579. doi:10.1162/jocn.2008.20047

Janson C (1985) Aggressive competition and individual food-consumption in wild brown capuchin monkeys (Cebus-Apella). Behav Ecol Sociobiol 18(2):125–138. doi:10.1007/Bf00299041

Karau SJ, Williams KD (1993) Social loafing: a meta-analytic review and theoretical integration. J Pers Soc Psychol 65(4):681–706. doi:10.1037/0022-3514.65.4.681

Kennerley SW, Wallis JD (2009) Reward-dependent modulation of working memory in lateral prefrontal cortex. J Neurosci 29(10):3259–3270. doi:10.1523/JNEUROSCI.5353-08.2009

Kim S, Cai X, Hwang J, Lee D (2012) Prefrontal and striatal activity related to values of objects and locations. Front Neurosci 6:108. doi:10.3389/fnins.2012.00108

Kobayashi S, Lauwereyns J, Koizumi M, Sakagami M, Hikosaka O (2002) Influence of reward expectation on visuospatial processing in macaque lateral prefrontal cortex. J Neurophysiol 87 (3):1488–1498

Leboeuf BJ (1974) Male-male competition and reproductive success in elephant seals. Am Zool 14 (1):163–176

Maier A, Wilke M, Aura C, Zhu C, Ye FQ, Leopold DA (2008) Divergence of fMRI and neural signals in V1 during perceptual suppression in the awake monkey. Nat Neurosci 11 (10):1193–1200. doi:10.1038/nn.2173

Marsh AA, Blair KS, Jones MM, Soliman N, Blair RJ (2009) Dominance and submission: the ventrolateral prefrontal cortex and responses to status cues. J Cogn Neurosci 21(4):713–724. doi:10.1162/jocn.2009.21052

McCabe K, Houser D, Ryan L, Smith V, Trouard T (2001) A functional imaging study of cooperation in two-person reciprocal exchange. Proc Natl Acad Sci U S A 98 (20):11832–11835. doi:10.1073/pnas.211415698

Mogami T, Tanaka K (2006) Reward association affects neuronal responses to visual stimuli in macaque te and perirhinal cortices. J Neurosci 26(25):6761–6770. doi:10.1523/JNEUROSCI.4924-05.2006

Pan X, Sawa K, Tsuda I, Tsukada M, Sakagami M (2008) Reward prediction based on stimulus categorization in primate lateral prefrontal cortex. Nat Neurosci 11(6):703–712. doi:10.1038/nn.2128

Persichetti AS, Aguirre GK, Thompson-Schill SL (2015) Value is in the eye of the beholder: early visual cortex codes monetary value of objects during a diverted attention task. J Cogn Neurosci 27(5):893–901. doi:10.1162/jocn_a_00760

Pooresmaeili A, FitzGerald TH, Bach DR, Toelch U, Ostendorf F, Dolan RJ (2014) Cross-modal effects of value on perceptual acuity and stimulus encoding. Proc Natl Acad Sci U S A 111 (42):15244–15249. doi:10.1073/pnas.1408873111

Pusey A, Williams J, Goodall J (1997) The influence of dominance rank on the reproductive success of female chimpanzees. Science 277(5327):828–831. doi:10.1126/science.277.5327.828

Reeve J, Olson BC, Cole SG (1985) Motivation and performance: two consequences of winning and losing in competition. Motiv Emot 9(3):291–298

Rilling JK, Sanfey AG, Aronson JA, Nystrom LE, Cohen JD (2004) The neural correlates of theory of mind within interpersonal interactions. NeuroImage 22(4):1694–1703. doi:10.1016/j.neuroimage.2004.04.015

Robbins MM (1999) Male mating patterns in wild multimale mountain gorilla groups. Anim Behav 57:1013–1020. doi:10.1006/anbe.1998.1063

Roesch MR, Olson CR (2003) Impact of expected reward on neuronal activity in prefrontal cortex, frontal and supplementary eye fields and premotor cortex. J Neurophysiol 90(3):1766–1789

Rubenstein DI, Nuñez CM (2009) Sociality and reproductive skew in horses and zebras. Reproductive skew in vertebrates: proximate and ultimate causes. Cambridge University Press, Cambridge, pp. 196–226

Rumbaugh DM, Richardson WK, Washburn DA, Savage-Rumbaugh ES, Hopkins WD (1989) Rhesus monkeys (*Macaca mulatta*), video tasks, and implications for stimulus-response spatial contiguity. J Comp Psychol 103(1):32–38

Sanfey AG, Rilling JK, Aronson JA, Nystrom LE, Cohen JD (2003) The neural basis of economic decision-making in the Ultimatum Game. Science 300(5626):1755–1758. doi:10.1126/science.1082976 300/5626/1755 [pii]

Serences JT (2008) Value-based modulations in human visual cortex. Neuron 60(6):1169–1181. doi:10.1016/j.neuron.2008.10.051

Shmuel A, Augath M, Oeltermann A, Logothetis NK (2006) Negative functional MRI response correlates with decreases in neuronal activity in monkey visual area V1. Nat Neurosci 9 (4):569–577. doi:10.1038/nn1675

Tauer JM, Harackiewicz JM (2004) The effects of cooperation and competition on intrinsic motivation and performance. J Pers Soc Psychol 86(6):849–861. doi:10.1037/0022-3514.86.6.849 [doi] 2004–14304-005 [pii]

Tricomi E, Rangel A, Camerer CF, O'Doherty JP (2010) Neural evidence for inequality-averse social preferences. Nature 463(7284):1089–1091. doi:10.1038/nature08785

Tutin CEG (1979) Mating patterns and reproductive strategies in a community of wild chimpanzees (Pan-Troglodytes-Schweinfurthii). Behav Ecol Sociobiol 6(1):29–38. doi:10.1007/Bf00293242

Wallis JD, Anderson KC, Miller EK (2001) Single neurons in prefrontal cortex encode abstract rules. Nature 411(6840):953–956. doi:10.1038/35082081

Washburn DA, Hopkins WD, Rumbaugh DM (1990) Effects of competition on video-task performance in monkeys (*Macaca mulatta*). J Comp Psychol 104(2):115–121

Watanabe M (1989) The appropriateness of behavioral responses coded in post-trial activity of primate prefrontal units. Neurosci Lett 101(1):113–117 doi:0304-3940(89)90450-3 [pii]

Watanabe M (1996) Reward expectancy in primate prefrontal neurons. Nature 382 (6592):629–632. doi:10.1038/382629a0

Watanabe M, Sakagami M (2007) Integration of cognitive and motivational context information in the primate prefrontal cortex. Cereb Cortex 17(Suppl 1):i101–i109. doi:10.1093/cercor/bhm067

Watanabe M, Hikosaka K, Sakagami M, Shirakawa S (2002) Coding and monitoring of motivational context in the primate prefrontal cortex. J Neurosci 22(6):2391–2400 doi:22/6/2391 [pii]

White IM, Wise SP (1999) Rule-dependent neuronal activity in the prefrontal cortex. Exp Brain Res 126(3):315–335

Yoshida K, Saito N, Iriki A, Isoda M (2011) Representation of others' action by neurons in monkey medial frontal cortex. Curr Biol 21(3):249–253. doi:10.1016/j.cub.2011.01.004

Yoshida K, Saito N, Iriki A, Isoda M (2012) Social error monitoring in macaque frontal cortex. Nat Neurosci 15(9):1307–1312. doi:10.1038/nn.3180

Zahn R, Moll J, Paiva M, Garrido G, Krueger F, Huey ED, Grafman J (2009) The neural basis of human social values: evidence from functional MRI. Cereb Cortex 19(2):276–283. doi:10.1093/cercor/bhn080

Zink CF, Tong Y, Chen Q, Bassett DS, Stein JL, Meyer-Lindenberg A (2008) Know your place: neural processing of social hierarchy in humans. Neuron 58(2):273–283. doi:10.1016/j.neuron.2008.01.025

# Chapter 10
# Self-Recognition Process in the Human Prefrontal Cortex

**Ken Yaoi, Mariko Osaka, and Naoyuki Osaka**

**Abstract** "What is *self*?" or "Where is *self*?"—for a long time, many academics have been discussing various problems associated with the complex and ambiguous existence of the "self." In recent years, a number of neuroimaging studies have investigated the neural basis of the self-recognition processes. In this chapter, we classify the self into two types, "bodily self" and "mental self," and review the studies on the brain regions associated with self-recognition. First, we pick up sense of ownership and sense of agency as the bodily self. These abilities are assumed to be based on the brain regions that process the visual-perceptual information about body parts and that match the visual and proprioceptive information with the movement information in premotor cortex. Second, self-representation, which is one of the main parts of the mental self, is known to be involved with medial prefrontal cortex and cortical midline structure. Additionally, to maintain the sense of self, we continuously have to integrate information about ourselves without the distinction between bodily and mental self. In the "integrated self," the brain regions that receive and process sensory input from the body and the prefrontal cortex that carries out higher-order cognitive functions have an important role.

**Keywords** Self-recognition • Bodily self • Mental self • Sense of ownership • Sense of agency • Self-representation • Self-reference effect • Medial prefrontal cortex (MPFC) • Cortical midline structure

---

K. Yaoi (✉) • N. Osaka
Department of Psychology, Graduate School of Letters, Kyoto University, Yoshida-Honmachi, Sakyo-ku, Kyoto 606-8501, Japan
e-mail: fin.psy@gmail.com

M. Osaka
Center for Information and Neural Networks, National Institute of Information and Communication Technology, Osaka, Japan



M. Watanabe (ed.), *The Prefrontal Cortex as an Executive, Emotional, and Social Brain*, DOI 10.1007/978-4-431-56508-6_10

## 10.1 A "Self" Is a "Self"

In our daily lives, we rarely wonder about the fact that a "self" is a "self." The information captured by our sensory organs such as eyes and ears is recognized as what we see and what we hear by ourselves. Our arms and legs belong to ourselves, and we think that our thoughts and behaviors are mostly controlled by ourselves, consciously or subconsciously. Of course, we are "unconscious" while asleep, except perhaps when dreaming, and we cannot consciously control our behaviors. Even in such a state, our "self" is not lost. Most people would wake up or react in some ways if shaken or water is poured on them. This indicates that stimulation upon ourselves can be processed even when we are unconscious. It is extremely rare that someone would experience amnesia upon awakening one morning and lose the entire personal memories such as the name, living situation, and childhood experiences, asking "Where am I? Who am I?" (in psychiatry, this condition is classified as a dissociative disorder called "generalized amnesia"). Indeed, our "self" is properly maintained while we are asleep.

## 10.2 Self-Recognition and Its Relationship with Knowing Other

Self-recognition is thought to enable us to process information about ourselves at various levels, from "bodily" to "mental," and to distinguish self from nonself. We rarely experience sudden loss or change of self-recognition in our lives, and, thus, the self-recognition ability may seem commonplace. Our self-recognition is, however, not necessarily absolute and permanent. As with generalized amnesia, self-recognition could be lost or impaired due to various psychiatric disorders or brain injuries or could be partially altered by a simple factor such as an illusion. A number of studies have been conducted in cases of impaired self-recognition, in which one cannot recognize self in a picture, the most studied example being Alzheimer's disease (Biringer and Anderson 1992; Bologna and Camp 1997; Hehman et al. 2005). In his work *Altered Egos* (2001), the American neurologist Feinberg described many remarkable cases of lost or altered self-recognition due to various psychiatric disorders and brain injuries. Feinberg's work vividly illustrates that our "self-recognition" could be fragile and vulnerable as opposed to being inherent and permanent as we often assume.

On the other hand, as previously mentioned, we have to discriminate between self and other person to recognize ourselves. Seen in this light, it can be said that the self-recognition processes is one of the social brain's basic functions. Actually, many cognitive neuroscience research have indicated a strong link between social cognition and self-recognition processes, especially in the prefrontal cortex (Bergström et al. 2014). For example, medial prefrontal cortex (MPFC) shows greater activation when participants performed social and cognitive perspective-taking tasks such as theory of mind (ToM) (Saxe and Wexler 2005; Völlm et al. 2006). And some studies that investigated the neural correlates of the feeling of agency also point to the implication of the MPFC (Vinogradov et al. 2006). Furthermore, it has also been suggested that the lateral part of prefrontal cortex might hold different perspectives in working memory and inhibit the interference from one's own perspective (Gallagher and Frith 2003; Ruby and Decety 2003; Easton et al. 2009).

## 10.3 Where Does the "Self" *Reside*?

Self-recognition is an ability by which we recognize ourselves as self, consciously or otherwise. It is established by processing information about ourselves and, therefore, can be viewed as a cognitive function of humans and other animals. What is the underlying information processing system that gives rise to self-recognition? In humans and other animals, the brain is the center of the processing of vast information. Accordingly, at least a large part of our self-recognition is likely produced by information processing within the brain. In fact, injuries to or diminished function of certain brain regions often result in loss or impaired self-recognition ability, while functions essential for sustaining life or other cognitive activities are mostly unaltered. In their studies of patients with dementia caused by loss or impaired function in the frontal lobe or the temporal lobe (frontotemporal dementia), Miller and colleagues reported that atrophies in the right frontal lobe were strongly associated with dramatic changes in the personality of the affected individuals (Miller et al. 2001). Patients with this disorder often show sudden changes in personality and exhibit strange behaviors, such as abruptly leaving in the middle of a conversation. However, the patients typically lack insight into the changes caused by the disorder. The loss or impaired self-recognition was also mostly caused by stroke, cerebral thrombosis, or brain injuries in the patients described in the abovementioned work by Feinberg, *Altered Egos*.

What is the brain function that underlies the self-recognition ability? Is there a specific region of the brain where the "self" resides? The authors argue that there is no "specific brain region" that produces "self." Rather, distinct brain regions appear to play specific roles in establishing self-recognition, and also consciousness, in various physical and mental aspects. As we discuss later in this chapter, distinct brain regions are reportedly activated during different types of activities, such as seeing our own faces and movements of our arms or contemplating our own personality. Thus, coordinated activities of multiple brain regions are likely required to process various types of information needed to establish self-recognition.

The new methods developed in recent years have enabled researchers to indirectly quantify brain "activities" without invasive measures. These technological advances have facilitated investigations into brain activities associated with the formation of self-recognition. Human brain activities are now commonly studied (or, more precisely, indirect indicators of the neuronal activities are obtained) using electroencephalogram (EEG), positron emission tomography (PET), functional magnetic resonance imaging (fMRI), and magnetoencephalography (MEG). Below, we will discuss several studies that applied these technologies to investigate the brain regions critical for self-recognition.

## 10.4 Bodily Self and Its Neurological Basis in the Brain

### *10.4.1 Bodily Self and Mental Self*

In reviewing the studies on the brain regions associated with self-recognition, we will classify self-recognition into two types: (1) the recognition of the bodily self and (2) the recognition of the mental self. Although this distinction is provided primarily to simplify the discussion and is not an absolute classification, it seems reasonable as the processes of bodily and mental recognition greatly differ from each other and are likely associated with different brain regions.

Bodily self is the recognition of our body parts (i.e., face, hands, etc.) and their movements and is established based on the information provided by the sensory systems, such as vision, proprioception, and the somatosensory input. For instance, we can usually recognize ourselves when we see the image of ourselves in a mirror. This is because we continually acquire the information about external characteristics of ourselves, such as faces, physical appearances, and their movements. We can use the information to cross-reference and correctly identify the image in the mirror. A slightly different process is involved when we move a cursor on the display of a personal computer using a mouse. The cursor is located away from the hand, which we can move at will. However, we have the understanding that the cursor moves according to the movement of the hand (or the mouse in the hand), and therefore we can move it to an intended location on the computer display as well. This process requires the ability to recognize that the distance and direction of the movements of the hand and arm correspond to the movements of the cursor on the display.

From these examples, we can predict that the key brain regions for recognition of our bodily self are the regions that process external information provided by the sensory systems, such as the visual and somatosensory system. Below, we will discuss the specific brain regions associated with bodily self-recognition.

### *10.4.2 Self in the Mirror: The Face and the Body*

The "face" is the most unique among all the body parts. The face is what distinguishes us from others and is deeply intertwined with our self-identity. Naturally, brain regions activated by face recognition have drawn attention of researchers, and a number of studies have been conducted using fMRI and other brain imaging techniques mentioned above. Sugiura and colleagues (2005) compared the brain activities of subjects as they looked at their own faces, faces of their close acquaintances (friends), or faces of strangers. The results revealed that the right inferior frontal gyrus (also known as frontal operculum) and the left fusiform gyrus were activated when the subjects looked at their own faces. Many others reported similar findings (Platek et al. 2004; Uddin et al. 2005; Sugiura et al. 2006 and others). Platek and colleagues (2008) concluded that these studies consistently showed activation in the left fusiform gyrus[1], the bilateral middle frontal gyrus and inferior frontal gyrus, and the right precuneus, when subjects looked at their own faces (Fig. 10.1). Based on these findings, Platek expanded the hypothesis by Northoff and colleagues (2006) and proposed that these brain regions form a hierarchical information processing system consisting of three distinct levels. According to this hypothesis, the fusiform gyrus first identifies the information about the facial features obtained by the visual system. The precuneus then performs a cross-reference analysis of the incoming information about the face of oneself. And finally, the middle and inferior frontal gyri perform a higher-level processing to distinguish self from others. It should be noted that these brain regions are active not *only* when one sees the "face of self" but also when one performs various cognitive tasks. For instance, it is well documented that a part of the fusiform gyrus is activated when one sees a face regardless to whom it belongs (Kanwisher et al. 1997). The region is, therefore, referred to as the fusiform face area (FFA). Currently, there is no clear explanation on *how* the fusiform gyrus and other brain regions respond more strongly to the face of self than to the face of others.

Let us consider how we recognize the body parts other than the face. We can recognize our own hands and feet and distinguish them from those of others, although not as readily as the face. Recent studies indicate that the brain region called the extrastriate body area (EBA) within the extrastriate cortex, which is a part of the visual cortex in the occipital lobe, plays a critical role in bodily self-recognition other than the face (David et al. 2007, 2009; Myers and Sowden 2008; Vocks et al. 2010).[2] The name "body area" was previously provided because, as with the FFA in face recognition, this region was strongly activated when the subject saw images or movements of body parts, regardless of self or nonself (Downing et al. 2001, 2006). To further explain the activity of the EBA in recognizing one's own body parts, Myers and Sowden postulated the presence of a distinct neural subpopulation within the EBA that responds selectively to body

[1]A recent report suggested that the left and right fusiform gyri have distinct functions in recognizing the face of self (Ma and Han 2012).

[2]Another report showed that other brain regions involved in the perception of body parts within the fusiform gyrus (fusiform body area) are activated more strongly upon seeing one's own body parts than the body parts of others (Vocks et al. 2010).

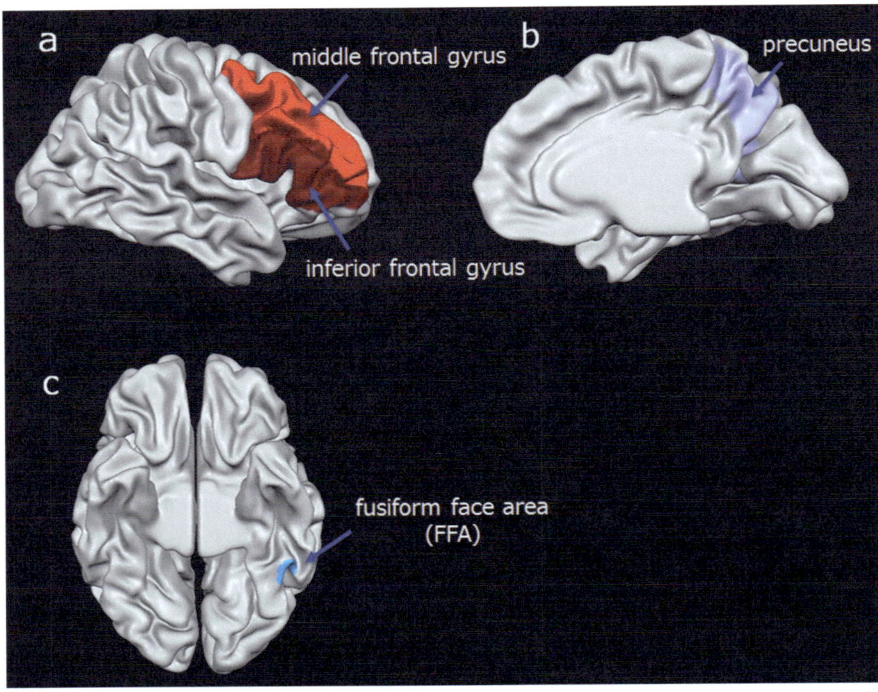


**Fig. 10.1** Brain regions activated by looking at the face of one's self (the images were generated based on the data by Platek et al. (2008) using the BrainVoyager Brain Tutor software (http://www.brainvoyager.com/products/braintutor.html). (**a**) *Middle* frontal gyrus and inferior frontal gyrus. (**b**) *Right* precuneus (image shows the medial side of the right hemisphere). (**c**) *Left* FFA (Image shows a ventral view of the cerebrum looking upward)

parts of self or nonself (Myers and Sowden 2008). It is of interest to note that neighboring or overlapping brain regions are involved in the perceptive recognition of bodily self and nonself. It will likely provide insight into the relationship between self-recognition and recognition of others.

### *10.4.3 Whose "Hand" Is This?: Sense of Ownership and Sense of Agency*

We have discussed so far the brain regions involved in the recognition of bodily self with a special focus on the perceptive recognition obtained by "looking" at our own face and body. Of course, there are other types of bodily self beyond what is visible. The two more fundamental aspects of bodily self are the ability to develop the sense that the body parts "belong to me" (this is called the "feeling of ownership") and the ability to sense that body movements are "controlled by myself" (this is called the "sense of agency") (Gallagher 2000; Newen and Vogeley 2003). These senses of bodily self are so "inherent" to us, and we may not feel that we "recognize" these abilities. However the feeling of ownership and the sense of agency can be lost due to brain injuries caused by accidents or disease. In the condition called asomatognosia, the patient loses the feeling of ownership and acquires a strange belief that a part of the body does not belong to oneself. The most frequently reported type of asomatognosia is called the alien hand syndrome, in which the patients complain that their left arm moves on its own against their will. In this disorder, both the feeling of ownership and the sense of agency have presumably been lost, and the patients often claim that their body parts belong to someone else and sometimes personify the body parts by naming them. This condition is often associated with an impairment in the right parietal lobe, especially in the region called the inferior and superior parietal lobules (*The Tell-Tale Brain*, Ramachandran 2011).

Other conditions are not as localized and permanent as asomatognosia and however may cause temporary confusion to the feeling of ownership. A well-documented example is the recently reported phenomenon called the rubber hand illusion (Botvinick and Cohen 1998). This phenomenon was demonstrated in an experiment, in which a subject is seated in front of a hand model made of rubber with the subject's own hand hidden from view. When the rubber hand and the subject's hand are *simultaneously* stroked by a paintbrush, the subject begins to think the rubber hand as the subject's own hand. The rubber hand illusion can be described as a phenomenon in which the subjects develop the feeling of ownership for the rubber hand that could not be their own hand in real life. Using this phenomenon, the mechanism that gives rise to the feeling of ownership can be investigated by comparing the brain activities of the subjects when they feel the rubber hand as their own and when they do not. Accordingly, the rubber hand illusion has been used in studies on the neural basis of the feeling of ownership. Ehrsson and colleagues (2004, 2005) demonstrated that the premotor cortex, which is in the posterior region of the frontal lobe, is activated as subjects experience the rubber hand illusion, and the potency of the activation was proportional to the strength of the sense of illusion. Studies in primates showed that the premotor cortex receives major inputs from the parietal lobe where visual and tactile information converge, and some of the neurons in the premotor cortex are activated in response to both visual and tactile stimulation of a certain body part (Graziano et al. 1997, 2004). These findings indicate that the rubber hand illusion occurs when tactile stimulation in a body part (a stroke of a paintbrush on the hand) is connected with visual stimulation (seeing a paintbrush stroking a rubber hand) by the activity of the neurons in the premotor cortex. Taken together, these studies suggest that, to establish the feeling of ownership, the input from the tactile sense and proprioceptive sense (the sense of the position and movement of the body obtained by the sensory system in the skin and muscle) must be synchronized (coincide temporally and spatially) with the visual input (Shimada 2009).

Let us now discuss the sense of agency and its neural basis. Consider the two situations: you raise your arm, and someone else lifts your arm. In both situations, you would recognize the arm is "yours" by the feeling of ownership established by the visual and tactical input. However, you would not feel that someone is lifting your arm when you raise your arm and vice versa. This indicates that the sense of agency is functioning properly. What is the mechanism of the sense of agency, and what is the underlying neural basis? A prerequisite to establish the sense of agency is the sensation that "the body moves (is moving) according to my will." You would not have the sense of agency if your arm moves downward against your will when you are trying to raise it (you would feel an external force pushing the arm downward). Thus, it is evident that the sense of agency is closely associated with the ability to control the voluntary movement of the body, as well as the proprioception that provides the information of the position and movement of the body. In fact, studies on the neural basis of the sense of agency suggest that the motor cortex in the frontal lobe and the association cortex in the parietal lobe play critical roles in establishing the sense of agency (Farrer et al. 2003; Leube et al. 2003; Schnell et al. 2007; Yomogida et al. 2010). These brain regions process the information about the own body and spatial context of the external environment and control body movement. How is the information processed in the motor cortex and the parietal lobe to establish the sense of agency? When we intend to move our own body, the motor cortex in the brain sends the signal to the muscle, and a "copy" of the signal is sent to the parietal lobe (this is called the "efference copy") (von Holst 1954; Andersen et al. 1997). The efference copy makes it possible to monitor and predict the movements of the body according to the movement signal sent to the muscle. This information is then matched with the proprioceptive input generated by the movement (Andersen et al. 1997; Murata 2009). In order to establish the sense of agency, the information about the result of the movement, which is the "feedback" sensory input obtained by the visual and proprioceptive systems, must coincide temporally with the command for the movement, which is the information on the movement provided by the efference copy (de Vignemont and Fourneret 2004; David et al. 2008). The object that generates the feedback input need not be a part of our body and can be applied to an object that moves synchronously with the movement of our body. For instance, when we move the cursor on the computer display, the movements of our arm are monitored by the efference copy, and the feedback input is generated by looking at the movements of the cursor. We develop the sense of agency that we are moving the cursor by temporally matching the visual input with the information about the movement of the arm monitored by the efference copy. The sense of agency, which is the mechanism by which we distinguish the movements of ourselves and others, also reportedly involves the "mirror neurons." These neurons are activated by seeing movements of *both* self and others (Shimada 2009; Murata 2009). The details are beyond the scope of this review; however, it is noteworthy that these neurons do not distinguish the movements of self and others and yet participate in controlling our own movements. This suggests that these neurons are required for the comprehension of actions of others through sensing the actions of ourselves.

In this section we have reviewed only a portion of the studies that investigated bodily self-recognition and its underlying neural basis. These studies have revealed that the brain regions that play critical roles in establishing bodily self-recognition are the regions that process the visual-perceptual information about body parts (fusiform gyrus and EBA) and the regions that match the visual and proprioceptive information with the movement information (premotor cortex and the parietal lobe). However, as mentioned above, these brain regions are not only involved in self-recognition but also, to some extent, in recognizing body parts and movements of others. Thus, in discussing self-recognition and its neural basis, we cannot simply describe, "XY brain region performs self-recognition." These findings also imply that we connect with others by recognizing others through the understanding of our own self (using the self as a model). Perhaps, the opposite should be considered as well; we may recognize our own self through the understanding of others.

## 10.5 Mental Self and Its Neural Basis

### *10.5.1 Who Am I?*

We will discuss mental self in this section. Simply put, mental self is established by a series of information we have about ourselves (memories). The information spans past, present, and (predictable) future and forms the basis of self-identity and self-concept, such as "what type of person I am" (this of course includes the mental images of the physical features of the face and body of ourselves; the classification of bodily and mental self is not absolute in part because of this). For example, when we introduce ourselves to someone we meet for the first time, we often tell the person our name, age, occupation, and the place of origin. In some instances, the person may ask us about something we did not mention, and we would respond to the question. Thus, in order to communicate with others about ourselves, we need to access (or recollect) the self-representation (also called self-image) established by the mental self. In the opposite situation, others may provide you with information about us. For instance, when someone tells you, "You are a modest person," you acquire the information about yourself and register the word "modest" in your memory (regardless of whether the person was sincere or not). In this case, the information acquired from an external source needs to be connected with the mental self. Thus, our mental self is closely associated with autobiographical memory, which is formed based on our own experiences and deeply meaningful to ourselves. Development of the mental self likely involves a cognitive process in which exogenous information is integrated into the autobiographical memory, and then information is retrieved by searching the autobiographical memory.

### 10.5.2 Self-Reference Effect

Before reviewing the studies that investigated the brain regions associated with the recognition of the mental self, we will discuss the unique cognitive nature of self-representation as the background of these studies. Self-representation has a unique nature distinct from "other-representation," which is based on the information about the "nature of existence of others (in ourselves)," or from other types of representation. A representative example of the unique aspects of self-representation is the self-reference effect (also known as self-association effect). Self-reference effect can be observed in a task (reference task) in which a subject is asked to judge the applicability of an object. When the subject refers to (or accesses) one's own self-representation in the task, the information provided is memorized better than in a task that requires a simpler process. In the self-reference effect, the memory is formed primarily through spontaneous learning (acquisition of memory without being explicitly instructed) (Rogers et al. 1977; Symons and Johnson 1997). To give a specific example, a subject is asked to answer two questions, "Are you *generous*?" and "Does *stubborn* mean the same as obstinate?" The subject will memorize the word *generous* through spontaneous learning and remember better than the word *stubborn*. The origin of self-reference effect remains unclear, and several theories exist. One hypothesis attributes to the unique nature of self-representation with the highly abundant and well-integrated information, compared to other-representation (Klein and Loftus 1988; Horiuchi 1995). One caveat of this example may be the dramatically different levels of cognitive processing in a task to judge one's own personality and a task to answer a meaning of a word. Let us consider another example in which the reference subject is another person instead of one's self. Several studies suggest that the memory performance in a similar reference task generally diminishes if the reference subject was a distant other (such as a celebrity) compared to a task requiring self-reference (Bower and Gilligan 1979; Kuiper and Rogers 1979; and others). As mentioned earlier, in a task that requires reference to the concept of oneself, the subject will need to access self-representation. Similarly, in a task that requires reference to another person (within one's self), the subject would need to access the representation of that person. Because the stimulation of the two types resulted in different memory performances, the cognitive nature of self-representation appears distinct from distant other-representations. However, it remains controversial whether self-representation is "unique" and clearly distinguishable from all other types of representation. For example, varying results have been reported in reference task experiments in which the subjects were given a task to judge close others (friends or spouses), and memory performances in response to the stimuli were compared to those obtained in a self-reference task (Symons and Johnson 1997; Heatherton et al. 2006). Much more information would exist regarding close others, such as friends or spouses, compared to distant others, and close other-representations likely have a similar nature as self-representation, which may have influenced the experiments.

Further investigation is warranted to determine whether there is a difference in the cognitive nature between self-representation and close other-representation.

### 10.5.3 Is the Mental Self "Special"?

The introduction above will provide some background information for studies of the brain regions associated with the metal self. Many studies reported experiments with reference tasks similar to the ones discussed above. Kelley and colleagues (Kelley et al. 2002) measured the brain activities of subjects by fMRI during three judgment tasks. The subjects were shown a word (adjective) on the display and were asked whether it applied to themselves (self-reference condition) or whether it applied to someone not close to them (former US President Bush) (other-reference condition). They were also asked whether the word was in capital letters or small letters. The brain imaging data indicated that the ventral part of medial prefrontal cortex (MPFC), which is the inner region of the prefrontal cortex, was more active in the self-reference condition compared to the other-reference condition. Based on these findings, Kelley and colleagues conclude that the MPFC plays a critical role in the cognitive activity of accessing or judging self-representation. Several other studies, such as the one by Craik (Craik et al. 1999), used similar experimental techniques. In most of these studies, the results indicated the activation of MPFC in a task requiring self-reference (Johnson et al. 2002; Fossati et al. 2003). Based on these series of studies, Northoff and Bermpohl (2004) proposed the term "cortical midline structure (CMS)" for the region encompassing the prefrontal cortex and the posterior cingulate cortex (PCC). They postulated that the CMS plays a critical role in processing the information related to mental self, such as self-representation, self-consciousness, and autobiographical memory. They divided the CMS into four regions with distinct functions. The dorsomedial prefrontal cortex (DMPFC), which is the upper (dorsal) area of the MPFC, has an important role in the evaluation of the information about self. The area below (ventral to) the DMPFC is the orbital and adjacent medial prefrontal cortex, so-called ventromedial prefrontal cortex (VMPFC). This area is reportedly associated with the representation of stimuli that have been connected to self by reference tasks. The anterior cingulate cortex (ACC) is the front portion of the cingulate gyrus, which is the area encompassing the corpus callosum that connects the left and right hemispheres. The ACC is associated with cognitive functions, such as selection and suppression of self-responses, and is thought to monitor or control the internal information about the self. The PCC and the neighboring area, called the precuneus, are strongly implicated in autobiographical memory and the process to integrate information about the self into the self in accordance with various contexts. Does this mean that the MPFC and other areas are the "special" areas for the mental self?

In the studies by Kelley and Craik, the others compared to the self were famous people who were not close to the subjects. Both studies found that distinct neural bases exist for the reference process depending on the types of representation, which supports the idea of "uniqueness of self-representation" suggested by the self-reference effect. Interestingly, however, results are more variable for the neural basis of the reference processes involving self-representation and representation of close others. This is reminiscent of the conflicting results to the question on memory performance in response to the reference stimuli involving self and close others. For instance, Heatherton and colleagues conducted a study on reference tasks involving the subjects themselves and their friends. The results indicated that MPFC was active in the self-reference condition (Heatherton et al. 2006), which led them to propose a critical role for MPFC in self-reference. In contrast, Schmitz et al. (2004), Vanderwal et al. (2008), and others used similar experimental designs and found MPFC activation in both self-reference and friend-reference conditions. Hence, it remains undetermined whether the MPFC has a "unique" role in reference tasks that require access to self-representation.

What might be the reason for the conflicting results in these studies? If the neural basis for the self-reference process was "unique," experiments with similar designs are expected to yield consistent data. An issue that complicates this problem is that in most studies, only one object (person) was used as the "other" that was compared to the self. Experiments comparing self to only one other are probably inadequate in answering questions such as: whether the MPFC plays a critical role in processes that specifically require access to self-representation, whether the MPFC is also involved in accessing to other-representation but only "weakly," and whether there is a difference between "close others" and "distant others." The authors, therefore, designed an experiment, in which the reference task required access to person representation for three individuals: (1) self (study subject), (2) a close other (friend), and (3) a distant other (prime minister). The brain activities were measured by fMRI (Yaoi et al. 2009). Other experimental settings were similar to previous studies. The subjects were shown a word (an adjective describing personality such as "laid-back" or "kind"; Table 10.1) and were asked how applicable the word was to the person in a scale of 1–4 or how many letters there were in the word. The fourth task was used to obtain the baseline (control) for the other three tasks that required references to person representation. Counting the number of letters does not require referencing to person representation; however, the actions to "look at the adjective" and to "respond by pressing the button" are shared with the reference tasks.

While the subjects performed these tasks, their brain activities were measured using an fMRI scanner (Fig. 10.2). The areas in the images marked in red showed significantly higher activities in each reference task compared to the baseline activities in the letter-count task. Our data clearly indicate that there is no significant difference in active areas among the three tasks with the self-, friend-, and

**Table 10.1** Examples of the adjectives used in the reference tasks

| | | | |
|---|---|---|---|
| Kind | Cheerful | Honest | Diligent |
| Docile | Obedient | Poor talker | Fast talker |
| Selfish | Lethargic | Cold-hearted | Irresponsible |

Yaoi et al. (2009)

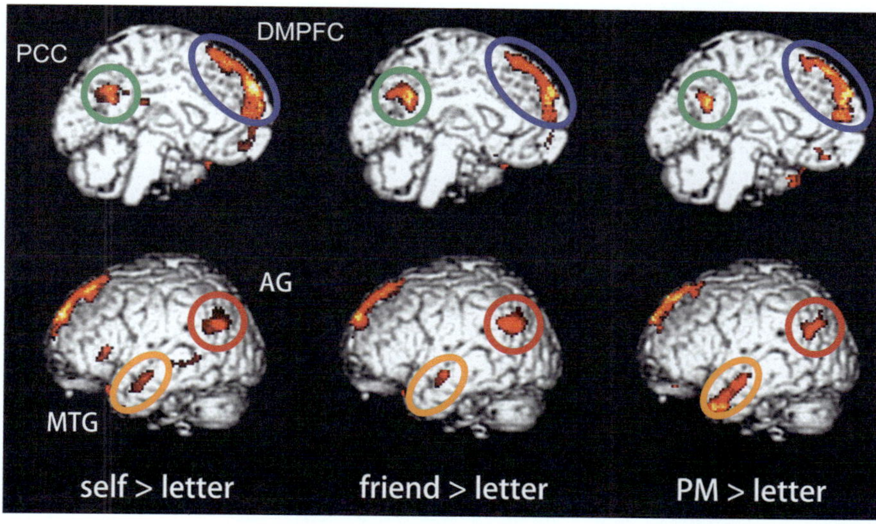


**Fig. 10.2** The regions activated in the three types of person-reference tasks compared to the letter-count task. DMPFC and several other areas were activated in all three (*DMPFC* dorsomedial prefrontal cortex; *PCC* posterior cingulate cortex, *AG* angular gyrus, *MTG* middle temporal gyrus) (From Yaoi et al. (2009) with modifications). The *upper images* are the medial side of the hemisphere

prime minister-reference conditions. In all tasks, the following regions were activated: DMPFC, PCC, the left middle temporal gyrus (MTG) located in the middle of the temporal lobe, and the angular gyrus (AG) located in the posterior region of the parietal lobe. No difference was detected even when we directly compared the self-, friend-, and prime minister-reference tasks. What do these results mean? Let us first discuss the activities in the three areas, PCC, left MTG, and AG. Several studies demonstrated that these areas play important roles in episodic memory including autobiographical memory (memories of experiences with defined time and space) and in retrieval and encoding of verbal memories (Ojemann et al. 2004; Wagner et al. 2005; and others). When judging how an adjective is applicable to a person (including self), various information about the individual, including the person representation, needs to be searched and retrieved. The activation of these regions likely reflects these common cognitive activities during the reference tasks. How about the results that MFPC (DMPFC) was also activated in the self-, friend-, and prime minister-reference tasks?

By a simple comparison to previous studies, our results disagree with Heatherton's claim that MPFC activation is unique to self-reference and are consistent with the results by Schmitz and colleagues that MPFC is activated in both self- and friend-reference tasks. Although we cannot determine specific processes in the judgment tasks associated with MPFC in this experiment, our data do not support the notion that the self-reference process requires activities in specific brain regions compared to other-reference process. Rather, our data indicate a neural basis for the process of accessing the person representation within the episodic memory and for making judgment about the person representation. In fact, our results are consistent with a meta-analysis of similar studies (meta-analysis is an analysis of a large number of previous studies to identify common factors, brain activities, etc. associated with a specific phenomenon). The meta-analysis performed by Legrand and Ruby (Legrand and Ruby 2009) revealed that an expansive neural network encompassing the MPFC, precuneus, temporoparietal junction, and the temporal pole. This network is not only involved in the judgment about information related to self but also involved in inference about a wide range of represented information, including other-representation, and in memory retrieval. The authors named the network the "E-network" (E stands for "evaluation"). The brain activities common in self- and other-reference that we found (shown in Fig. 10.2) support the existence of the E-network.[3]

It is unclear whether the E-network could explain every aspect of the neural basis of self- and other-reference processes. Why did previous studies, including ours, yield seemingly conflicting results about the activation of MPFC, even though the experimental paradigm was the same? The authors postulate that the MPFC is *not* specifically associated only with the reference process involving self-representation, but functions in accessing internal representation of self and others and possibly many other objects. This may not be limited to MPFC. The aforementioned cortical midline structure (CMS) appears to process various types of internal information, rather than to evaluate and integrate the information only about mental self. As mentioned earlier, self-representation incorporates a greater volume of information compared to representation of others. Self-reference tasks, therefore, would typically require deeper access into internal representation. If the activity of MPFC reflects the *difference in degree* of access into internal representation, it would be expected that MPFC shows stronger activation in self-reference tasks (in this regard, self could be considered unique). To corroborate this point, our recent data suggested that VMPFC was differentially activated in self- and other-reference tasks only when the word was complex and required longer time to judge (Yaoi et al. 2013). It is also possible that the process of accessing self- or other-representation, and its neural basis, may be influenced by how the self or others are internally recognized by the subjects. Perhaps there are individual and cultural variations, resulting in seemingly conflicting MPFC activation data that could or could not be interpreted as unique to self-reference. Thus, conflicting results in the previous studies with a similar experimental paradigm could have resulted from the difference in the nature of subject populations (a meta-analysis would likely disregard these variables). This possibility has been suggested by several studies in cultural psychology. Cousins (1989) reported that there are qualitative differences in the patterns of self-recognition between Japanese and Americans. In recent

[3]On the other hand, Denny et al. (2012) performed a meta-analysis of neuroimaging studies and reported that there is a ventral-to-dorsal gradient in MPFC for self- to other judgments from a direct comparison between both. Furthermore, other reviews or meta-analyses reached approximately the same conclusion (van der Meer et al. 2010; Qin and Northoff 2011; Wagner et al. 2012).

years, there have been interesting social neuroscience studies that corroborate this point (Han et al. 2008; Han and Northoff 2008). Zhu and colleagues (2007) compared the brain activities of Chinese and Caucasians in judgment tasks about the subjects themselves and their mothers. Interestingly, the MPFC was activated only in the self-judging task in Caucasians, while the MPFC was activated in both self- and mother-judging tasks in Chinese. These results provide evidence that cultural differences exist in the self- and other-reference processes and its neural basis. Very few studies in cultural neuropsychology are available at this time, and it is still unclear whether the cultural difference truly exists and, if so, to what extent. Future development is warranted in this important area of cultural neuroscience.

## 10.6 Integrated Self

We have so far reviewed the brain regions that play critical roles in the recognition of bodily self and mental self separately. Further questions arise as we reflect on the sense of "self." We continuously process the information about ourselves as needed without the distinction between bodily and mental and maintain the sense of "I." The incoming information through our body is continuously processed by the sensory and cognitive systems in response to stimuli. Part of the information will draw attention and surface to the consciousness, and other information will be given (special) meanings within the consciousness and eventually integrated into the mental self. We do not perceive any gaps in the information processing systems, and it seems as if a higher-level process exists so that various types of information about ourselves are shaped and integrated into distinct "self."

What is the system that creates the "integrated self"? The system would process the vast information about self, and it would be difficult to explain by the activities of the brain regions that we have discussed, each of which processes a portion of the information pertinent to self-recognition. To address the question of integrated self-recognition, we would need to explore the relationship between the "consciousness" and the "brain," which would form the ultimate foundation of self-recognition. Assume there are neurons that are electrochemically activated when we look at our own faces. How do these "activities" create the "feeling" in our mind that tells us "this is my face"? This is the so-called hard problem of the consciousness (Chalmers 1995). Although the problem may seem very difficult, several theories have provided hints for an answer. In his book *The Feeling of What Happens* (1999), Damasio proposed a hierarchical cognitive neural structure as a mechanism that continuously creates the feeling of an integrated self. According to his hypothesis, this structure encompasses a range of systems, from the sensory and cognitive systems to the higher-order self-recognition system. The brain regions that form this network include the brain stem and other regions that receive and process sensory input from the body and the prefrontal cortex (or medial prefrontal cortex) that carries out higher-order cognitive functions required for self-recognition (the brain stem lies between the cerebrum and spinal cord, consisting of the diencephalon, midbrain, pons, and medulla oblongata; and it controls vital functions for life). More recently, Feinberg (2011) proposed a similar hierarchical structure as the foundation of the creation of self-recognition, including consciousness. These theories share the view that self-recognition is a cognitive function consisting of a complex hierarchical system, in which sensory and cognitive information about our body and movements are processed at a lower, subconscious level, and then seamlessly integrated into the information processed at a higher, conscious level that generates internal memories and self-image.

## 10.7 Concluding Remarks: You Are "Not" Alone

In this chapter, we classified the self into two types, "bodily self" and "mental self," and reviewed the studies on the brain regions associated with self-recognition. First, we picked up sense of ownership and sense of agency as the bodily self. These abilities are assumed to be based on the brain regions that process the visual-perceptual information about body parts and that match the visual and proprioceptive information with the movement information. Second, self-representation, which is one of the main parts of the mental self, is known to be involved with medial prefrontal cortex and cortical midline structure.

Taking particular note of the role of MPFC, this region also has been known to be one of the important parts of the neural network that activates when the person "does nothing," so-called default mode network (DMN) or resting state network (Gusnard et al. 2001; Buckner et al. 2008; Spreng and Grady 2010; Salomon et al. 2014). Without going into detail, recent studies discovered that DMN is a group of areas in the human brain characterized by functions of a self-referential nature (Sheline et al. 2009; Buuren et al. 2010). On the other hand, recent social neuroscience research has emphasized the link between self- and other-representation in the MPFC. As shown above, self-recognition process is one of the social brain's basic functions that interface self and others, and actually researchers have indicated a strong link between social cognition and self-recognition processes. Some researchers argue that this overlap is not coincidence and that our "default mode" is wanting to become continuous with others (Lieberman 2013). In any case, the MPFC is not specifically associated only with the reference process involving self-representation, but functions in accessing internal representation of self and others. MPFC would support both our mental self and a wide variety of social activities by managing internal representation and make us social beings.

The studies on self-recognition discussed in this chapter explore only certain parts of self. It would be extremely difficult to answer how various aspects of "self," from bodily senses and cognition to internal representation, are integrated to create our self-recognition. We are only at the stage of searching and accumulating knowledge regarding individual processes at this time. Indeed, the field of research has just begun to explore the neural basis of the fundamental ability for social life of humans and other animals, such as the recognition and understanding of self and others (so-called social brain). As we learn about the cognitive systems that allow us to recognize the self as the self, and how we coexist with others and relate to the external world, we will begin to answer the fundamental question of "what is *self*."

## References


Andersen RA, Snyder LH, Bradley DC, Xing J (1997) Multimodal representation of space in the posterior parietal cortex and its use in planning movements. Annu Rev Neurosci 20:303–330. doi:10.1146/annurev.neuro.20.1.303

Bergström ZM, Vogelsang DA, Benoit RG, Simons JS (2014) Reflections of oneself: neurocognitive evidence for dissociable forms of self-referential recollection. Cereb Cortex 25:2648–2657. doi:10.1093/cercor/bhu063

Biringer F, Anderson JR (1992) Self-recognition in Alzheimer's disease: a mirror and video study. J Gerontol 47:385–388

Bologna SM, Camp CJ (1997) Covert versus overt self-recognition in late stage Alzheimer's disease. J Int Neuropsychol Soc 3:195–198

Botvinick M, Cohen J (1998) Rubber hands "feel" touch that eyes see. Nature 391:756. doi:10.1038/35784

Bower G, Gilligan S (1979) Remembering information related to one's self. J Res Pers 13:420–432

Buckner RL, Andrews-Hanna JR, Schacter DL (2008) The brain's default network: anatomy, function, and relevance to disease. Ann N Y Acad Sci 1124:1–38. doi:10.1196/annals.1440.011

Chalmers DJ (1995) Facing up to the problem of consciousness. J Conscious Stud 2:200–219

Cousins SD (1989) Culture and self-perception in Japan and United States. J Pers Soc Psychol 56:124–131

Craik FIM, Moroz TM, Moscovitch M et al (1999) In search of the self: a positron emission tomography study. Psychol Sci 10:26–34

Damasio AR (1999) The feeling of what happens: body and emotion in the making of consciousness. Harcourt Brace & Company, New York

David N, Cohen MX, Newen A et al (2007) The extrastriate cortex distinguishes between the consequences of one's own and others' behavior. NeuroImage 36:1004–1014. doi:10.1016/j.neuroimage.2007.03.030

David N, Newen A, Vogeley K (2008) The "sense of agency" and its underlying cognitive and neural mechanisms. Conscious Cogn 17:523–534. doi:10.1016/j.concog.2008.03.004

David N, Jansen M, Cohen MX et al (2009) Disturbances of self-other distinction after stimulation of the extrastriate body area in the human brain. Soc Neurosci 4:1–9. doi:10.1080/17470910801938023

de Vignemont F, Fourneret P (2004) The sense of agency: a philosophical and empirical review of the "Who" system. Conscious Cogn 13:1–19. doi:10.1016/S1053-8100(03)00022-9

Denny BT, Kober H, Wager TD, Ochsner KN (2012) A meta-analysis of functional neuroimaging studies of self- and other judgments reveals a spatial gradient for mentalizing in medial prefrontal cortex. J Cogn Neurosci 24:1742–1752. doi:10.1162/jocn_a_00233

Downing PE, Jiang Y, Shuman M, Kanwisher N (2001) A cortical area selective for visual processing of the human body. Science 293:2470–2473. doi:10.1126/science.1063414

Downing PE, Peelen MV, Wiggett AJ, Tew BD (2006) The role of the extrastriate body area in action perception. Soc Neurosci 1:52–62. doi:10.1080/17470910600668854

Easton S, Blanke O, Mohr C (2009) A putative implication for fronto-parietal connectivity in out-of-body experiences. Cortex 45:216–227. doi:10.1016/j.cortex.2007.07.012

Ehrsson HH, Spence C, Passingham RE (2004) That's my hand! Activity in premotor cortex reflects feeling of ownership of a limb. Science 305:875–877. doi:10.1126/science.1097011

Ehrsson HH, Holmes NP, Passingham RE (2005) Touching a rubber hand: feeling of body ownership is associated with activity in multisensory brain areas. J Neurosci 25:10564–10573. doi:10.1523/JNEUROSCI.0800-05.2005
Farrer C, Franck N, Georgieff N et al (2003) Modulating the experience of agency: a positron emission tomography study. NeuroImage 18:324–333. doi:10.1016/S1053-8119(02)00041-1
Feinberg TE (2001) Altered egos – how the brain creates the self. Oxford University Press, Oxford
Feinberg TE (2011) The nested neural hierarchy and the self. Conscious Cogn 20:4–15. doi:10.1016/j.concog.2010.09.016
Fossati P, Hevenor SJ, Graham SJ et al (2003) In search of the emotional self: an fMRI study using positive and negative emotional words. Am J Psychiatry 160:1938–1945
Gallagher I (2000) Philosophical conceptions of the self: implications for cognitive science. Trends Cogn Sci 4:14–21
Gallagher I, Frith CD (2003) Functional imaging of 'theory of mind'. Trends Cogn Sci 7:77–83
Graziano MSA, Hu XT, Gross CG (1997) Visuospatial properties of ventral premotor cortex. J Neurophysiol 77:2268–2292
Graziano MSA, Gross CG, Taylor CSR, Moore T (2004) A system of multimodal areas in the primate brain. In: Spence C, Driver J (eds) Crossmodal space and crossmodal attention. Oxford University Press, Oxford, pp 51–67
Gusnard DA, Akbudak E, Shulman GL, Raichle ME (2001) Medial prefrontal cortex and self-referential mental activity: relation to a default mode of brain function. Proc Natl Acad Sci U S A 98:4259–4264. doi:10.1073/pnas.071043098
Han S, Northoff G (2008) Culture-sensitive neural substrates of human cognition: a transcultural neuroimaging approach. Nat Rev Neurosci 9:646–654. doi:10.1038/nrn2456
Han S, Mao L, Gu X et al (2008) Neural consequences of religious belief on self-referential processing. Soc Neurosci 3:1–15. doi:10.1080/17470910701469681
Heatherton TF, Wyland CL, Macrae CN et al (2006) Medial prefrontal activity differentiates self from close others. Soc Cogn Affect Neurosci 1:18–25. doi:10.1093/scan/nsl001
Hehman JA, German TP, Klein SB (2005) Impaired self – recognition from recent photographs in a case of late – stage Alzheimer's Disease. Soc Cogn 23:118–124
Johnson SC, Baxter LC, Wilder LS et al (2002) Neural correlates of self-reflection. Brain 125:1808–1814. doi:10.1093/brain/awf181
Kanwisher N, McDermott J, Chun MM (1997) The fusiform face area: a module in human extrastriate cortex specialized for face perception. J Neurosci 17:4302–4311
Kelley WM, Macrae CN, Wyland CL et al (2002) Finding the self? An event-related fMRI study. J Cogn Neurosci 14:785–794. doi:10.1162/08989290260138672
Klein SSB, Loftus J (1988) The nature of self-referent encoding: the contributions of elaborative and organizational processes. J Pers Soc Psychol 55:5–11. doi:10.1037/0022-3514.55.1.5
Kuiper NA, Rogers TB (1979) Encoding of personal information: self-other differences. J Pers Soc Psychol 37:499–514. doi:10.1037/0022-3514.37.4.499
Legrand D, Ruby P (2009) What is self-specific? Theoretical investigation and critical review of neuroimaging results. Psychol Rev 116:252–282. doi:10.1037/a0014172
Leube D, Knoblich G, Erb M et al (2003) The neural correlates of perceiving one's own movements. NeuroImage 20:2084–2090. doi:10.1016/j.neuroimage.2003.07.033
Lieberman, MD (2013) Social: why our brains are wired to connect. Crown, New York.
Ma Y, Han S (2012) Functional dissociation of the left and right fusiform gyrus in self-face recognition. Hum Brain Mapp 33:2255–2267. doi:10.1002/hbm.21356
Miller BL, Seeley WW, Mychack P et al (2001) Neuroanatomy of the self: evidence from patients with frontotemporal dementia. Neurology 57:817–821
Myers A, Sowden PT (2008) Your hand or mine? The extrastriate body area. NeuroImage 42:1669–1677. doi:10.1016/j.neuroimage.2008.05.045
Newen A, Vogeley K (2003) Self-representation: searching for a neural signature of self-consciousness. Conscious Cogn 12:529–543. doi:10.1016/S1053-8100(03)00080-1

Northoff G, Bermpohl F (2004) Cortical midline structures and the self. Trends Cogn Sci 8:102–107. doi:10.1016/j.tics.2004.01.004

Northoff G, Heinzel A, de Greck M et al (2006) Self-referential processing in our brain – a meta-analysis of imaging studies on the self. NeuroImage 31:440–457. doi:10.1016/j.neuroimage.2005.12.002

Ojemann GA, Schoenfield-McNeill J, Corina D (2004) Different neurons in different regions of human temporal lobe distinguish correct from incorrect identification or memory. Neuropsychologia 42:1383–1393. doi:10.1016/j.neuropsychologia.2004.01.008

Platek SM, Keenan JP, Gallup GG, Mohamed FB (2004) Where am I? The neurological correlates of self and other. Cogn Brain Res 19:114–122. doi:10.1016/j.cogbrainres.2003.11.014

Platek SM, Wathne K, Tierney NG, Thomson JW (2008) Neural correlates of self-face recognition: an effect-location meta-analysis. Brain Res 1232:173–184. doi:10.1016/j.brainres.2008.07.010

Qin P, Northoff G (2011) How is our self related to midline regions and the default-mode network? NeuroImage 57:1221–1233. doi:10.1016/j.neuroimage.2011.05.028

Ramachandran VS (2011) The tell-tale brain: a neuroscientist's quest for what makes us human. W. W. Norton & Company, New York

Rogers TB, Kuiper NA, Kirker WS (1977) Self-reference and the encoding of personal information. J Pers Soc Psychol 35:677–688

Ruby P, Decety J (2003) What you believe versus what you think they believe: a neuroimaging study of conceptual perspective-taking. Eur J Neurosci 17:2475–2480

Salomon R, Levy DR, Malach R (2014) Deconstructing the default: cortical subdivision of the default mode/intrinsic system during self-related processing. Hum Brain Mapp 35:1491–1502. doi:10.1002/hbm.22268

Saxe R, Wexler A (2005) Making sense of another mind: the role of the right temporo-parietal junction. Neuropsychologia 43:1391–1399

Schmitz TW, Kawahara-Baccus TN, Johnson SC (2004) Metacognitive evaluation, self-relevance, and the right prefrontal cortex. NeuroImage 22:941–947. doi:10.1016/j.neuroimage.2004.02.018

Schnell K, Heekeren K, Schnitker R et al (2007) An fMRI approach to particularize the frontoparietal network for visuomotor action monitoring: detection of incongruence between test subjects' actions and resulting perceptions. NeuroImage 34:332–341. doi:10.1016/j.neuroimage.2006.08.027

Sheline YI, Barch DM, Price JL et al (2009) The default mode network and self-referential processes in depression. Proc Natl Acad Sci U S A 106:1942–1947. doi:10.1073/pnas.0812686106

Spreng RN, Grady CL (2010) Patterns of brain activity supporting autobiographical memory, prospection, and theory of mind, and their relationship to the default mode network. J Cogn Neurosci 22:1112–1123. doi:10.1162/jocn.2009.21282

Sugiura M, Watanabe J, Maeda Y et al (2005) Cortical mechanisms of visual self-recognition. NeuroImage 24:143–149. doi:10.1016/j.neuroimage.2004.07.063

Sugiura M, Sassa Y, Jeong H et al (2006) Multiple brain networks for visual self-recognition with different sensitivity for motion and body part. NeuroImage 32:1905–1917. doi:10.1016/j.neuroimage.2006.05.026

Symons CS, Johnson BT (1997) The self-reference effect in memory: a meta-analysis. Psychol Bull 121:371–394

Uddin LQ, Kaplan JT, Molnar-Szakacs I et al (2005) Self-face recognition activates a frontoparietal "mirror" network in the right hemisphere: an event-related fMRI study. NeuroImage 25:926–935. doi:10.1016/j.neuroimage.2004.12.018

van der Meer L, Costafreda S, Aleman A, David AS (2010) Self-reflection and the brain: a theoretical review and meta-analysis of neuroimaging studies with implications for schizophrenia. Neurosci Biobehav Rev 34:935–946. doi:10.1016/j.neubiorev.2009.12.004

Vanderwal T, Hunyadi E, Grupe DW et al (2008) Self, mother and abstract other: an fMRI study of reflective social processing. NeuroImage 41:1437–1446. doi:10.1016/j.neuroimage.2008.03.058

Vinogradov S, Luks TL, Simpson GV et al (2006) Brain activation patterns during memory of cognitive agency. NeuroImage 31:896–905. doi:10.1016/j.neuroimage.2005.12.058

Vocks S, Busch M, Grönemeyer D et al (2010) Differential neuronal responses to the self and others in the extrastriate body area and the fusiform body area. Cogn Affect Behav Neurosci 10:422–429. doi:10.3758/CABN.10.3.422

Völlm BA, Taylor ANW, Richardson P et al (2006) Neuronal correlates of theory of mind and empathy: a functional magnetic resonance imaging study in a nonverbal task. NeuroImage 29:90–98. doi:10.1016/j.neuroimage.2005.07.022

von Holst E (1954) Relations between the central Nervous System and the peripheral organs. Br J Anim Behav 2:89–94. doi:10.1016/S0950-5601(54)80044-X

Wagner AD, Shannon BJ, Kahn I, Buckner RL (2005) Parietal lobe contributions to episodic memory retrieval. Trends Cogn Sci 9:445–453. doi:10.1016/j.tics.2005.07.001

Wagner DD, Haxby JV, Heatherton TF (2012) The representation of self and person knowledge in the medial prefrontal cortex. Wiley Interdiscip Rev Cogn Sci 3:451–470. doi:10.1002/wcs.1183

Yaoi K, Osaka N, Osaka M (2009) Is the self special in the dorsomedial prefrontal cortex? An fMRI study. Soc Neurosci 4:455–463. doi:10.1080/17470910903027808

Yaoi K, Osaka M, Osaka N (2013) Medial prefrontal cortex dissociation between self and others in a referential task: an fMRI study based on word traits. J Physiol 107:517–525

Yomogida Y, Sugiura M, Sassa Y et al (2010) The neural basis of agency: an fMRI study. NeuroImage 50:198–207. doi:10.1016/j.neuroimage.2009.12.054

Zhu Y, Zhang L, Fan J, Han S (2007) Neural basis of cultural influence on self-representation. NeuroImage 34:1310–1316. doi:10.1016/j.neuroimage.2006.08.047

# Chapter 11
# Shared Attention and Interindividual Neural Synchronization in the Human Right Inferior Frontal Cortex

**Norihiro Sadato**

**Abstract** During a dyadic social interaction, two individuals can share visual attention through gaze, directed to each other (eye contact) or to a third person or an object (joint attention). Eye contact and joint attention are tightly coupled to generate the state of shared attention across individuals. Hyperscanning fMRI conducted with pairs of adults during joint attention tasks showed interindividual neural synchronization in the right inferior frontal gyrus. To explore how joint attention generates the state of shared attention, and whether its memory trace persists during a subsequent eye-contact condition, two-day hyperscanning fMRI study was conducted, in which pairs of participants performed a real-time mutual gaze task followed by a joint attention task on the first day and mutual gaze tasks several days later. The joint attention task enhanced eye-blink synchronization, which is a behavioral index of shared attention. When the same participant pairs underwent mutual gaze without joint attention on the second day, enhanced eye-blink synchronization persisted, which was positively correlated with interindividual neural synchronization within the right inferior frontal gyrus. Neural synchronization was also positively correlated with enhanced eye-blink synchronization during the previous joint attention task session. These results indicate that shared attention is represented and retained by pair-specific neural synchronization during mutual gaze that cannot be reduced to the individual level. This interbrain effect highlights the role of the right inferior frontal gyrus in the execution and learning of attentional coordination and sharing attention between self and others.

**Keywords** Inferior frontal gyrus • Hyperscanning functional MRI • Joint attention • Mutual gaze • Eye blink • Social cognition • Action representation • Hebbian association

---

N. Sadato (✉)
National Institute for Physiological Sciences, 38 Nishigonaka, Myodaiji, Okazaki, Aichi 444-8585, Japan
e-mail: sadato@nips.ac.jp



M. Watanabe (ed.), *The Prefrontal Cortex as an Executive, Emotional, and Social Brain*, DOI 10.1007/978-4-431-56508-6_11

## 11.1 Shared Attention Critical for Social Interaction

### *11.1.1 Social Interaction, Social Attention, and Shared Attention*

Humans can detect another person's focus of attention, orient own attention to the same location, and draw inferences regarding the other's goals using mainly through eye-gaze information (Allison et al. 2000; Calder et al. 2007; Nummenmaa and Calder 2009). This capability, known as social attention, is particularly important when during direct interaction with others. Face-to-face social interaction has three prominent characteristics (Schilbach et al. 2013). First, different roles for the interacting individuals emerge, such as initiator and responder. Second, sharing of attention, intention, and motivation are created de novo within an interaction and are critical for the interaction itself. Finally, there is a context for the interaction based on past events and experience. Shared attention, or coordinated visual attention during face-to-face interaction, such as mutual gaze and joint attention (Emery 2000), is a typical and fundamental process that fulfills the above three characteristics (Koike et al. 2016).

### *11.1.2 Mutual Gaze*

Mutual gaze provides a communicative link between humans by sharing the message "I am attending to you" (Farroni et al. 2002; Schilbach 2015). As gaze direction explicitly indicates the attentional target, mutual gaze is regarded as shared attention directed toward one another. Human infants and adults interact with one another dyadically by looking, touching, smiling, and vocalizing toward each other in turn-taking sequences, called protoconversations, during which infants gaze into the eyes of the partner (Hobson 2002; Trevarthen 1979, 1993). This face-to-face visual engagement, mutual gaze, is a universal feature of adult–infant interactions that represents mutual attentiveness and enhances positive emotional states (Keller et al. 1988). Thus, mutual gaze is implicated in the sharing of various psychological states.

### *11.1.3 Joint Attention*

Joint attention (JA) coordinates attention between partners to share an awareness of objects or events (Mundy et al. 1986). The importance of JA in the development of social cognition, as well as development of language, has been stressed (Mundy and Newell 2007). As an early-onset interactive process that leads to various kinds of social learning (Mundy and Newell 2007), it emerges as early as 6–12 months of age (Corkum and Moore 1998). Two types of joint attention are known: Initiating JA (IJA) is the ability to create spontaneously a shared point of reference using mutual gaze, and by alternating gaze between objects and other individuals, and responding JA (RJA) is the ability to follow the direction of the initiator's gaze to share attention toward an object (Mundy et al. 2009).

### *11.1.4 Link Between Mutual Gaze and Joint Attention*

The shared space of the common psychological ground resulting from mutual gaze may provide a communicative context. An adult's initial eye contact or mutual gaze prior to looking at an object is a critical cue that can establish joint attention with infants as young as 9 months old (Striano and Reie 2006). Mutual gaze therefore provides a communicative context for joint attention (Farroni et al. 2002). Thus IJA, RJA, and mutual gaze are tightly linked (Emery 2000; Perrett and Emery 1994) and function to share attention within a dyad or toward a third object.

## 11.2 Neural Substrates of Social Attention

### *11.2.1 Isolated Brain Approaches*

#### Eye Gaze

The neural substrates of social attention have been studied extensively, particularly using eye-gaze paradigms. Bilateral removal of the superior temporal sulcus (STS) region in macaques impairs perception of gaze direction without affecting perception of facial identity (Heywood and Cowey 1992). Recent human functional MRI (fMRI) studies have identified the involvement of the posterior STS (pSTS) in social perception through eye movement (Allison et al. 2000). Gaze processing extends beyond the STS to include the amygdala (George et al. 2001; Kawashima et al. 1999) and the inferior temporal (Wicker et al. 1998), parietal (Calder et al. 2007; Hoffman and Haxby 2000; Wicker et al. 1998; Mosconi et al. 2005; Hooker et al. 2003), medial prefrontal and anterior cingulate cortices (Calder et al. 2002), and other frontal regions (Mosconi et al. 2005; Hooker et al. 2003; Bristow et al. 2007). These different regions seem to process different aspects of the visual and social properties of gaze. Other regions of relevance include temporal areas implicated in face perception, frontoparietal attention regions, and areas implicated in emotion and social cognition (Nummenmaa and Calder 2009). For example, Sato et al. (2009) showed that automatic attentional shifts triggered by gaze, gestures, and symbols commonly activated the pSTS, the inferior parietal lobule, the inferior frontal gyrus, and the occipital cortices in the right hemisphere. This evidence indicated that the pSTS is related to the attentional shift per se. Recently, a combined fMRI–diffusion tensor imaging study by Ethofer et al. (2011) showed that dynamic gaze shifts toward an observer enhanced functional connectivity between the right pSTS and right anterior insula which, abutting the lower inferior frontal gyrus, plays a crucial role as part of the ventral attention system that is recruited by salient stimuli (Corbetta et al. 2008). Calder et al. (2002) reported that without making an explicit judgment, effects were observed in the anterior rostral portion of the medial prefrontal cortex (arMFC), when comparing an averted eye-gaze condition with a direct-gaze condition. They suggest that the activation of the arMFC is related to the implicit interpretation of averted gaze regarding the shift of an avatar's attention, which is a process that recruits the theory of mind module postulated by Baron-Cohen (1995). Thus, research in gaze processing is now outlining the neural basis of social attention, the cognitive components of which include the directing of social attention, attention shifting, processing of emotional reactions, and attribution of mental states (Saito et al. 2010).

#### Joint Attention

There are several neuroimaging studies of joint attention. Williams et al. (2005) used an RJA task that focused on the sharing of attention toward objects. In a joint attention condition, an avatar's gaze and the movement of a set of dot stimuli was concordant, whereas it was discordant in a non-joint attention condition. Corresponding regions of brain activation were in the anterior and posterior cingulate cortices. Laube et al. (2011) showed that the right pSTS and the right fusiform gyrus were involved in both processing of head- and eye-gaze direction during RJA. Using live interaction joint attention tasks, Redcay et al. (2010, 2012) showed activation in the ventromedial prefrontal cortex for RJA and intraparietal sulcus and middle frontal gyrus for IJA. Overlap for both IJA and RJA was observed in the dorsal medial prefrontal cortex, right inferior frontal gyrus, and right pSTS. Utilizing a virtual reality paradigm and functional MRI, Schilbach et al. (2010) showed that IJA and RJA are accompanied by activation of overlapping but partially independent neural networks. Unique activation for IJA was reported in the ventral striatum bilaterally, and unique activation for RJA was reported within the ventral medial prefrontal cortex. The extent to which the neural substrates of IJA and RJA are functionally segregated remains controversial.

### *11.2.2 Need for Hyperscanning to Identify the Neural Substrates of Shared Attention*

Until recently, the neural substrates of cross subject sharing of attention were hardly known. This is because much of the previous work on social attention has measured the responses of an individual brain to shared attention stimuli. However, since shared attention arises from the dynamic interaction of two agents, simultaneous measurement of the brain activities of two persons engaged in actual eye contact and joint attention is critical, because shared attention is an interactively constituted phenomenon which cannot be reduced to responses at the individual level (Konvalinka and Roepstorff 2012; Schilbach 2015). There have been several studies that investigated the flow of information between the brains of two partners by scanning participants one after another during offline interactions (pseudo hyperscanning, Anders et al. 2011; Konvalinka and Roepstorff 2012; Schippers et al. 2010; Stephens et al. 2010, for review). However, this technique cannot capture mutual influence *during* the interaction, which may be represented by interindividual neural synchronization (Astolfi et al. 2010; Cui et al. 2012; Dumas et al. 2010; Jiang et al. 2012; Muller et al. 2013; Osaka et al. 2014; Saito et al. 2010; Sänger et al. 2013; Tanabe et al. 2012; Yun et al. 2012).

The brain could be conceptualized as a discrete functional system, with external factors modulating rather than determining the operation of the system. In contrast, another way to conceptualize the brain is that it is an input–output system primarily driven by interaction with the external world (Fox et al. 2007). Support for the intrinsic perspective on brain function comes from studies of brain activity present even in the absence of task performance or stimuli, called intrinsic brain activity. This intrinsic brain activity is not random noise but is specifically correlated between related neurons (Tsodyks et al. 1999) and cortical columns (Kenet et al. 2003) and within widely distributed neuroanatomical systems (Biswal et al. 1995; Fox et al. 2005; Greicius et al. 2003; Hampson et al. 2002; Lowe et al. 1998). Given this spatial organization at multiple levels, intrinsic brain activity might have an important role in coordination of neuronal processing within the brain (Fox et al. 2007). Expanding this concept, intersubject synchronization might represent the intersubjective sharing of psychological states during eye contact. To evaluate intersubject synchronization, however, it is critical to exclude the possibility that the observed neural synchronization simply reflects similarity in their behavior (Konvalinka and Roepstorff 2012).

## *11.2.3 Interindividual Neural Synchronization at Right IFG During JA*

### Hyperscanning fMRI with JA

Based on this conceptualization and caveat, Saito et al. (2010) conducted hyperscanning fMRI of paired subjects (Montague et al. 2002) while they were engaged in joint attention tasks, with eye contact as the baseline. Saito et al. (2010) reported neural synchronization by intersubject correlation of the "innovations" which are the residual time courses of the neural activities obtained by modeling out the task-related effects and other confounding effects (Fig. 11.1). Given the linear addition of task-related activity on top of the persistent resting spontaneous activity

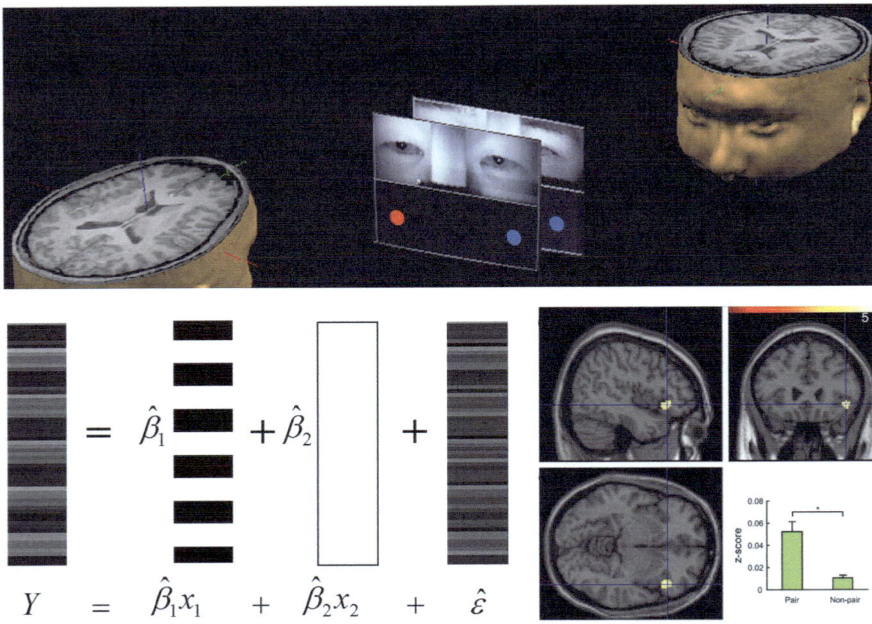


**Fig. 11.1** (*Top*) schematic diagram of the "hyperscan." Double-video systems implemented on two MRI scanners captured video images of each participant's eyes and eyebrows, which were transferred to the screen splitter that bound them to the computer-generated visual stimuli. The combined images were projected onto the screen in front of the counterpart through the projector. The other participant's eyes were presented on the upper half of the screen, and computer-generated images of balls were displayed at both ends of the screen in the lower half. The timing of the MRI scanning and the stimulus presentation were synchronized by the pulse signal from the controller of the double-video system to the two MRI scanners and the PC for the presentation of visual stimuli. (*Bottom left*) general linear model showing that the observed time series of the BOLD signal in the given voxel (Y) is the linear sum of the task-related activities (x1), constant term (x2), and the innovation (ε). (*Bottom right*) significant positive correlations of the innovation between the paired subjects who had been "face-to-face" during fMRI compared with the non-paired subjects. Images are superimposed on three orthogonal sagittal, transaxial, and coronal sections of T1-weighted high-resolution MR images. The *blue lines* in each section cross in the right IFG (44, 26, –6). The *color scale* indicates the *t* values. Standardized correlation value (z-score) of the pair and non-pair group. *Error bars* indicate the standard error of the mean (Modified from Saito et al. 2010)

(Fox et al. 2006), the adequate removal of task-induced variance in functional connectivity should yield a correlation profile similar to "continuous" resting-state data. Fair et al. (2007) speculated that interregional correlations might be altered during task states such that correlated spontaneous neural activity is strengthened in task-induced areas and weakened in non-task-induced areas. Thus, the innovation is a useful source of intrinsic dynamic information within different brain regions (Riera et al. 2004; Fair et al. 2007). Saito et al. (2010) made use of the correlations between pairs of innovations at voxels in different brain regions to construct measures that reflect interpersonal influences. As the baseline of JA is the eye-contact condition, the interpersonal correlation of the innovations of the paired persons, compared with those of the non-paired persons, represented the eye-contact effect. The advantage of the innovation approach is that it eliminates the effect of task, and therefore any remaining intersubjective correlation of the innovation data is not caused by task similarity across partners.

#### Interindividual Neural Synchronization at Right IFG During Joint Attention

A comparison of "pair" and "non-pair" correlations of the innovations showed that the correlation of the right IFG ($x = 44$, $y = 26$, $z = -6$, MNI coordinate) between the two brains was more prominent in the "paired" group than in a dummy "non-paired" group (Fig. 11.1 bottom right). The peak correlation corresponds to Brodmann area 47 adjacent to the anterior insular cortex (Saito et al. 2010). As mutual gaze implies the sharing of the intentional relation from the self to the agent ("I look at you"), and from the agent to the self ("You look at me"), the between-subject correlation of the innovation suggests that the right IFG is related to the between-subject sharing of the intentional relation. This sharing might create a context that enhances the detection of the communicative intent emitted with the eye movement (Frith and Frith 2006), making possible collaborative activities with shared goals (in the case of joint attention, looking at the same objects). The neural synchronization of the right IFG might represent the innate self-other equivalence in intention in action (Meltzoff 2007), which in turn provides a "like-me" framework. Within the "like-me" framework, it has been argued that internal representations of actions are shared between the self and others (shared action representations) and that this integration of information about one's own actions, and those of others, might involve the IFG. Recently, de Vignemont and Haggard (2008) argued that shared action representations are represented within the motor system. Within the hierarchical model of motor control, shared action representations involve intentional representations of action prior to the dispatch of a motor command. Shared action representations allow the observer to internalize someone else's actions as if he or she were the agent, and not just an external witness, providing the first-person perspective (de Vignemont and Haggard 2008). Saito et al. (2010) concluded that the right IFG may be the site of the neural representation of the "shared space of common psychological ground" mediated by eye gaze.

### *11.2.4 Hebbian Association Causes Synchronization of the Right IFG*

#### What Does the Neural Synchronization Represent?

In the context of shared action representations, interindividual neural synchronization can be understood based on the premise that the perceptual system of one brain can become coupled to the motor system of another (Dumas et al. 2010;

Jacob 2009; Schippers and Keysers 2011) through Hebbian association. This Hebbian account was previously invoked to explain automatic mimicry (Keysers and Perrett 2004; Del Giudice et al. 2009; Sasaki et al. 2012). That is, the basis of automatic mimicry is the process by which motor and perceptual action representations become tightly linked in such a way that perceiving another person's action activates the same representations as performing the action. It has been argued that action representations, or perceptuo-motor common representations, can be formed as an internal model through Hebbian associations trained during motor execution (Keysers and Perrett 2004; Del Giudice et al. 2009). Given that we continuously monitor our own actions, their sensory consequences are systematically and synchronously paired with motor commands. This predicts the emergence of Hebbian connections that link motor programs to sensory consequences (forward internal models), and vice versa (inverse internal models), even during social interaction (Wolpert et al. 2003; Treur 2011). In social Hebbian connections, one's own motor programs are linked to the sensory consequences provided by another's actions. Koike et al. (2016) applied this motor-perceptual common representation account to attention control. Their hypothesis was that the training of joint attention causes a social Hebbian association between initiating and responding joint attention, IJA and RJA, respectively. This is because the control of directing attention toward a third object for initiating JA is temporally linked to sensory consequences of the partner's response of directing attention to the same object, that is, RJA. Thus, social Hebbian association could link the neural activities induced by IJA to those by induced by RJA of the partner, resulting in neural synchronization. If this is true, then both IJA and RJA should activate the right IFG, and this synchronization should be retained as social memory after the JA experience.

## Behavioral Markers of Shared Attention: Blink

To quantify interpersonal aspects of the social interaction such as shared attention, finding behavioral markers is critical (Schilbach 2014). Attentional coordination during shared attention is in the spatial domain. Less explicitly included in the shared attention is the common "time window" of the attention directed to each other during mutual gaze that precedes the JA. To perform a JA task, the initiator is required to confirm that the partner is attending to the initiator during a preceding eye-contact condition, and the responder is required to attend to the initiator's eye movements. Thus, they are to share an attentional temporal window (Koike et al. 2016). Eye blinks are known to define the attentional temporal window. Demands for attentional resources modulate the rate of eye blinks (Bentivoglio et al. 1997; Shultz et al. 2011), and the timing of eye blinks is associated with implicit (Herrmann 2010) and explicit (Orchard and Stern 1991) attentional pauses in task content. Eye blinks of participants are synchronized while viewing the same video stories (Nakano et al. 2009) and between listener and speaker in face-to-face conversation (Nakano and Kitazawa 2010). Considering that blinks define the attentional "window," synchronization of eye blinks between face-to-face interactants can be taken as an index of shared attention. Once a Hebbian association is established, the initiation of eye contact between the previously trained pair will induce the control–response linkage in the attentional domain that can be measured via eye-blink synchronization (Koike et al. 2016).

### Hyperscanning fMRI with Mutual Gaze of Pre- and Post-JA Task

#### Hypothesis

Koike et al. (2016) hypothesized that shared attention during a JA task would be represented by blink synchronization and retained as a social memory and that this social memory would be represented by enhanced interindividual neural synchronization in the right IFG. Based on the Hebbian account, they also expected the right IFG to be activated by both RJA and IJA.

#### Experimental Setup

To test these hypotheses, Koike et al. (2016) conducted hyperscanning fMRI during a JA task and during mutual eye gaze both before and after the JA task (Fig. 11.2a). Three fMRI experiments were carried out. In Experiment 1, 34 (17 pairs) participants performed real-time mutual gaze (MG1 condition, Fig. 11.2a) followed by the JA tasks (Figs. 11.2b–d) on day 1; on day 2, participants again underwent the real-time mutual gaze condition (MG2 condition, Fig. 11.2a). There was a control condition in which participants believed that they were performing real-time interaction using eye contact, but in actuality, they watched a video recorded on day 1 (VIDEO condition, Fig. 11.2a). Experiment 2 was a 2-day hyperscanning fMRI study with 30 participants consisting of the real-time mutual gaze task without JA on day 1. In Experiment 3, 32 participants completed the MG1 and JA tasks as in Experiment 1 on day 1, but on day 2, they performed the real-time mutual gaze task with a new partner.

#### Eye-Blink Synchronization

In Experiment 1, the mutual gaze condition on day 1 (MG1) with an unknown partner did not elicit significant eye-blink synchronization (Fig. 11.3a). On day 2, during the mutual gaze condition (MG2), eye-blink synchronization was significant (Fig. 11.3a), and eye-blink synchronization in MG2 was significantly more prominent than during MG1 (Fig. 11.3a). Without online interaction between participants (VIDEO condition, watching the video recorded during the MG1), eye-blink synchronization was not significant (Fig. 11.3a). The difference in eye-blink synchronization between the MG2 and VIDEO conditions was also statistically significant (Fig. 11.3a).

There was a significant interindividual eye-blink synchronization even during JA tasks between paired partners (real pair) compared with eye-blink

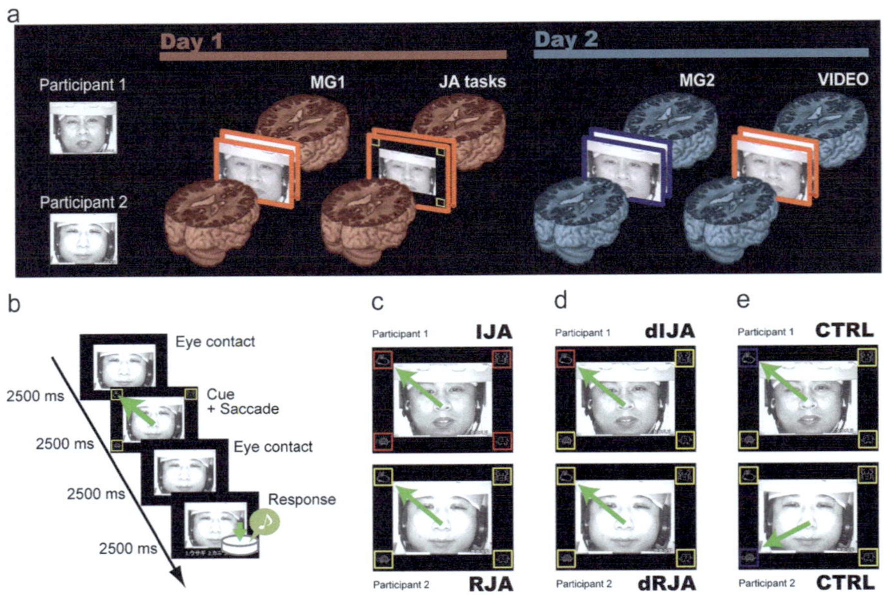


**Fig. 11.2** (**a**) Time line of Experiment 1. Image of the brain schematically indicates fMRI data obtained in day 1 (*orange*) during real-time eye contact through video (*orange frame*, MG1) and during joint attention task (*red frame*, JA tasks). In day 2, fMRI data (*blue brains*) was obtained during real-time eye contact through video (*blue frame*, MG2) and during watching the face video of the partner on day 1 (*orange frame*, VIDEO). (**b**) Time course of JA tasks. (**c**) Settings of IJA/RJA. The "all-four red" cue prompted the participant 1 to freely select one of the objects and shift his/her gaze on it. At the same time, identical objects with yellow frame were presented to the counterpart, participant 2. This "all-four-yellow" cue prompted the participant 2 to shift his/her gaze to the object that participant A attended to (*Green arrows*). Once the objects disappear, the participants are required to return back to the eye-contact situation for 2,500 ms. Then the names of four objects were presented under the live image of partner's face. Using a *button*, participants were required to select the name of the object that they had watched. Sound effect feedback was used to inform whether or not they successfully shared their attention to one object. (**d**) Designated-choice IJA/RJA (dIJA/dRJA). One red frame object and three yellow frame objects were presented to participant 1 who has to shift eye gaze toward the red target. (**e**) Control (CTRL) task. Both of them have to shift eye gaze individually toward the blue target without caring about partner's eye movement (Modified from Koike et al. 2016)

synchronization between randomly selected participants (pseudo pair, Fig. 11.3b). The strength of eye-blink synchronization during JA was positively correlated with enhanced eye-blink synchronization during MG2 compared with MG1 (Fig. 11.3c).

## Neural Synchronization

During the mutual gaze condition on day 1 (MG1), interindividual neural synchronization was found in the middle occipital gyrus and MTG (Fig. 11.3d) adjacent to the right EBA (white outline in Figs. 11.3d–f). During the mutual gaze condition on

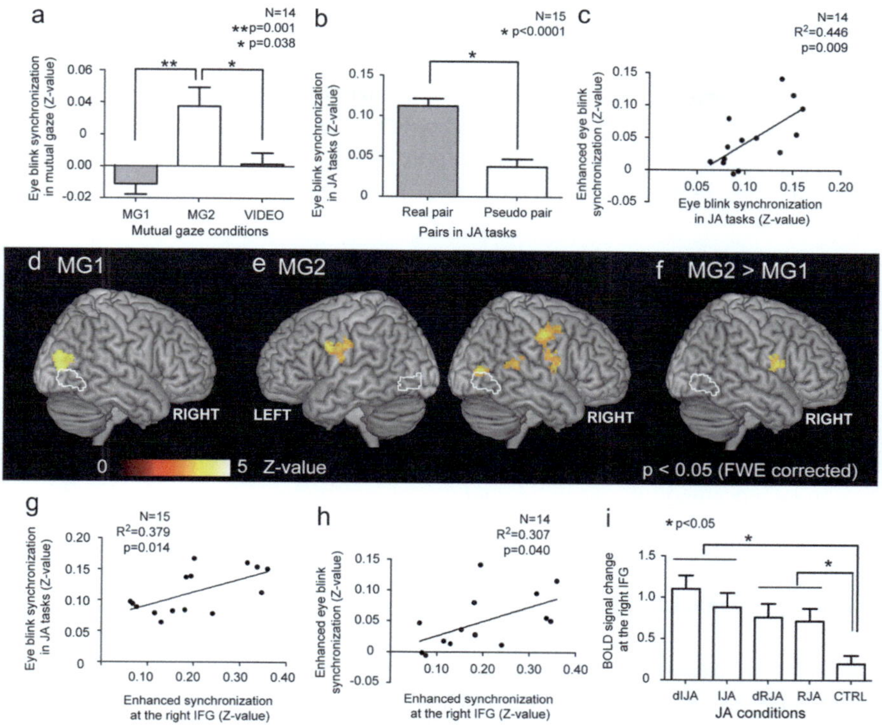


**Fig. 11.3** Interindividual eye-blink and neural synchronization. (**a**) The eye-blink synchronization between paired participants during the MG1, MG2, and VIDEO conditions. (**b**) Eye-blink synchronization in the JA task between paired participants (real pair) and two participants who were not paired but performed JA tasks with the same temporal parameters (pseudo pair). (**c**) Correlation between eye-blink synchronization during JA tasks and enhancement of eye-blink synchronization from MG1 to MG2. (**d**) Interindividual neural synchronizations before (MG1), **e** after JA task (MG2), and (**f**) their increment were superimposed on the 3D surface of a template brain. *White contour* indicates functionally defined extrastriate body area (EBA). The enhancement of neural synchronization at the right IFG cluster defined by MG2-MG1 (**f**) was correlated with (**g**) eye-blink synchronization during JA tasks and with (**h**) enhanced eye-blink synchronization. (**i**) Task-related activation of the right IFG during JA task. *Error bars*, standard error of the mean (s.e.m.) (Modified from Koike et al. 2016)

day 2 (MG2), interindividual synchronization extended anteriorly to the right posterior superior temporal sulcus, bilateral IFG, and ventral premotor cortex (Fig. 11.3e). The enhancement in interindividual synchronization during MG2 compared with MG1 was statistically significant only in the right IFG (Fig. 11.3f).

### Relationship Between Neural and Behavioral Synchronization

The enhancement of interindividual neural synchronization in the right IFG was significantly correlated with eye-blink synchronization during JA tasks (Fig. 11.3g) and with the enhancement of eye-blink synchronization (Fig. 11.3h). Consistent with the social Hebbian learning hypothesis, the right IFG was activated by both IJA and RJA, while no activation was found during the control condition (Fig. 11.3i).

### The Learning Effect Was Task and Pair Specific

Without JA experience (Experiment 2), no enhancement of behavioral synchronization was observed. Even following JA (Experiment 3), synchronization was not enhanced when the partner was swapped. In parallel with the behavioral data, there was no enhancement of interindividual neural synchronization in the right IFG.

## Eye-Blink Synchronization During JA

The JA task caused blink synchronization. To successfully conduct the task, participants had to coordinate the timing of opening and closing their window of attention with their partner's, resulting in eye-blink synchronization (task effect). Consistent with the task effect, significant eye-blink synchronization was also observed in the pseudo pair (Fig. 11.3b). As the task design was identical across the pairs, this indicates that the JA task aligned the attentional window within the dyad. Therefore, any difference in blink synchronization between real and pseudo pairs (Fig. 11.3b) constitutes a pair-specific effect. There was no eye-blink synchronization during first mutual gaze (MG1), reflecting no commonly shared task that can provide cues for eliciting similar behavior. Thus, eye-blink synchronization during mutual gaze which emerged after the JA task (Fig. 11.3a) does not reflect a task effect. The lack of eye-blink synchronization in the VIDEO condition confirms the importance of online mutual interaction for the emergence of eye-blink synchronization during MG2. Furthermore, the strength of eye-blink synchronization during JA was positively correlated with enhanced eye-blink synchronization during MG2 (Fig. 11.3c). Given constant task effects in synchronization during JA, this correlation indicates that blink synchronization during MG2 is affected by the pair-specific effect of blink synchronization during JA. In other words, the shared attention induced by JA was retained as a pair-specific "social" memory and represented by enhanced synchronization during mutual gaze.

## Enhanced Neural Synchronization in IFG During Mutual Gaze After JA

Across the whole brain, only the right IFG showed enhanced neural synchronization following JA (MG2–MG1, Fig. 11.3f), whereas no synchronization was observed during VIDEO. Enhanced synchronization in the right IFG was positively correlated with eye-blink synchronization during JA tasks and with the enhancement of eye-blink synchronization. Finally, the right IFG was activated by both IJA and RJA. These findings indicate that the right IFG is related to the generation of shared attention through social Hebbian association during JA and to its retention that is evoked by mutual gaze.

## The Role of Right IFG in Shared Attention

In general, the IFG is linked to several executive processes of the social stimuli, such as controlling, overriding, or inhibiting behavioral and emotional responses (Aron et al. 2004; Dillon and Pizzagalli 2007; Mitchell 2011), as well as mirroring (Leslie et al. 2004), empathizing (Schulte-Rüther et al. 2007), or imitating the behavior of another individual (Lee 2006). The IFG is also related to a unification of different types of sensory information to perform these executive processes (Frühholz and Grandjean 2013). The right IFG is an interface between self and other, especially during social situations. The right IFG is involved in unconscious incorporation of facial information of one's partner (Leslie et al. 2004) and in distinguishing self-related facial information from that of others (Sugiura et al. 2005). Furthermore, the right IFG is involved in the release of attention that is linked to spontaneous eye blinks (Nakano et al. 2013). The release of attention activates the default-mode network that is associated with internal processing while suppressing the dorsal attentional network (Nakano et al. 2013). As the right IFG and adjacent anterior insula switch between central-executive and default-mode networks (Sridharan et al. 2008), neural synchronization in the right IFG may represent synchronized shifting of attention toward self and others (Pfeiffer et al. 2013).

The right IFG was activated by both responding and initiating JA (Fig. 11.3i), consistent with previous studies (Redcay et al. 2012; Saito et al. 2010; Williams et al. 2005). Furthermore, neural synchronization of the right IFG occurred spontaneously during MG2. These findings are in line with the notion that mirror neuron properties of the right IFG and ventral premotor cortex (Gallese et al. 1996; Rizzolatti et al. 1996; Keysers et al. 2010) are caused by social Hebbian learning (Keysers and Perrett 2004; Wolpert et al. 2003) which binds self-derived behavior to that of others through online interaction (Mundy and Newell 2007; Treur 2011). The present study suggests that the right IFG was affected by social Hebbian association which binds self-derived directed attention (Tomasello and Carpenter 2007) to that of others.

Enhanced interindividual neural synchronization in the right IFG was statistically significant even after the removal of the eye-blink-related activation (Koike et al. 2016). Thus, the neural synchronization of the right IFG is not related to the blink per se but represents learned shared attention. Considering shared attention is to be understood as a complementary action due to its social salience, relevance in initiating communication, and joint action (Pfeiffer et al. 2013), the present finding is consistent with a previous study by Newman-Norlund et al. (2007) who showed the right IFG is more active during complimentary as compared to imitative actions.

## 11.3 Neural Synchronization in the "Social Default Mode"

Koike et al. (2016) showed enhanced synchronization of eye blinks within a dyad that was not attributable to similarity in their behavior but was instead due to the pair-specific relation (Konvalinka and Roepstorff 2012). Regarding the interindividual functional connectivity by means of neural correlation, Koike et al. (2016) treated the two brains as a spontaneous "two-in-one" system during the mutual gaze condition that can be regarded as a "social default mode," as the activity of an individual brain consists of spontaneously organized networks during the resting state (Fox et al. 2005, 2006). Right middle temporal gyrus (MTG) showed significant and consistent interindividual synchronization during mutual gaze (Fig. 11.3d, e). Unlike the right IFG, there was no learning effect (Fig. 11.3f). As no interindividual neural synchronization occurred during the VIDEO condition, MTG synchronization should have emerged as a result of online mutual interaction during mutual gaze. The EBA are known to receive both sensory inputs of others' body information (Downing et al. 2001) and efference copies (Astafiev et al. 2004; Orlov et al. 2010); thus, the adjacent MTG may conceivably receive information about self and other's eye blinks. Consistent with this notion, MTG has a role in detecting contingency between own and partner's behavior (Redcay et al. 2010). Given that the summation of inputs to the MTG region is identical between the two participants, even pairs of new partners synchronize their visual area activation (Koike et al. 2016).

In contrast, in the right IFG, interindividual connectivity became more conspicuous after partners became familiar with one another, i.e., after the JA training (Fig. 11.3f), and the connectivity profiles showed pair specificity. Thus, the property of the two-in-one system during the social default mode reflects the relationship between two participants, as the property of an intra-brain network reflects the mental state during a no-task condition or default mode (Yan et al. 2009). Mutual gaze underlies almost all face-to-face social interactions. Therefore, the effect of mutual gaze should be carefully considered to explore interindividual networks involved in face-to-face communication. Further investigation of this two-in-one system, during minimum task constraints, i.e., mutual gaze, might help to reveal the functional roles of interindividual neural synchronization, as default-mode network studies in the resting state have shed light on task-related brain networks (Fox et al. 2005, 2006).

## 11.4 Conclusion

The enhancement of behavioral and interindividual neural synchronization of the right IFG during mutual gaze after a JA task represents a pair-specific construct of shared attention that cannot be reduced to the individual level. As default-mode network studies on the resting state have shed light on state-related brain activities (Fox et al. 2005), further investigation of interindividual neural interaction will help to reveal the neural underpinnings of the state of interacting persons (Konvalinka and Roepstorff 2012).

## References

Allison T, Puce A, McCarthy G (2000) Social perception from visual cues: role of the STS region. Trends Cogn Sci 4:267–278

Anders S, Heinzle J, Weiskopf N, Ethofer T, Haynes J-D (2011) Flow of affective information between communicating brains. NeuroImage 54:439–446

Aron AR, Robbins TW, Poldrack R a (2004) Inhibition and the right inferior frontal cortex. Trends Cogn Sci 8:170–177

Astafiev SV, Stanley CM, Shulman GL, Corbetta M (2004) Extrastriate body area in human occipital cortex responds to the performance of motor actions. Nat Neurosci 7:542–548

Astolfi L, Toppi J, De Vico Fallani F, Vecchiato G, Salinari S, Mattia D, Cincotti F, Babiloni F (2010) Neuroelectrical hyperscanning measures simultaneous brain activity in humans. Brain Topogr 23:243–256

Baron-Cohen S (1995) Mindblindness: an essay on autism and theory of mind. The MIT Press, Cambridge

Bentivoglio AR, Bressman SB, Cassetta E, Carretta D, Tonali P, Albanese A (1997) Analysis of blink rate patterns in normal subjects. Mov Disord 12:1028–1034

Biswal B, Yetkin FZ, Haughton VM, Hyde JS (1995) Functional connectivity in the motor cortex of resting human brain using echo-planar MRI. Magn Reson Med 34:537–541

Bristow D, Rees G, Frith CD (2007) Social interaction modifies neural response to gaze shifts. Soc Cogn Affect Neurosci 2:52–61

Calder AJ, Lawrence AD, Keane J, Scott SK, Owen AM, Christoffels I, Young AW (2002) Reading the mind from eye gaze. Neuropsychologia 40:1129–1138

Calder AJ, Beaver JD, Winston JS, Dolan RJ, Jenkins R, Eger E, Henson RNA (2007) Separate coding of different gaze directions in the superior temporal sulcus and inferior parietal lobule. Curr Biol 17:20–25

Corbetta M, Patel G, Shulman GL (2008) The reorienting system of the human brain: from environment to theory of mind. Neuron 58:306–324

Corkum V, Moore C (1998) The origins of joint visual attention in infants. Dev Psychol 34:28

Cui X, Bryant DM, Reiss AL (2012) NIRS-based hyperscanning reveals increased interpersonal coherence in superior frontal cortex during cooperation. Neuroimage 59:2430–2437

de Vignemont F, Haggard P (2008) Action observation and execution: what is shared. Soc Neurosci 3:421–433

Del Giudice M, Manera V, Keysers C, Del Giudice M (2009) Programmed to learn – the ontogeny of mirror neurons. Dev Sci 12:350–363
Dillon DG, Pizzagalli DA (2007) Inhibition of action, thought, and emotion: a selective neurobiological review. Appl Prev Psychol 12:99–114
Downing PE, Jiang Y, Shulman M, Kanwisher N (2001) A cortical area selective for visual processing of the human body. Science 293:2470–2473
Dumas G, Nadel J, Soussignan R, Martinerie J, Garnero L (2010) Inter-brain synchronization during social interaction. PLoS One 5:e12166
Emery NJ (2000) The eyes have it: the neuroethology, function and evolution of social gaze. Neurosci Biobehav Rev 24:581–604
Ethofer T, Gschwind M, Vuilleumier P (2011) Processing social aspects of human gaze: a combined fMRI-DTI study. NeuroImage 55:411–419
Fair DA, Schlaggar BL, Cohen AL, Miezin FM, Dosenbach NU, Wenger KK, Fox MD, Snyder AZ, Raichle ME, Petersen SE (2007) A method for using blocked and event-related fMRI data to study "resting state" functional connectivity. NeuroImage 35:396–405
Farroni T, Csibra G, Simion F, Johnson MH (2002) Eye contact detection in humans from birth. Proc Natl Acad Sci U S A 99:9602–9605
Fox MD, Snyder AZ, Vincent JL, Corbetta M, Van Essen DC, Raichle ME (2005) The human brain is intrinsically organized into dynamic, anticorrelated functional networks. Proc Natl Acad Sci U S A 102:9673–9678
Fox MD, Snyder AZ, Zacks JM, Raichle ME (2006) Coherent spontaneous activity accounts for trial-to-trial variability in human evoked brain responses. Nat Neurosci 9:23–25
Fox MD, Snyder AZ, Vincent JL, Raichle ME (2007) Intrinsic fluctuations within cortical systems account for intertrial variability in human behavior. Neuron 56:171–184
Frith CD, Frith U (2006) The neural basis of mentalizing. Neuron 50:531–534
Frühholz S, Grandjean D (2013) Processing of emotional vocalizations in bilateral inferior frontal cortex. Neurosci Biobehav Rev 37:2847–2855
Gallese V, Fadiga L, Fogassi L, Rizzolatti G (1996) Action recognition in the premotor cortex. Brain 119(Pt 2):593–609
George N, Driver J, Dolan RJ (2001) Seen gaze-direction modulates fusiform activity and its coupling with other brain areas during face processing. NeuroImage 13:1102–1112
Greicius MD, Krasnow B, Reiss AL, Menon V (2003) Functional connectivity in the resting brain: a network analysis of the default mode hypothesis. Proc Natl Acad Sci U S A 100:253–258
Hampson M, Peterson BS, Skudlarski P, Gatenby JC, Gore JC (2002) Detection of functional connectivity using temporal correlations in MR images. Hum Brain Mapp 15:247–262
Herrmann A (2010) The interaction of eye blinks and other prosodic cues in German Sign Language. Sign Lang Linguist 13:3–39
Heywood CA, Cowey A (1992) The role of the "face-cell" area in the discrimination and recognition of faces by monkeys. Philos Trans R Soc L B Biol Sci 335:31–38
Hobson RP (2002) The cradle of thought. Pan Macmillan, London
Hoffman EA, Haxby JV (2000) Distinct representations of eye gaze and identity in the distributed human neural system for face perception. Nat Neurosci 3:80–84
Hooker CI, Paller KA, Gitelman DR, Parrish TB, Mesulam M, Reber PJ (2003) Brain networks for analyzing eye gaze. Cogn Brain Res 17:406–418
Jacob P (2009) The tuning-fork model of human social cognition: a critique. Conscious Cogn 18:229–243
Jiang J, Dai B, Peng D, Zhu C, Liu L, Lu C (2012) Neural synchronization during face-to-face communication. J Neurosci 32:16064–16069
Kawashima R, Sugiura M, Kato T, Nakamura A, Hatano K, Ito K, Fukuda H, Kojima S, Nakamura K (1999) The human amygdala plays an important role in gaze monitoring. A PET study. Brain 122(Pt 4):779–783
Keller H, Scholmerich A, Eibl-Eibesfeldt I (1988) Communication patterns in adult-infant interactions in Western and Non-Western cultures. J Cross-Cult Psychol 19:427–445

Kenet T, Bibitchkov D, Tsodyks M, Grinvald A, Arieli A (2003) Spontaneously emerging cortical representations of visual attributes. Nature 425:954–956

Keysers C, Perrett DI (2004) Demystifying social cognition: a Hebbian perspective. Trends Cogn Sci 8:501–507

Keysers C, Kaas JH, Gazzola V (2010) Somatosensation in social perception. Nat Rev Neurosci 11:417–428

Koike T, Tanabe HC, Okazaki S, Nakagawa E, Sasaki AT, Shimada K, Sugawara SK, Takahashi HK, Yoshihara K, Bosch-Bayard J, Sadato N (2016) Neural substrates of shared attention as social memory: a hyperscanning functional magnetic resonance imaging study. NeuroImage 125:401–412

Konvalinka I, Roepstorff A (2012) The two-brain approach: how can mutually interacting brains teach us something about social interaction. Front Hum Neurosci 6:215

Laube I, Kamphuis S, Dicke PW, Thier P (2011) Cortical processing of head- and eye-gaze cues guiding joint social attention. NeuroImage 54:1643–1653

Lee T-W (2006) Imitating expressions: emotion-specific neural substrates in facial mimicry. Soc Cogn Affect Neurosci 1:122–135

Leslie KR, Johnson-Frey SH, Grafton ST (2004) Functional imaging of face and hand imitation: towards a motor theory of empathy. NeuroImage 21:601–607

Lowe MJ, Mock BJ, Sorenson JA (1998) Functional connectivity in single and multislice echoplanar imaging using resting-state fluctuations. NeuroImage 7:119–132

Meltzoff AN (2007) "Like me": a foundation for social cognition. Dev Sci 10:126–134

Mitchell DGV (2011) The nexus between decision making and emotion regulation: a review of convergent neurocognitive substrates. Behav Brain Res 217:215–231

Montague PR, Berns GS, Cohen JD, McClure SM, Pagnoni G, Dhamala M, Wiest MC, Karpov I, King RD, Apple N, Fisher RE (2002) Hyperscanning: simultaneous fMRI during linked social interactions. NeuroImage 16:1159–1164

Mosconi MW, Mack PB, McCarthy G, Pelphrey KA (2005) Taking an "intentional stance" on eye-gaze shifts: a functional neuroimaging study of social perception in children. NeuroImage 27:247–252

Muller V, Sanger J, Lindenberger U (2013) Intra- and inter-brain synchronization during musical improvisation on the guitar. PLoS One 8:e73852

Mundy P, Newell L (2007) Attention, joint attention, and social cognition. Curr Dir Psychol Sci 16:269–274

Mundy P, Sigman M, Ungerer J, Sherman T (1986) Defining the social deficits of autism: the contribution of non-verbal communication measures. J Child Psychol Psychiatry 27:657–669

Mundy P, Sullivan L, Mastergeorge AM (2009) A parallel and distributed processing model of joint attention, social-cognition and autism. Autism Res 2:2–21

Nakano T, Kitazawa S (2010) Eyeblink entrainment at breakpoints of speech. Exp Brain Res 205:577–581

Nakano T, Yamamoto Y, Kitajo K, Takahashi T, Kitazawa S (2009) Synchronization of spontaneous eyeblinks while viewing video stories. Proc Biol Sci 276:3635–3644

Nakano T, Kato M, Morito Y, Itoi S, Kitazawa S (2013) Blink-related momentary activation of the default mode network while viewing videos. Proc Natl Acad Sci U S A 110:702

Newman-Norlund RD, van Schie HT, van Zuijlen AMJ, Bekkering H (2007) The mirror neuron system is more active during complementary compared with imitative action. Nat Neurosci 10:817–818

Nummenmaa L, Calder AJ (2009) Neural mechanisms of social attention. Trends Cogn Sci 13:135–143

Orchard LN, Stern JA (1991) Blinks as an index of cognitive activity during reading. Integr Physiol Behav Sci 26:108–116

Orlov T, Makin TR, Zohary E (2010) Topographic representation of the human body in the occipitotemporal cortex. Neuron 68:586–600

Osaka N, Minamoto T, Yaoi K, Azuma M, Osaka M (2014) Neural synchronization during cooperated humming: a hyperscanning study using fNIRS. Procedia Soc Behav Sci 126:241–243

Perrett D, Emery N (1994) Understanding the intentions of others from visual signals: neurophysiological evidence. Curr Psychol Cogn 13:683–694

Pfeiffer UJ, Vogeley K, Schilbach L (2013) From gaze cueing to dual eye-tracking: novel approaches to investigate the neural correlates of gaze in social interaction. Neurosci Biobehav Rev 37:2516–2528

Redcay E, Dodell-Feder D, Pearrow MJ, Mavros PL, Kleiner M, Gabrieli JDE, Saxe R (2010) Live face-to-face interaction during fMRI: a new tool for social cognitive neuroscience. NeuroImage 50:1639–1647

Redcay E, Kleiner M, Saxe R (2012) Look at this: the neural correlates of initiating and responding to bids for joint attention. Front Hum Neurosci 6:169

Riera J, Bosch J, Yamashita O, Kawashima R, Sadato N, Okada T, Ozaki T (2004) fMRI activation maps based on the NN-ARx model. NeuroImage 23:680–697

Rizzolatti G, Fadiga L, Gallese V, Fogassi L (1996) Premotor cortex and the recognition of motor actions. Cogn Brain Res 3:131–141

Saito DN, Tanabe HC, Izuma K, Hayashi MJ, Morito Y, Komeda H, Uchiyama H, Kosaka H, Okazawa H, Fujibayashi Y, Sadato N (2010) "Stay tuned": inter-individual neural synchronization during mutual gaze and joint attention. Front Integr Neurosci 4:127

Sänger J, Muller V, Lindenberger U (2013) Directionality in hyperbrain networks discriminates between leaders and followers in guitar duets. Front Hum Neurosci 7:234

Sasaki AT, Kochiyama T, Sugiura M, Tanabe HC, Sadato N (2012) Neural networks for action representation: a functional magnetic-resonance imaging and dynamic causal modeling study. Front Hum Neurosci 6:236

Sato W, Kochiyama T, Uono S, Yoshikawa S (2009) Commonalities in the neural mechanisms underlying automatic attentional shifts by gaze, gestures, and symbols. NeuroImage 45:984–992

Schilbach L (2014) On the relationship of online and offline social cognition. Front Hum Neurosci 8:278

Schilbach L (2015) Eye to eye, face to face and brain to brain: novel approaches to study the behavioral dynamics and neural mechanisms of social interactions. Curr Opin Behav Sci 3:130–135

Schilbach L, Wilms M, Eickhoff SB, Romanzetti S, Tepest R, Bente G, Shah NJ, Fink GR, Vogeley K (2010) Minds made for sharing: initiating joint attention recruits. J Cogn Neurosci 22:2702–2715

Schilbach L, Timmermans B, Reddy V, Costall A, Bente G, Schlicht T, Vogeley K (2013) Toward a second-person neuroscience. Behav Brain Sci 36:393–414

Schippers MB, Keysers C (2011) Mapping the flow of information within the putative mirror neuron system during gesture observation. NeuroImage 57:37–44

Schippers M, Roebroeck A, Renken R, Nanetti L, Keysers C (2010) Mapping the information flow from one brain to another during gestural communication. Proc Natl Acad Sci U S A 107:9388–9393

Schulte-Rüther M, Markowitsch HJ, Fink GR, Piefke M (2007) Mirror neuron and theory of mind mechanisms involved in face-to-face interactions: a functional magnetic resonance imaging approach to empathy. J Cogn Neurosci 19:1354–1372

Shultz S, Klin A, Jones W (2011) Inhibition of eye blinking reveals subjective perceptions of stimulus salience. Proc Natl Acad Sci U S A 108:21270–21275

Sridharan D, Levitin DJ, Menon V (2008) A critical role for the right fronto-insular cortex in switching between central-executive and default-mode networks. Proc Natl Acad Sci U S A 105:12569–12574

Stephens GJ, Silbert LJ, Hasson U (2010) Speaker-listener neural coupling underlies successful communication. Proc Natl Acad Sci U S A 107:14425–14430

Striano T, Reid VM (2006) Social cognition in the first year. Trends Cogn Sci 10:471–476

Sugiura M, Watanabe J, Maeda Y, Matsue Y, Fukuda H, Kawashima R (2005) Cortical mechanisms of visual self-recognition. NeuroImage 24:143–149

Tanabe HC, Kosaka H, Saito DN, Koike T, Hayashi MJ, Izuma K, Komeda H, Ishitobi M, Omori M, Munesue T, Okazawa H, Wada Y, Sadato N (2012) Hard to "tune in": neural mechanisms of live face-to-face interaction with high-functioning autistic spectrum disorder. Front Hum Neurosci 6:268

Tomasello M, Carpenter M (2007) Shared intentionality. Dev Sci 10:121–125

Treur J (2011) A computational agent model for hebbian learning of social interaction. Lect Notes Comput Sci 7062:9–19

Trevarthen C (1979) Communication and cooperation in early infancy: a description of primary intersubjectivity. In: Bullowa M (ed) Before speech the beginning of interpersonal communication. Cambridge University Press, Cambridge, pp 321–347

Trevarthen C (1993) The function of emotions in early communication and development. In: Nadel J (ed) New perspectives in early communicative development. Routledge, New York, pp 48–81

Tsodyks M, Kenet T, Grinvald A, Arieli A (1999) Linking spontaneous activity of single cortical neurons and the underlying functional architecture. Science 286:1943–1946

Wicker B, Michel F, Henaff MA, Decety J (1998) Brain regions involved in the perception of gaze: a PET study. NeuroImage 8:221–227

Williams JHG, Waiter GD, Perra O, Perrett DI, Whiten A (2005) An fMRI study of joint attention experience. NeuroImage 25:133–140

Wolpert DM, Doya K, Kawato M (2003) A unifying computational framework for motor control and social interaction. Philos Trans R Soc B 358:593–602

Yan C, Liu D, He Y, Zou Q, Zhu C, Zuo X, Long X, Zang Y (2009) Spontaneous brain activity in the default mode network is sensitive to different resting-state conditions with limited cognitive load. PLoS One 4:e5743

Yun K, Watanabe K, Shimojo S (2012) Interpersonal body and neural synchronization as a marker of implicit social interaction. Sci Rep 2:959

# Part IV
# Default Mode of Brain Activity and the Prefrontal Cortex

# Chapter 12
# Default Mode of Brain Activity Observed in the Lateral, Medial, and Orbital Prefrontal Cortex in the Monkey

**Masataka Watanabe**

**Abstract** Human neuroimaging studies have demonstrated the presence of a "default system" that consists mainly of the medial prefrontal and medial parietal areas and shows cognitive task-induced deactivation. The default activity is thought to be concerned with internal thought processes; however, there have been few attempts to demonstrate a default system that shows task-induced deactivation in nonhuman primates. Recently, a positron emission tomography (PET) study demonstrated working memory (WM) task-induced deactivations in the medial prefrontal (MPFC), lateral prefrontal (LPFC), orbital prefrontal (OFC), and medial posterior parietal areas during rest, suggesting the existence of internal thought processes in the monkey. In humans, activities of the executive system (such as LPFC) and default system (such as MPFC) are generally anticorrelated, and the OFC is rarely activated during rest. The rest-related activity in the monkey LPFC and OFC may be associated more with emotional or motivational aspects of internal thought processes. Dopamine in the prefrontal cortex plays important roles in cognitive operations, and a previous monkey microdialysis study revealed an increase of dopamine release in the LPFC during the WM task compared with that during rest. A recent study indicated a decrease in dopamine release in the anterior default system (MPFC/ACC) during the WM task compared with that during rest, indicating a rest-related increase in dopamine release in the anterior default system. As dopamine release in the LPFC contributes to cognitive operations such as WM, dopamine release in the MPFC/ACC during rest may be associated with other kinds of cognitive operations, such as internal thought.

**Keywords** Default mode of brain activity • Monkey • Rest • Working memory • Lateral prefrontal cortex • Medial prefrontal cortex • Orbital prefrontal cortex • Dopamine

---

M. Watanabe (✉)
Department of Physiological Psychology, Tokyo Metropolitan Institute of Medical Science, 2-1-6 Kamikitazawa, Tokyo 156-8506, Japan
e-mail: watanabe-ms@igakuken.or.jp



M. Watanabe (ed.), *The Prefrontal Cortex as an Executive, Emotional, and Social Brain*, DOI 10.1007/978-4-431-56508-6_12

## 12.1 Default Mode of Brain Activity in the Human

### *12.1.1 Higher Cerebral Blood Flow or Metabolic Activity During Rest*

Since the beginning of human neuroimaging studies using radioactive ligands, it has been noticed that there are certain brain areas that show higher cerebral blood flow (rCBF), compared with that in other areas, during the resting period. Ingvar (1979) reported that rCBF measured with the intra-arterial 133 Xenon clearance technique was significantly (20–40%) higher in the frontal than in the postcentral, occipital, and temporal regions and called this higher frontal activity "hyperfrontal activity." He considered the high frontal activity in the resting conscious state that is unaccompanied by movements, speech, or behavioral reactions, to be concerned with an anticipatory "simulation of behavior," since the activity was low in the sensory-gnostic cortical areas, while there should be internal processes related to anticipatory programming of several alternative behaviors. Technical advances in positron emission tomography (PET) methodology have allowed examination of activities in whole brain areas during the resting period. Raichle et al. (2001) used O $^{15}$-labeled radiopharmaceuticals as a ligand and showed higher rCBF in the medial cortical areas, including the medial prefrontal cortex (MPFC) and posterior cingulate cortex (PCC)/precuneus, than in other areas. More recently, Rilling et al. (2007) conducted a [$^{18}$F] fluorodeoxyglucose (FDG) – PET study on both humans and chimpanzees. In the human resting brain, they found higher metabolic activity in the MPFC, lateral prefrontal cortex (LPFC), premotor cortex, and PCC/precuneus. In these studies, it was revealed that there are brain areas that show higher activity during the resting period.

### *12.1.2 Regions with Task-Induced Deactivation Revealed by PET and fMRI*

In most human neuroimaging studies using PET or fMRI (functional magnetic resonance imaging), a subtractive methodology is commonly used, in which functional images acquired in a control state are subtracted from images acquired in a task state. The subtraction data are usually expressed as activations caused by the task requirement. Raichle and his colleagues, on the other hand, payed attention to the "deactivations" obtained by the "reverse subtraction" (Raichle et al. 2001). Thus, Shulman et al. (1997) compared average brain activity at rest with average activity during a task by conducting meta-analysis on 132 human neuroimaging studies and obtained a set of regions showing decreased activity during task performance. Such "task-induced deactivation" has been observed in the "default system" or "default mode network" of the brain that mainly involves the MPFC,

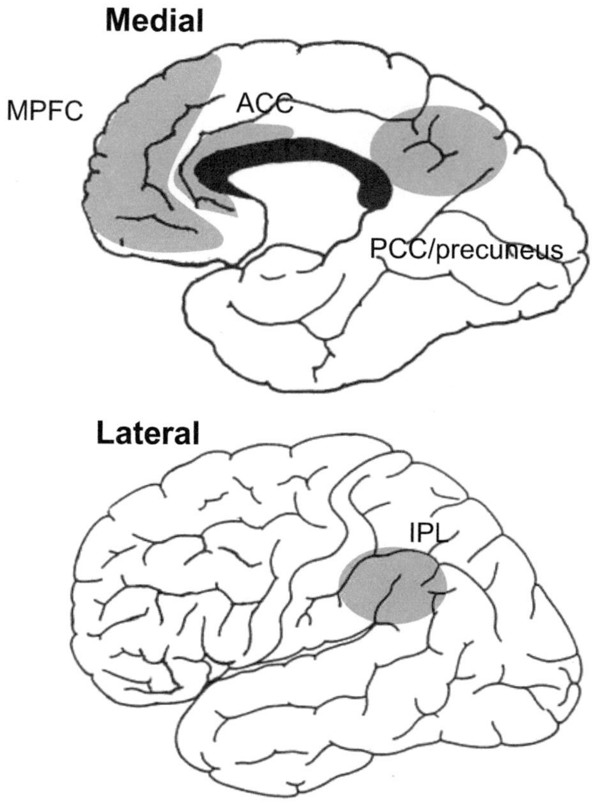


**Fig. 12.1** Regions with default mode of brain activity in humans. Medial and lateral indicate medial and lateral aspects of the human brain, respectively. Abbreviations, *ACC* anterior cingulate cortex, *IPL* inferior parietal lobule, *MPFC* medial prefrontal cortex, *PCC* posterior cingulate cortex

anterior cingulate cortex (ACC), PCC including the retrosplenial cortex, precuneus, and inferior parietal lobule (IPL) (Fig. 12.1).

The fact that some brain regions were more active at rest than during task performance led to the hypothesis that the brain remained active in an organized fashion during the resting state, and the regions that routinely exhibit a decrease in activity during task performance were thought to mediate processes that are important for the resting state (Fox and Raichle 2007). Importantly, rest-related activity increases are observed irrespective of whether the subject is resting in either visual fixation or with eyes closed. Furthermore, regardless of the task under investigation, the rest-related activity increase almost always includes the same default brain system (Gusnard and Raichle 2001).

### 12.1.3 Regions with Task-Induced Suppression of High-Frequency EEG Power

It is not easy to examine EEG activity in the human default system since the default system consists mainly of medial cortical areas that are difficult to access using the surface EEG recording. However, there are a few rare studies in which EEG was recorded from the default system in epileptic patients in whom depth-EEG electrodes were implanted for presurgical examination of epileptic foci. Miller et al. (2009) found diminished neural activity of high-frequency power (76–200 Hz) in the MPFC and precuneus, while the subject was engaged in a task from resting-state levels. Jerbi et al. (2010) also observed a task-induced decrease of gamma (>50 Hz) power, which is known to be associated with attention demanding cognitive task performance, in the human MPFC and PCC. Interestingly, the high gamma suppression found in the default system co-occurred with task-related enhancement outside the default system. The authors suggested that gamma modulations represent an electrical correlate of blood oxygen level-dependent (BOLD) signal (measured by fMRI) modulations (Jerbi et al. 2010).

### 12.1.4 Correlated Low-Frequency Spontaneous BOLD Activity Revealed by fMRI During Rest

It has consistently been observed that regions with similar functionality tend to show a correlation in their spontaneous BOLD activity. Correlations in spontaneous BOLD activity of very slow frequencies (around 0.1 Hz) are commonly observed across areas within the default system (Fox et al. 2005; Fransson 2005). Furthermore, regions with apparently opposing functionality have been found to be negatively correlated or anticorrelated in their spontaneous activity. Thus, in relation to attention demanding cognitive task performance, regions with activity increases are anticorrelated with a set of regions showing activity decreases (Fox et al. 2005; Buckner et al. 2008). Among studies of the default mode of brain activity, the functional connectivity analysis using this measure (correlations in spontaneous BOLD activity of very slow frequencies) is most commonly used.

Importantly, MPFC, ACC, and PCC that constitute the "core" default system are always included in all measures, while it is also important to note that different measures demonstrate different default systems. Furthermore, during the anesthetized condition, the task cannot be trained, and thus no task-induced deactivation can be obtained, whereas default activity can be demonstrated by correlated spontaneous low-frequency BOLD activity (Greicius et al. 2008; Vincent et al. 2007).

## 12.2 Functional Significance of Default Mode of Brain Activity

### *12.2.1 Developmental and Age-Related Changes of Default Mode of Brain Activity*

According to Thomason et al. (2008), children of 7–11 years old showed reliable reductions in BOLD response in the default system in response to increasing task demands or task difficulty as adults did. However, in children, task-induced deactivation was also observed in the somatosensory area and insular cortex, suggesting that the default system is not well differentiated. In functional connectivity analysis, Fair et al. (2008) found that the default systems are only sparsely functionally connected at early school age (7–9 years old), whereas these regions are integrated into a cohesive, interconnected network during development. In elderly people, Grady et al. (2006) observed a linear decrease in memory task-induced deactivation with age in the default system, as well as a decrease in activation in task-related areas (e.g., the dorsal part of LPFC) with age. The default system of Alzheimer disease (AD) patients appears to be impaired, since AD patients show reduced task-induced deactivation in the MPFC while showing task-induced activation, instead of deactivation in the PCC/precuneus (Lustig et al. 2003).

Abnormality in the default mode of brain activity was also investigated in relation to developmental disorders such as autism and ADHD. Kennedy et al. (2006) found that the autism group failed to demonstrate a task-induced deactivation effect. Furthermore, there was a strong correlation between a clinical measure of social impairment and activity within the ventral MPFC, a part of the default system. According to Fassbender et al. (2009), although both ADHD and control groups displayed a pattern of increasing deactivation of the MPFC with increasing task difficulty, the ADHD group was significantly less deactive than controls. Peterson et al. (2009) indicated that youths with ADHD showed significantly less prominent deactivation in the ACC and PCC when off medication, whereas when on stimulant medication, they showed prominent deactivation comparable to control group levels in these areas.

### *12.2.2 Abnormality of the Default Mode of Brain Activity in Psychiatric Patients*

Task-induced deactivation in the default system has also been examined in schizophrenic patients in which the default system has often been reported to be overactive (Harrison et al. 2007). Garrity et al. (2007) showed that positive symptoms of the disease (hallucinations, delusions, and thought confusions) correlated with increased task-induced deactivation in the default system.

Abnormality of default activity is also observed in depressed patients. For example, Grimm et al. (2009) indicated that depressive patients showed significantly reduced task-induced deactivation in several regions of the default system. Decreased deactivations were correlated with depression severity and feelings of hopelessness.

### *12.2.3 Functional Significance of the Default Mode of Brain Activity*

Functional roles of the default mode of brain activity could be deduced from neuroimaging studies where relative activity increases in response to task manipulations can be obtained in the regions of the default system as well as from clinical studies on patients with disturbance of the region in the default system. According to Gusnard and Raichle (2001), the functional significance of the ventral MPFC is related to a continuous process of online monitoring of associations between sensory information, responses, and outcomes under changing circumstances. The dorsal MPFC is related to monitoring or reporting one's own mental state, such as self-generated thoughts, intended speech, and emotions, and attributing mental states to others (explicit representation of states of the self). The ACC, which is known to be involved in error detection and outcome monitoring, is considered to be a center of high-level contextual integration of different kinds of information including socially and empathy-related information (Lavin et al. 2013).

Based on these studies, the default activity has been thought to be associated with internal thought processes (Christoff et al. 2004) such as the recall of autobiographical episodic memories (Mazoyer et al. 2001), self-referential processing (Kelley et al. 2002), conceptual processing (Binder et al. 1999), spontaneous semantic processing (Mckiernan et al. 2003), mind-wandering (Mason et al. 2007), and monitoring of the external environment, body image, and emotional state (Gusnard et al. 2001).

## 12.3 Default Mode of Brain Activity Observed in the Monkey

### *12.3.1 Default Mode of Brain Activity Observed in the Nonhuman Primate*

To consider the functional significance of the default mode of brain activity, it would be interesting to determine whether default activity is similarly observed, or how it is modulated (when it is modulated), during sleep (Horovitz et al. 2009) or anesthesia (Greicius et al. 2008) in humans. It would also be interesting to know whether similar default activity is observed in the nonhuman primate without linguistic ability. If the default mode of brain activity is observed in the nonhuman primate, it is possible that it is a manifestation of a more basic aspect of functional organization of the brain that may be conserved throughout animal evolution (Welberg 2007).

A recent [$^{18}$F] FDG-PET study in the chimpanzee demonstrated that the highest level of metabolic activity during rest occurred in the MPFC, LPFC, and medial and lateral parietal areas (Rilling et al. 2007). A single-photon emission computed tomography (SPECT) study on rhesus monkeys, which was not directly concerned with examining the default activity, suggested that there is relatively higher metabolism in both the MPFC and LPFC than in the other cortical areas in the resting monkey (Machado et al. 2008).

In the anesthetized monkey, low-frequency, spontaneous activity measured by fMRI showed a correlation between the medial parietal and medial prefrontal areas (Vincent et al. 2007). A recent study demonstrated task-induced suppression of neuronal activity in the monkey posterior cingulate cortex (Hayden et al. 2009). However, there are very few PET or fMRI studies in the nonhuman primate indicating the presence of a default system showing task-induced deactivation.

### *12.3.2 Default Mode of Brain Activity Revealed by Task-Induced Deactivation in the Monkey*

As I described before, areas with task-induced deactivation are similar but often differ from areas with the highest level of metabolic activity or from areas with correlated low-frequency spontaneous BOLD activity during rest. Thus, it was not clear whether regions showing the highest level of metabolic activity during rest or regions with correlated spontaneous low-frequency BOLD activity within the medial brain areas in nonhuman primates corresponded to regions showing task-induced deactivation.

Kojima et al. (2009) conducted a PET study on unanesthetized monkeys with [$^{15}$O]$H_2O$ as a ligand. Monkeys were first trained on four kinds of tasks: spatial working (WM) task, nonspatial WM task, spatial non-WM task, and nonspatial non-WM task. Spatial delayed response task was employed as the spatial WM task. In the nonspatial WM task, the monkey was required to respond to the right or left depending on what pattern stimulus had been presented. During the non-WM tasks, there were no delay periods and thus there were no memory requirements. Following training, monkeys underwent PET scanning in a non-illuminated experimental room during all task conditions. Tilting the PET camera gantry (parallel to the orbito-meatal line) allowed the monkey to sit in an upright position, making it possible for the monkey to view the CRT display and perform the task. Monkeys were also scanned during rest, while they sat quietly on the monkey chair without task performance.

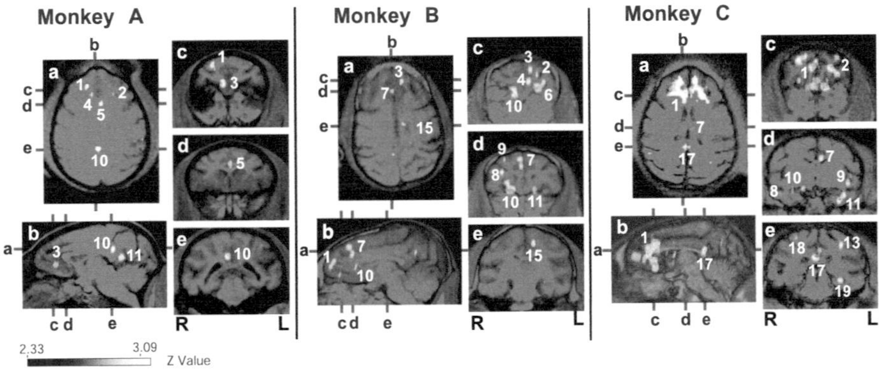


**Fig. 12.2** Regions with higher activity during rest than during the spatial WM task. Subtraction images are shown separately for each monkey (**A–C**). *Upper left* (*a*), *lower left* (*b*), and *right panels* (*c, d* and *e*) indicate transverse, sagittal, and coronal brain sections of each monkey, respectively. *Vertical line b* in the *upper left panel* indicates the L-R line corresponding to the sagittal section pictured in the *lower left panel* (*b*). *Horizontal line a* in panel (*b*) indicates the top-bottom line corresponding to the transverse section pictured in panel (*a*). *Lines c*, *d*, and *e* in panels (*a*, *b*) indicate the A-P line corresponding to the coronal sections pictured in the *right panels* (*c*, *d*, and *e*) (From Kojima et al. 2009 with permission)

To compare the brain activity of monkeys during the resting period with that during one of the four kinds of tasks, subtraction images (images acquired during rest minus images acquired during task) were made for each monkey for each task. In all the three monkeys studied, higher activities were observed during rest than during the spatial WM task in the LPFC, MPFC, ACC, OFC, and PCC/precuneus (Fig. 12.2). Notably, there were almost no differences in subtraction images among different task situations; not only during the nonspatial WM task but also during the spatial and nonspatial controls tasks, task-induced deactivation was observed in areas similar to those induced by spatial WM task. Furthermore, the magnitude of the task-induced deactivation did not differ depending on the task condition since there was no significant difference in the number of voxels showing task-induced deactivation among the four kinds of rest-related activity for all monkeys.

## 12.4 Functional Significance of Default Mode of Brain Activity Observed in the Lateral and Orbital Prefrontal Cortex Compared with That in the Medial Prefrontal Cortex in the Monkey

### *12.4.1 Default Mode of Brain Activity Observed in the Monkey MPFC*

Human default activity has been observed predominantly in medial parts of the brain (the anterior medial prefrontal and posterior medial parietal areas) (Raichle et al. 2001). Similar to the human default system, all monkeys showed higher rest-related activity in the MPFC, ACC, and PCC/precuneus. Functional roles of the MPFC and ACC may differ between the monkey and human. However, in the monkey, MPFC is involved in monitoring behavioral outcomes, especially in social contexts (Rushworth et al. 2007). The ACC is proposed to support the goal-directed response selection by generating a response plan based on the expectation of reward conditions associated with the responses (Matsumoto et al. 2003). Thus, there are some functional similarities between the MPFC and ACC in humans (Gusnard et al. 2001) and those in monkeys. It has been proposed that activity in the human default system is related to internal thought processes (Christoff et al. 2004). According to Kennedy et al. (2006), the lack of deactivation in the MPFC in the autism group is suggested to indicate abnormality in internal thought processes at rest. Thus, the results of Kojima et al. (2009)'s study also demonstrating default activity in the MPFC suggest that there might be internal thought processes in the monkey. Of course, it is not known what the exact nature of the internal thought might be, given the lack of linguistic ability in monkeys. Future studies are needed to conduct PET or fMRI studies to examine what kind of cognitive operations are associated with increased activity in the monkey MPFC/ACC.

### *12.4.2 Default Mode of Brain Activity Observed in the Monkey LPFC*

It is notable that Kojima et al. (2009) observed consistently higher rest-related activity in the LPFC and OFC besides the medial brain areas in the monkey. The LPFC is thought to be concerned with higher executive control (Miller and Cohen 2001). However, Rilling et al. (2007) found rest-related higher metabolic activity in the LPFC besides medial cortical areas in chimpanzees. Mantini et al. (2011) performed a meta-analysis of fMRI data collected in ten awake monkeys to reveal areas that show task-induced deactivation. They found that not only medial cortical areas but also LPFC were more active during rest than during task performance.

Thus, it appears that higher rest-related LPFC activity is rather the rule than the exception in the nonhuman primate.

Furthermore, previous human neuroimaging studies also reported rest-related activation in the LPFC. Ingvar (1979), who reported "hyperfrontal activity" during rest, considered the high frontal activity in the resting conscious state to be concerned with an anticipatory "simulation of behavior." Rilling et al. (2007) also found that in the human resting brain, metabolic activity was higher not only in medial cortical areas but also in the LPFC. Christoff et al. (2009) conducted an fMRI study in relation to mind-wandering. They observed activations not only in the default system but also in the executive system including the LPFC. Interestingly, neural recruitment in both default and executive systems was strongest when subjects were unaware of their own mind-wandering, suggesting that mind-wandering is most pronounced when it lacks meta-awareness. The executive and default systems have been assumed to be functionally anticorrelated (Fox et al. 2005; Buckner et al. 2008). The parallel recruitment of the two systems is suggested to indicate that mind-wandering may evoke a unique mental state that may allow otherwise opposing systems to work in cooperation. Thus, the LPFC appears to be coactive with the default system during rest when the human, and possibly nonhuman, is in a certain kind of mental state (Koshino et al. 2014).

### 12.4.3 Default Mode of Brain Activity Observed in the OFC in the Monkey

The OFC is known to be associated with processing reward information (Thorpe et al. 1983). During the PET scan experiment in Kojima et al. (2009), the monkeys were allowed to perform the task and obtain the liquid rewards during each of the task conditions, but not during rest. It is notable that the OFC and ventral striatum that are activated in relation to reward processing in the human (Porcelli and Delgado 2009) did not show higher activity during the task performance than during rest in the monkey. The monkeys were fluid restricted and therefore must have been eager to perform the tasks in order to obtain their reward. Thus, the monkey's desire for the liquid reward might have been very strong during rest. Monkey OFC neurons are related to both reward delivery and reward expectancy (Schultz and Tremblay 2006). Human neuroimaging studies indicate that the OFC is concerned not only with receiving the reward but also with reward expectancy (O'Doherty et al. 2002). It is speculated that the magnitude of reward expectancy-related activity might have been larger than that of reward-delivery-related activity. On the other hand, the resting condition may have been frustrating to the monkey that was not allowed to perform the task to obtain the reward. Considering that the OFC is also related to processing aversive stimuli (Thorpe et al. 1983), the resting activity in the OFC may be related to the monkey's frustration associated with the resting condition. Thus, reward expectancy and/or frustration during rest might have induced higher rest-related than task-related activity in this brain area.

### 12.4.4 *Spontaneous Thought Processes and Default Mode of Brain Activity Observed in the LPFC and OFC*

Human LPFC is shown to be activated in relation to mind-wandering (Christoff et al. 2009). Higher rest-related activity observed in the monkey LPFC may also be concerned with a certain kind of internal thought that could be a monkey's mind-wandering. The OFC is known to be involved in processing reward and aversive stimuli (Thorpe et al. 1983). Neuronal activities associated with processing appetitive and aversive stimuli have been observed not only in the OFC but also in the LPFC (Hikosaka and Watanabe 2000; Kobayashi et al. 2006). Human OFC and LPFC are both activated in relation to reward processing (Thut et al. 1997). The rest-related activity observed in the regions of the LPFC and OFC in Kojima et al. (2009)'s study, which do not constitute the human default system, may also be related to internal thought processes, which may involve more motivational and emotional contents.

Interestingly, Rilling et al. (2007) suggested that the resting state of chimpanzees involves emotionally laden episodic memory retrieval and some level of mental self-projection. While some researchers have argued that monkeys may not have a Theory of Mind (Povinelli et al. 1991), there are data indicating that the monkey has a degree of social intelligence, as exemplified by their capacity for altruistic behavior (Burkart et al. 2007) and fairness judgments (Brosnan and De Waal 2003). Furthermore, a recent study suggested that monkeys demonstrate Theory of Mind abilities when tested in more ecologically relevant situations (Flombaum and Santos 2005). Considering that monkeys live within a complex social structure and, thus, are required to process self and others in the context of society, it is not implausible to consider that internal thought processes exist not only in chimpanzees but also in monkeys.

## 12.5 Default Mode of Brain Activity and Dopamine Release in the Prefrontal Cortex

### 12.5.1 *Increase of Dopamine Release in the Lateral Prefrontal Cortex in Relation to WM Task Performance*

Dopamine (DA) plays important roles in executive control in the LPFC (Cools and D'Esposito 2011; Robbins and Arnsten 2009). An increase in DA release was previously reported in the monkey LPFC during a WM task compared with that during a non-WM control task and rest periods (Watanabe et al. 1997). However, no previous study investigated the release of DA, which plays important roles in cognitive operations, in the default system during a WM task and during rest. Kodama et al. (2015) used a microdialysis technique to examine changes in DA release in the monkey anterior default system (MPFC and ACC) during the WM task compared with those during rest.

### *12.5.2 Changes in Dopamine Release in Relation to WM Task Performance in the Anterior Default System*

In Kodama et al. (2015)'s study, monkeys were trained to perform a spatial WM (delayed alternation) task. Sampling was conducted each day from 4 to 6 points in the MPFC and ACC while the monkey was performing a WM task and during the resting period when the monkey sat at the primate chair without task performance and without juice delivery. Sampling was also done when an unpredictable juice reward was delivered.

The DA concentration in the perfusate was determined by high-performance liquid chromatography and electrochemical detection. The concentrations of DA at each sampling point on each day were measured for each behavioral situation (WM task, resting period, and unpredictable juice delivery period). Then, DA concentrations at each sampling position during the WM task and unpredictable juice delivery period were expressed as percent increases or decreases over basal rest values (mean values of the resting periods in each daily session) to normalize the variations between animals and between different experimental days.

In the MPFC and ACC, an increase in DA release was observed during rest compared with that during the WM task (Fig. 12.3, left) (Kodama et al. 2015). Increased DA release in the LPFC during the WM task is considered to play important roles for WM task performance (Watanabe et al. 1997). The mental operations performed in the MPFC during rest may have been supported by increased DA release. Considering that internal thought processes may exist in monkeys (Watanabe 2011), DA in the monkey anterior default system may be related to some form of internal thought process.

### *12.5.3 Changes in Dopamine Release in Relation to Unpredictable Reward Delivery in the Anterior Default System*

A significant increase in DA release was also observed during rest compared with that during the unpredictable reward delivery period in the MPFC (Kodama et al.

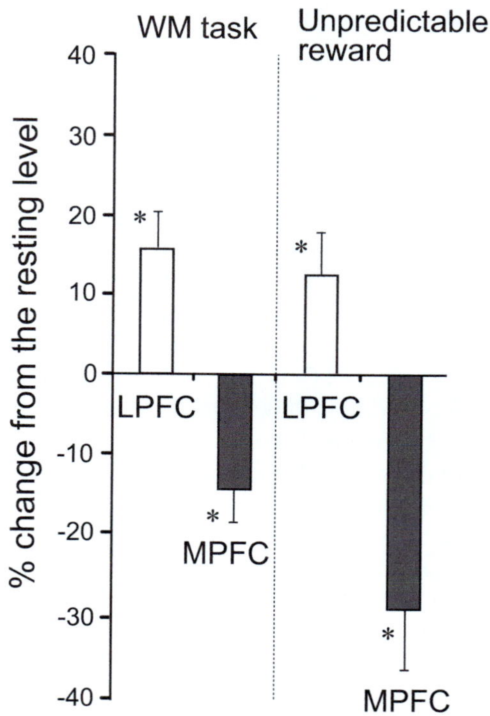


**Fig. 12.3** Changes in dopamine release from the basal resting level during the WM task (*left side*) and during the unpredictable reward period (*right side*) in the LPFC and MPFC. The data of the MPFC (*shaded graphs*) during the WM task and the unpredictable reward period are from Kodama et al. (2015). The datum of the LPFC during the WM task is from Watanabe et al. (1997). The datum of the LPFC during the unpredictable reward period is from Kodama et al. (2014). [*] indicates that the change is statistically significant (two-tailed t test, $p < 0.05$). *Error bars* indicate SEM

2015) (Fig. 12.3, right). In the human, attentive processes that suppress internal thought processes are associated with a decrease in default mode activity, and the default mode activity is related to watchfulness, and a passive, low-level monitoring of the external environment for unexpected events in conditions when active attention is relaxed (Buckner et al. 2008). Kodama et al. (2014) had observed an increase in DA in the LPFC with unpredictable reward delivery compared with that during rest (Fig. 12.3, right). Because monkeys could not predict the time of reward delivery, they were considered to be continuously attentive to the reward delivery. The higher level of attention caused by unpredictable reward delivery, which can inhibit self- and other-referential processing, may have induced an increase and a decrease in DA release in the LPFC and MPFC/ACC, respectively.

A recent study on healthy subjects indicated that oral DA administration facilitates self-awareness and autonoetic metacognition, which constitute internal thought processes, concomitant with increased gamma power in the MPFC measured by magnetoencephalography (Joensson et al. 2015). Garrity et al. (2007) showed that positive symptoms of schizophrenia, such as hallucinations and delusions, were correlated with increased task-induced deactivation in the default system. Considering that hallucinations and delusions are a form of heightened internal thought processes, the increase in DA release in the MPFC during rest in the present study may be related to internal thought processes in monkeys.

## 12.6 Conclusion

While the monkey was at rest, besides MPFC, ACC, and PCC/precuneus that are core areas of the default system, LPFC and OFC were more active than when the monkey was performing a WM task (Kojima et al. 2009). In both chimpanzee PET and monkey fMRI studies, rest-related activity increases were also reported in the LPFC. Some human neuroimaging studies also indicated an increase of LPFC activity in relation to certain mental operations such as mind-wandering (Christoff et al. 2009). Thus, the LPFC appear to co-work with default system in relation to certain mental operations both in humans and nonhumans. On the other hand, rest-related activity increase in the OFC was observed only in Kojima et al. (2009)'s study. The OFC is known to be highly susceptible to fMRI signal dropouts (Weiskopf et al. 2006), and thus reliable rest-related activity could not be detected in previous fMRI studies. In most neuroimaging studies on default mode of brain activity, subjects (both human and nonhuman) were not food or liquid restricted, and thus they must not be highly motivated for food or liquid. Considering that the OFC plays important roles in processing reward information, the rest-related increase observed in OFC may be caused by the heightened motivation for liquid in the subject in this study. Since LPFC is also concerned with processing reward information (Hikosaka and Watanabe 2000; Kobayashi et al. 2006; Thut et al. 1997), rest-related increase in the LPFC may also be related to a heightened motivational state of the monkey.

As dopamine release in the LPFC is important for WM task performance, that in the MFC/ACC may be important for mental operations, such as internal thought, during rest. It has been well documented that the human LPFC is activated during the WM task. An increase in rCBF in the monkey LPFC has also been reported during the WM task (Inoue et al. 2004). However, WM-related increases in the LPFC in those studies were obtained by the subtraction of the activity during the non-WM control task from the activity during the WM task. Although activities of executive system (such as LPFC) and default system are generally considered to be anticorrelated when measured by fMRI using low-frequency BOLD signal, there have been few studies indicating higher activity in the LPFC during the WM task than during rest. Indeed, Kojima et al. (2009) observed prominent activity in the LPFC during rest compared with that during WM task while there was no significant activity observed in the LPFC when rest-related activity was subtracted from WM-related activity.

There were rest-related increases in both rCBF and DA release in the anterior default system, while in the LPFC, there was an increase in rCBF, but a decrease in DA release, during rest. If DA plays important roles in cognitive operations, it is puzzling why DA is decreased, but not increased, when LPFC is active as exemplified by the increased rCBF during rest. It is speculated that DA release is more related to those mental operations, such as internal thought, than to retaining memory information. Further studies are needed to clarify the relationship between the rCBF and DA release in the PFC in relation to the default mode of brain activity.

## References

Binder JR, Frost JA, Hammeke TA, Bellgowan PSF, Rao SM, Cox RW (1999) Conceptual processing during the conscious resting state: a functional MRI study. J Cogn Neurosci 11:80–93

Brosnan SF, De Waal FB (2003) Monkeys reject unequal pay. Nature 425:297–299

Buckner RL, Andrews-Hanna JR, Schacter DL (2008) The brain's default network: anatomy, function, and relevance to disease. Ann N Y Acad Sci 1124:1–38

Burkart JM, Fehr E, Efferson C, van Schaik CP (2007) Other-regarding preferences in a non-human primate: common marmosets provision food altruistically. Proc Natl Acad Sci U S A 104:19762–19766

Christoff K, Ream JM, Gabrieli JD (2004) Neural basis of spontaneous thought processes. Cortex 40:623–630

Christoff K, Gordon AM, Smallwood J, Smith R, Schooler JW (2009) Experience sampling during fMRI reveals default network and executive system contributions to mind wandering. Proc Natl Acad Sci U S A 106:8719–8724

Cools R, D'Esposito M (2011) Inverted-U-shaped dopamine actions on human working memory and cognitive control. Biol Psychiatry 69:e113–e125

Fair DA, Cohen AL, Dosenbach NU, Church JA, Miezin FM, Barch DM, Raichle ME, Petersen SE, Schlaggar BL (2008) The maturing architecture of the brain's default network. Proc Natl Acad Sci U S A 105:4028–4032

Fassbender C, Zhang H, Buzy WM, Cortes CR, Mizuiri D, Beckett L, Schweitzer JB (2009) A lack of default network suppression is linked to increased distractibility in ADHD. Brain Res 1273:114–128

Flombaum JI, Santos LR (2005) Rhesus monkeys attribute perceptions to others. Curr Biol 15:447–452

Fox MD, Raichle ME (2007) Spontaneous fluctuations in brain activity observed with functional magnetic resonance imaging. Nat Rev Neurosci 8:700–711

Fox MD, Snyder AZ, Vincent JL, Corbetta M, Van Essen DC, Raichle ME (2005) The human brain is intrinsically organized into dynamic, anticorrelated functional networks. Proc Natl Acad Sci U S A 102:9673–9678

Fransson P (2005) Spontaneous low-frequency BOLD signal fluctuations: an fMRI investigation of the resting-state default mode of brain function hypothesis. Hum Brain Mapp 26:15–29

Garrity AG, Pearlson GD, McKiernan K, Lloyd D, Kiehl KA, Calhoun VD (2007) Aberrant "default mode" functional connectivity in schizophrenia. Am J Psychiatry 164:450–457

Grady CL, Springer MV, Hongwanishkul D, McIntosh AR, Winocur G (2006) Age-related changes in brain activity across the adult lifespan. J Cogn Neurosci 18:227–241

Greicius MD, Kiviniemi V, Tervonen O, Vainionpaa V, Alahuhta S, Reiss AL, Menon V (2008) Persistent default-mode network connectivity during light sedation. Hum Brain Mapp 29:839–847

Grimm S, Boesiger P, Beck J, Schuepbach D, Bermpohl F, Walter M, Ernst J, Hell D, Boeker H, Northoff G (2009) Altered negative BOLD responses in the default-mode network during emotion processing in depressed subjects. Neuropsychopharmacology 34:932–943

Gusnard DA, Raichle ME (2001) Searching for a baseline: functional imaging and the resting human brain. Nat Rev Neurosci 2:685–694

Gusnard DA, Akbudak E, Shulman GL, Raichle ME (2001) Medial prefrontal cortex and self-referential mental activity: relation to a default mode of brain function. Proc Natl Acad Sci U S A 98:4259–4264

Harrison BJ, Yucel M, Pujol J, Pantelis C (2007) Task-induced deactivation of midline cortical regions in schizophrenia assessed with fMRI. Schizophr Res 91:82–86

Hayden BY, Smith DV, Platt ML (2009) Electrophysiological correlates of default-mode processing in macaque posterior cingulate cortex. Proc Natl Acad Sci U S A 106:5948–5953

Hikosaka K, Watanabe M (2000) Delay activity of orbital and lateral prefrontal neurons of the monkey varying with different rewards. Cereb Cortex 10:263–271

Horovitz SG, Braun AR, Carr WS, Picchioni D, Balkin TJ, Fukunaga M, Duyn JH (2009) Decoupling of the brain's default mode network during deep sleep. Proc Natl Acad Sci U S A 106:11376–11381

Ingvar DH (1979) "Hyperfrontal" distribution of the cerebral grey matter flow in resting wakefulness; on the functional anatomy of the conscious state. Acta Neurol Scand 60:12–25

Inoue M, Mikami A, Ando I, Tsukada H (2004) Functional brain mapping of the macaque related to spatial working memory as revealed by PET. Cereb Cortex 14:106–119

Jerbi K, Vidal JR, Ossandon T, Dalal SS, Jung J, Hoffmann D, Minotti L, Bertrand O, Kahane P, Lachaux JP (2010) Exploring the electrophysiological correlates of the default-mode network with intracerebral EEG. Front Syst Neurosci 4:27

Joensson M, Thomsen KR, Andersen LM, Gross J, Mouridsen K, Sandberg K, Østergaard L, Lou HC (2015) Making sense: dopamine activates conscious self-monitoring through medial prefrontal cortex. Hum Brain Mapp 36:1866–1877

Kelley WM, Macrae CN, Wyland CL, Caglar S, Inati S, Heatherton TF (2002) Finding the self? An event-related fMRI study. J Cogn Neurosci 14:785–794

Kennedy DP, Redcay E, Courchesne E (2006) Failing to deactivate: resting functional abnormalities in autism. Proc Natl Acad Sci U S A 103:8275–8280

Kobayashi S, Nomoto K, Watanabe M, Hikosaka O, Schultz W, SakagamiM (2006) Influences of rewarding and aversive outcomes on activity in macaque lateral prefrontal cortex. Neuron 51:861–870

Kodama T, Hikosaka K, Honda Y, Kojima T, Watanabe M (2014) Higher dopamine release induced by less rather than more preferred reward during a working memory task in the primate prefrontal cortex. Behav Brain Res 266:104–107

Kodama T, Hikosaka K, Honda Y, Kojima T, Tsutsui K, Watanabe M (2015) Dopamine and glutamate release in the anterior default system during rest: a monkey microdialysis study. Behav Brain Res 294:194–197

Kojima T, Onoe H, Hikosaka K, Tsutsui K, Tsukada H, Watanabe M (2009) Default mode of brain activity demonstrated by positron emission tomography imaging in awake monkeys: higher rest-related than working memory-related activity in medial cortical areas. J Neurosci 29:14463–14471

Koshino H, Minamoto T, Yaoi K, Osaka M, Osaka N (2014) Coactivation of the default mode network regions and working memory network regions during task preparation. Sci Rep 4:5954

Lavin C, Melis C, Mikulan E, Gelormini C, Huepe D, Ibañez A (2013) The anterior cingulate cortex: an integrative hub for human socially-driven interactions. Front Neurosci 7:64

Lustig C, Snyder AZ, Bhakta M, O'Brien KC, McAvoy M, Raichle ME, Morris JC, Buckner RL (2003) Functional deactivations: change with age and dementia of the Alzheimer type. Proc Natl Acad Sci U S A 100:14504–14509

Machado CJ, Snyder AZ, Cherry SR, Lavenex P, Amaral DG (2008) Effects of neonatal amygdala or hippocampus lesions on resting brain metabolism in the macaque monkey: a microPET imaging study. NeuroImage 39:832–846

Mantini D, Gerits A, Nelissen K, Durand JB, Joly O, Simone L, Sawamura H, Wardak C, Orban GA, Buckner RL, Vanduffel W (2011) Default mode of brain function in monkeys. J Neurosci 31:12954–12962

Mason MF, Norton MI, Van Horn JD, Wegner DM, Grafton ST, Macrae CN (2007) Wandering minds: the default network and stimulus independent thought. Science 315:393–395

Matsumoto K, Suzuki W, Tanaka K (2003) Neuronal correlates of goal-based motor selection in the prefrontal cortex. Science 301:229–232

Mazoyer B, Zago L, Mellet E, Bricogne S, Etard O, Houde O, Crivello F, Joliot M, Petit L, Tzourio-Mazoyer N (2001) Cortical networks for working memory and executive functions sustain the conscious resting state in man. Brain Res Bull 54:287–298

McKiernan KA, Kaufman JN, Kucera-Thompson J, Binder JR (2003) A parametric manipulation of factors affecting task-induced deactivation in functional neuroimaging. J Cogn Neurosci 15:394–408
Miller EK, Cohen JD (2001) An integrative theory of prefrontal cortex function. Annu Rev Neurosci 24:167–202
Miller KJ, Weaver KE, Ojemann JG (2009) Direct electrophysiological measurement of human default network areas. Proc Natl Acad Sci U S A 106:12174–12177
O'Doherty JP, Deichmann R, Critchley HD, Dolan RJ (2002) Neural responses during anticipation of a primary taste reward. Neuron 33:815–826
Peterson BS, Potenza MN, Wang Z, Zhu H, Martin A, Marsh R, Plessen KJ, Yu S (2009) An FMRI study of the effects of psychostimulants on default-mode processing during Stroop task performance in youths with ADHD. Am J Psychiatry 166:1286–1294
Porcelli AJ, Delgado MR (2009) Reward processing in the human brain: insights from fMRI. In: Dreher JC, Tremblay L (eds) Handbook of reward and decision making. Academic, Oxford, pp 159–178
Povinelli DJ, Parks KA, Novak MA (1991) Do rhesus monkeys (*Macaca mulatta*) attribute knowledge and ignorance to others? J Comp Psychol 105:318–325
Raichle ME, MacLeod AM, Snyder AZ, Powers WJ, Gusnard DA, Shulman GL (2001) A default mode of brain function. Proc Natl Acad Sci U S A 98:676–682
Rilling JK, Barks SK, Parr LA, Preuss TM, Faber TL, Pagnoni G, Bremner JD, Votaw JR (2007) A comparison of resting-state brain activity in humans and chimpanzees. Proc Natl Acad Sci U S A 104:17146–17151
Robbins TW, Arnsten AF (2009) The neuropsychopharmacology of fronto-executive function: monoaminergic modulation. Annu Rev Neurosci 32:267–287
Rushworth MF, Buckley MJ, Behrens TE, Walton ME, Bannerman DM (2007) Functional organization of the medial frontal cortex. Curr Opin Neurobiol 2:220–227
Schultz W, Tremblay L (2006) Involvement of primate orbitofrontal neurons in reward, uncertainty, and learning. In: Zald DH, Rauch SL (eds) The orbitofrontal cortex. Oxford UP, Oxford, pp 173–198
Shulman GL, Fiez J, Corbetta M, Buckner R, Miezin FM, Raichle ME, Petersen S (1997) Common blood flow changes across visual task: II decreases in cerebral cortex. J Cogn Neurosci 9:648–663
Thomason ME, Chang CE, Glover GH, Gabrieli JD, Grecius MD, Gotlib IH (2008) Default mode function and task-induced deactivation have overlapping brain substrates in children. NeuroImage 41:1493–1503
Thorpe SJ, Rolls ET, Maddison S (1983) The orbitofrontal cortex: neuronal activity in the behaving monkey. Exp Brain Res 49:93–115
Thut G, Schultz W, Roelcke U, Nienhusmeier M, Missimer J, Maguire RP, Leenders KL (1997) Activation of the human brain by monetary reward. NeuroReport 8:1225–1228
Vincent JL, Patel GH, Fox MD, Snyder AZ, Baker JT, Van Essen DC, Zempel JM, Snyder LH, Corbetta M, Raichle ME (2007) Intrinsic functional architecture in the anaesthetized monkey brain. Nature 447:83–86
Watanabe M (2011) Are there internal thought processes in the monkey? Default brain activity in humans and nonhuman primates. Behav Brain Res 221:295–303
Watanabe M, Kodama T, Hikosaka K (1997) Increase of extracellular dopamine in primate prefrontal cortex during a working memory task. J Neurophysiol 78:2795–2798
Weiskopf N, Hutton C, Josephs O, Deichmann R (2006) Optimal EPI parameters for reduction of susceptibility-induced BOLD sensitivity losses: a whole-brain analysis at 3 T and 1.5 T. NeuroImage 33:493–504
Welberg L (2007) Unconscious activity in the monkey brain. Nat Rev Neurosci 8:407

# Chapter 13
# Coactivation of Default Mode Network and Executive Network Regions in the Human Brain

**Hideya Koshino**

**Abstract** The executive network (EN) is activated during higher-level cognition; however, the default mode network (DMN) is deactivated during cognitive tasks. Therefore, their relationship was described as anticorrelational. During the task period, the EN increases its activity and the DMN decreases its activity, whereas the DMN is activated and the EN is deactivated during the rest period. However, recent studies reported that the two networks are not necessarily anticorrelated, and instead they actually show coactivation during some cognitive tasks. In addition, recent research also reported that the relationship between the DMN and EN dynamically changes during different phases within a cognitive task. For example, the EN and DMN regions were coactivated during a task preparation period, whereas the EN was activated and the DMN regions were deactivated during a task execution period. The results suggest that the relationship between the EN and DMN could change dynamically depending on task demands.

**Keywords** Default mode network • Executive network • Brain imaging • Working memory • Resting state network • Cooperation between networks • Competition between networks • Functional heterogeneity

## 13.1 Introduction

Our understanding of brain functions has made dramatic progress in the last 20 years with the advent of brain imaging techniques, such as positron emission tomography (PET) and functional magnetic resonance imaging (fMRI). Early studies have focused on brain mapping to explore the correspondence between areas of the brain and their functions. This type of one-to-one correspondence between the brain regions and functions can be found in the sensory and motor

H. Koshino (✉)
Department of Psychology, California State University, 5500 University Parkway, San Bernardino, CA 92407, USA
e-mail: hkoshino@csusb.edu



M. Watanabe (ed.), *The Prefrontal Cortex as an Executive, Emotional, and Social Brain*, DOI 10.1007/978-4-431-56508-6_13

areas. However, many areas of the brain, especially the association cortex, tend to show activity during various cognitive tasks, and also higher-level cognitive functions, such as working memory, are associated with multiple brain regions. In other words, the correspondence between brain regions and their functions is not one to one, but rather many to many, especially for higher-level cognition. Therefore, brain regions tend to form networks with other brain regions in order to perform a certain task (e.g., Fuster 1997; Meslam 1990, 1998). Network approaches have become increasingly popular in recent years, including the human connectome (a map of neural connections in the brain, e.g., Seung 2013; Sporns 2012), functional connectivity (a measure of synchronization across brain regions, e.g., Friston 1994; Horwitz et al. 1998; McIntosh and Gonzalez-Lima 1994), and a graph theoretical approach (e.g., Bullmore and Sporns 2012; Menon 2011; Meunier et al. 2009; Power et al. 2011). The basic logic of functional connectivity is that if some regions are working together (functionally connected with each other), they should show similar activation pattern across the time course of certain tasks; and therefore, they should show significant correlations.

According to recent studies, brain regions are also synchronized at rest, forming resting state networks (RSN). When we are resting, or when we are tired, we might say that "my brain is not working," but in fact the brain is active. During particular tasks, brain networks depend also on current task demands. In contrast, networks are less dependent on specific tasks at rest, and therefore, the resting networks are considered intrinsic networks in the brain (e.g., Binder et al. 1999; Biswal et al. 1995; Damoiseaux et al. 2006; Deco and Corbetta 2011; De Luca et al. 2006; Menon 2011; Seeley et al. 2007; Smith et al. 2009; Toro et al. 2008; van den Heuvel and Hulshoff Pol 2010).

There are several resting state networks, including the dorsal attentional network (DAN), which includes the frontal eye fields (FEF) and superior parietal lobe (SPL), and these are closely related to spatial attention (e.g., Corbetta and Shulman 2002: Fox et al. 2006). The executive network (EN) includes the lateral prefrontal cortex (LPFC) and inferior parietal lobe (IPL), and is associated with various working memory and executive functions, including maintenance and manipulation of information, planning, shifting attention, and inhibition (e.g., Baddeley 2012; Dosenbach et al. 2007). The term EN is oftentimes used synonymously with the frontoparietal network (FPN) and the working memory network (WMN). The default mode network (DMN) typically involves the medial prefrontal cortex (MPFC) and the posterior cingulate cortex (PCC)/precuneus, and it is related to processing social and self-related information (e.g., Buckner et al. 2008; Greicius et al. 2003; Raichle 2015; Raichle et al. 2001). The salience network includes the anterior insula (AI) and the anterior cingulate cortex (ACC) and is associated with interoceptive awareness (e.g., Critchley et al. 2004), salience detection (e.g., Bressler and Menon 2010; Seeley et al. 2007; Sridharan et al. 2008), and maintenance of response set (e.g., Dosenbach et al. 2006). There are also other more local resting state networks in the visual areas, auditory areas, and somatosensory and motor areas (e.g., Biswal et al. 1995; Damoiseaux et al. 2006; Deco and Corbetta 2011; De Luca et al. 2006; van den Heuvel and Hulshoff Pol 2010). Recently a framework has been proposed to distinguish between the cognitive brain and the social brain (e.g., Lieberman 2013; Mars et al. 2012). The cognitive brain consists mainly of the lateral regions, such as the lateral PFC and parietal regions, whereas the social brain involves primarily the medial regions, including the medial PFC and posterior cingulate/precuneus. In this framework, the EN represents the cognitive brain and the DMN represents the social brain. Here, one critical question is concerned with relationships between the EN and the DMN. The DMN was once viewed as a task-negative network because it showed deactivation during various cognitive tasks, and the EN and the DMN are anticorrelated. However, recent research showed that the EN and the DMN are not necessarily anticorrelated, and instead they actually show coactivation during some cognitive tasks. In this chapter, dynamic relationships between the EN and the DMN, especially how they compete and collaborate with each other, will be discussed.

## 13.2 Executive Network

### *13.2.1 Executive Network and Working Memory*

The executive network (EN) is responsible for a variety of cognitive functions, especially higher-level cognition, purposeful goal-directed behavior, and cognitive control. The core regions typically include the lateral prefrontal cortex (LPFC) and the parietal lobe. The EN works with other brain regions in the temporal and occipital lobes that are more specialized for particular types of information processing to perform various mental activities depending on task demands. There are other networks that are active during cognitive tasks, and they are called task-positive networks. They include the frontoparietal network (FPN), typically consisting of DLPFC and IPL, that is active during many working memory tasks. The other task-positive networks include the dorsal attentional network (DAN) and the ventral attentional network (VAN) that are active during attention tasks.

The executive functions are the central components of working memory (WM), which is a system that temporarily stores information and operates on the contents of storage. There are several WM models, but the modal model is the one by Baddeley (e.g., Baddeley 2012; Baddeley and Hitch 1974). According to their model, there are two slave systems and one central executive system in WM. The slave systems are the phonological loop for verbal information processing and visuospatial sketchpad for the visuospatial information processing. The central executive acts as a supervisory system to control the flow of information among the slave systems. The phonological loop consists of a passive language buffer and active rehearsal processes. In the brain, the left hemisphere is more dominant in the processing of phonological information (e. g., Owen et al. 1998; Smith and Jonides 1999). For phonological rehearsal, the inferior frontal gyrus (IFG) and inferior parietal lobe (IPL) typically play major roles. The phonological representation would decay over time because the posterior brain regions do not have the maintenance function by themselves, and therefore, rehearsal is required to maintain information. The visuospatial sketchpad is for visual spatial information processing and is typically associated more with the right hemisphere. The rehearsal of spatial information in the visuospatial sketchpad is accomplished by allocating attention continuously to spatial representation (e. g., Jonides et al. 2005) and is associated with the frontal eye field (FEF) and the parietal lobe (mainly the superior parietal lobe). These are the same regions as those that control selective attention to the external environment (e.g., Jonides et al. 2005; Kastner and Ungerleider 2000). Recently, Baddeley added another working memory component named the episodic buffer, which binds information from the slave systems and from long-term memory to form a unitary episodic representation. The central executive is involved in controlling the slave systems in response to task demands, activation of representations in long-term memory, and distribution of attentional resources.

### *13.2.2 Executive Functions*

The central executive system in Baddeley's working memory model is similar to the concept of executive functions that has been used in cognitive neuroscience. Although there is still a debate about what aspects are included in executive functions, typical components are planning and task set formation (planning), updating of the stored contents (updating), allocation of attention (shifting), and suppression of irrelevant information and processes (inhibition) (e.g., Engle 2002; Miyake et al. 2000; Suchy 2009). Research suggests the prefrontal cortex (especially dorsolateral prefrontal cortex, ventrolateral prefrontal cortex, and the anterior cingulate gyrus) is deeply involved in executive system functions (e.g., Baddeley 2012; Miller and Cohen 2001; Niendam et al. 2012; Smith and Jonides 1999). Also, damage to the frontal regions may cause failure in various executive functions (e.g., Kolb and Whishaw 2003). Deficits in WM have also been implicated in various psychological and developmental disorders, including attention deficit/hyperactivity disorder (ADHD) (e.g., Klingberg et al. 2002, 2005), autism (e.g., Hill 2004; Just et al. 2004; Koshino et al. 2005, 2008), anxiety (e.g., Ashcraft and Kirk 2001; Beilock 2008), and alcoholism and drug addiction (e.g., Grenard et al. 2008).

With respect to planning, researchers often have used tasks such as the Tower of Hanoi and its variants. During the performance of the Tower of Hanoi task, activities in the DLPFC, VLPFC, ACC, and the parietal lobe have been reported (e.g., Dagher et al. 1999; Fincham et al. 2002). In updating, N-back tasks are widely used (e.g., Owen et al. 2005). In the N-back task, a series of stimuli (characters, objects, etc.) are visually or aurally presented, and the participants are told to judge whether the current stimulus is the same as the one presented N items before. The brain regions that are typically involved in N-back tasks include the ACC, DLPFC, VLPFC, frontal pole, and parietal lobe (see e.g., Owen et al. 2005 for review).

Shifting is a function that distributes attention or processing resources among a number of different cognitive processes. Dual tasking is a typical example of performance that requires shifting of attention. In this paradigm, participants are required to perform (two or more) different tasks simultaneously, successively (task switching), or over time (psychological refractory period). For example, WM span tasks, such as the reading span task (Daneman and Carpenter 1980) and the operation span task (Turner and Engle 1989), are dual tasks that include both maintenance and manipulation of WM contents. In the reading span task, a stimulus set consists of a sentence and a memory item (e.g., word or letter). Participants are typically asked to read and verify the sentence and then remember the memory item. Participants are told to continue with a number of stimulus set (typically from 2 to 6 items), then they are asked to recall the memory items. In the operation span task, a set consists of a math equation and a memory item, and participants are told to solve the math problem then to remember the memory item. The major brain regions related to the WM span task include the lateral prefrontal cortex and ACC (e.g., Bunge et al. 2000; Kondo et al. 2004; Smith et al. 2001).

The first fMRI research using simultaneous dual tasking was conducted by D'Esposito et al. (1995). They used a semantic classification task and a mental rotation task. In the semantic classification task, participants needed to judge whether a stimulus word belonged to a given category, for example, "banana belongs to fruit." In the mental rotation task, two rectangles were presented. Each rectangle contained double lines on one side and a small circle. Participants were asked to judge whether the position of the small circle relative to the double lines was the same between the two rectangles. The participants performed the semantic classification and mental rotation tasks alone (single-task condition) and simultaneously (dual-task condition). The results showed that activation in the DLPFC was higher in the dual-task condition. They concluded that the DLPFC was involved in task management, including the distribution of processing resources in the dual task. In the subsequent studies, some reported higher PFC activation in dual-task conditions (e.g., Dreher and Grafman 2003; Koechlin et al. 1999; Ramsey et al. 2004; Szameitat et al. 2002), but others have not found a unique contribution of the PFC in the dual-task conditions (e.g., Adcock et al. 2000; Bunge et al. 2000; Just et al. 2001; Klingberg 1998). For example, Klingberg (1998) asked participants to perform a 1-back auditory tone discrimination task and a 1-back visual brightness discrimination task simultaneously. They found activation in the middle frontal gyrus and ACC for both the single and dual task conditions, without any difference between the two conditions. Klingberg (1998) suggested that the PFC activation that was reported by D'Esposito et al. (1995) was due to a higher WM requirement but not to the dual-task coordination. Adcock et al. (2000) used an auditory verbal categorization task and visual mental rotation or face identification task. They showed that all brain areas that were active for the dual-task condition were also active during the component single task and that surplus activation within activated regions for the dual condition was accounted for by the addition of the second task. Nebel et al. (2005) also showed that there were no specific brain regions that were active for the dual-task condition compared to the single-task condition when the processing demands were equated between the two tasks. These results suggested that executive processes that are required for the dual-task coordination could be mediated by interaction among distinct systems for component tasks and may not require any particular areas dedicated to an executive system.

Task switching could be viewed as a different type of dual task, in which participants perform two tasks successively (e.g., Kim et al. 2012; Monsell 2003). The participants are typically required to form two task sets, maintain them in working memory, and switch between them according to instructions and cues. Thus, an attentional shift is required in this task. In behavioral studies, performance is impaired for task switching conditions compared to a no-switch condition. The scheduling between the two tasks also affects performance, in such a way that the participants' performance is worse when they are not given advance information about the task order (e.g., random order) than when they have foreknowledge about the order (e.g., fixed order). However, preparation time also affects performance when the task order is random, as is seen in a case like the task cueing paradigm (e.g., Sudevan and Taylor 1987; Meiran 1996), in which a cue indicating which task to perform is presented before each trial. If the participants are given enough time between a cue and a target (e.g., long stimulus-onset asynchrony SOA), their performance is typically better, compared to that for a short SOA.

Brain imaging research has found activation in the bilateral prefrontal and parietal regions, the lateral prefrontal cortex, cingulate gyrus, premotor, and posterior parietal regions for task switching conditions (e.g., Braver et al. 2003; Dove et al. 2000; Rushworth et al. 2002; Schubert and Szameitat 2003; Smith et al. 2001). For example, Dove et al. (2000) compared a switch condition with a no-switch condition. Participants were required to simply discriminate between "+" and "−" in the no-switch condition. They were asked to press a left-hand key for "+" and a right-hand key for "−" when the stimuli were presented in green. They were told to press the opposite keys when the stimuli were presented in red. Thus, the participants were required to switch their stimulus-response mapping depending on the color of the stimuli. The results showed bilateral activation in the lateral prefrontal cortex, premotor cortex, anterior insular cortex, left parietal lobe, supplementary motor area, and posterior cingulate cortex.

Another type of dual task consists of two tasks that are not performed simultaneously, but instead one task is started before the other with variable stimulus onset asynchronies (SOAs) between them. The shorter the SOA, the more the temporal overlap between the two tasks. Therefore, in this case, the main questions concern the brain functions associated with temporal order coordination. Also in cognitive psychology, these types of overlapping dual tasks have been studied under the terminology of the psychological refractory period (PRP, e.g., Pashler 1998; Pashler et al. 2001), which refers to the period of time during which the response to a second task is delayed because a first task is still being processed. The PRP effect is concerned with the two major views of attention: the central bottleneck and the attentional capacity. According to the attentional bottleneck view, the second stimulus cannot pass through the bottleneck while the first stimulus is being processed. Therefore, the response time (RT) for the second stimulus is delayed.

According to the capacity view, attentional resources are shared between the two tasks. The tasks can be performed simultaneously as long as the resource demands of the two tasks do not exceed the overall capacity. The capacity could either be a single pool (single-resource model, e. g., Norman and Bobrow 1975) or multiple pools of attentional resources (multiple-resource model, e. g., Wickens 2002), depending on stimulus modality (e.g., auditory or visual) and response modality (e.g., vocal or manual). Therefore, if two simultaneous tasks require different modalities, then they should have less interference with each other than if they require the same modalities. Studies that have investigated brain regions associated with PRP found greater right IFG activation for short than long SOAs (e.g., Herath et al. 2001; Jiang 2004). For example, Herath et al. (2001) manipulated the length of overlap between visual detection and somatosensory stimulus detection. When there was an overlap between the two tasks (SOA <300 ms), the right IFG was activated. Also, the increased activity of the right IFG was correlated with the PRP interference effect.

Inhibition is one of the most critical functions in cognition. Focusing attention on some processes is typically accompanied by inhibition of other processes. It takes a long time for inhibition to develop, and it is one of the functions that we seem to lose first as we get older (e.g., Braver and West 2008; Hasher and Zacks 1988; Reuter-Lorenz and Sylvester 2005). The tasks that are commonly used to study inhibition include the Stroop task (e.g., Stroop 1935; MacLeod 1991), Eriksen flanker tasks (e.g., Eriksen and Eriksen 1974), and the Simon task (Simon 1969; Hommel 2011). Studies have shown that ACC activation is observed in tasks requiring the overriding of a prepotent response, including Stroop tasks (e.g., Barch et al. 2001; MacLeod and MacDonald 2000 for review), flanker tasks (e.g., Botvinick et al. 1999; Bunge et al. 2000), Simon tasks (Peterson et al. 2002), global/local tasks (Lux et al. 2004; Weissman et al. 2003), and go/no-go tasks (e.g., Braver et al. 2001; Durston et al. 2002).

The executive functions might be hierarchically organized along the rostral-caudal axis of the prefrontal regions. For example, some researchers suggest that the frontal pole (BA10) is related to abstract planning, task-set formation, and monitoring, whereas the anterior DLPFC is associated with domain-general monitoring and the posterior DLPFC is related to domain-specific maintenance of information (e.g., Badre 2008; Christoff and Gabrieli 2000; Courtney 2004).

There is another point of view with respect to executive functions. Many brain imaging studies have emphasized the top-down role of the prefrontal cortex in executive functions. For example, as seen in the Baddeley's model, the prefrontal cortex is the control center of information processing. However, in recent years, it has been pointed out that executive functions might be accomplished by collaboration between top-down processing driven by the frontal regions and bottom-up processing driven by the posterior regions (e.g., Courtney 2004; Gruber and Goschke 2004; Postle 2006). In other words, executive functions could be emergent properties of distributed neural networks.

The emergence could appear in a system with hierarchies that contains feedback loops. The higher-level functions are generated by the interaction of the functions of the lower level and thus cannot be reduced to local features of the lower level. In other words, "the whole is more than the sum of its parts" (e.g., Kaufman 1995). From that point of view, the brain can be regarded as an emergent system. Although the brain is composed of neurons, understanding the structure and function of individual neurons is necessary but not sufficient to understand the higher-order cognitive processes, such as the execution functions.

## 13.3 DMN

### *13.3.1 Overview of DMN*

The default mode network (DMN) was discovered as a group of regions that seem to show higher activity during rest periods compared to task periods. It was considered that these regions form a network because they exhibit relatively common activity patterns in different cognitive tasks (e.g., Binder et al. 1999; Gusnard and Raichle 2001; Mazoyer et al. 2001; Shulman et al. 1997). The DMN typically includes the medial prefrontal cortex (MPFC), the posterior cingulate/precuneus (PCC/precuneus), the posterior parietal lobe (PPL), the lateral temporal cortex (LTC), and the hippocampal formation (HF) (e.g., Buckner et al. 2008). These regions tend to show strong functional connectivity at rest; therefore, they are considered to function together (e.g., Greicius et al. 2003; De Luca et al. 2006; Fransson 2005). The DMN was once thought as a task-negative network (e.g., Fox et al. 2005) that is active during rest and is deactivated during tasks compared to task-positive networks. The task-positive networks and the task-negative network (i.e., DMN) were suggested to be anticorrelated, in that the task-positive networks are activated, but the DMN is deactivated during a task period, whereas the task-positive networks are deactivated and the DMN is activated during a rest period (e.g., Fox et al. 2005). However, in recent years, the DMN is not necessarily viewed as a task-negative network because it shows activation during a variety of tasks including social and self-related functions (e.g., Spreng 2012). Research so far seems to suggest that DMN activities might be related to our self and social functions, such as those focusing on interest in the inner experience, theory of mind, social cognition, and episodic recall (e.g., Buckner et al. 2008; Smith et al. 2009).

### *13.3.2 DMN and Mind Wandering*

Activities of the DMN are related to interest in the inner experience (spontaneous cognitive activity that does not depend on the stimulus in the external environment), such as task-unrelated thoughts (TUT), daydreaming, and mind wandering (e.g., Christoff 2012; Christoff et al. 2009; Fox et al. 2015; Mason et al. 2007). Mind wandering is basically included in the task-unrelated thought (e.g., Christoff 2012). However, mind wandering is not an intentional activity, and we may not notice that it hinders the performance of tasks. For example, we might sometimes experience that when we are reading a book, our eyes are following the text, but we are not processing its meaning because of mind wandering. Such experiences might occur more often when we read what is necessity for our work when we are tired, but much less often when we read a book by our favorite authors. These would suggest how factors such as interests and motivation affect mind wandering. Also, current concerns and worries could increase mind wandering (Klinger 2009).

There are several methods to measure mind wandering. In one method, participants are asked to report mind wandering at the time it was noticed during a task (e.g., Smallwood and Schooler 2006). However, this method might add an extra process of constantly monitoring one's internal state during the experiment (Smallwood 2013). Another method uses a thought probe presented at random during the task, and participants are required to report whether or not they experienced mind wandering at that point. For example, in the experiment by Mason et al. (2007), the participants were given a novel task and a familiar task and asked whether they had task-unrelated thoughts. Results showed that they reported more mind wandering during the familiar task compared to the novel task. The results suggest that the familiar task can be performed with fewer processing resources; therefore, unused resources could be used for mind wandering. However, the novel task requires more processing resources, and therefore, there were not many resources left for mind wandering. Also, the participants who reported more mind wandering showed greater activity in the DMN.

One task that is often used in the study of mind wandering is the sustained attention to response task (SART). In this task, for example, a single digit from 1 to 9 is presented on a computer monitor one at a time, and participants are required to press a key for each digit except for "3". Mind wandering is frequently observed during such easy tasks. However, there are different types of mind wandering. For example, when we think about errors in previous blocks, this is not directly related to the current trial, but is still related to the current task itself. However, we might think about things that have nothing to do with the current task; for example, we might think about a fight we had with a friend a few days ago, or we might worry about a test we have to take next week. Stawarczyk et al. (2011) examined whether the DMN regions would show the same pattern of activation when task-unrelated activities and stimulus-independent activities are separated. There are four possibilities when task-unrelated activities and stimulus-independent activities are independently manipulated. These are (1) task-related and stimulus-dependent thoughts (i.e., focusing on current task); (2) task-related but stimulus-independent thoughts, such as thoughts about performance in previous trials; (3) task-unrelated but stimulus-/environment-related thoughts, such as thoughts about the MRI scanner noise, room temperature, brightness, and so on; and (4) task-unrelated and stimulus-independent thoughts, for example, daydreaming. A SART was used in this experiment, a thought probe was presented at the end of the block, and participants were required to report their thoughts after the probe. Results showed that the DMN activities associated with task-unrelated thought and stimulus-independent thoughts were additive. In other words, the medial prefrontal cortex (MPFC), posterior cingulate cortex (PCC), and the inferior frontal gyrus (IFG) showed activities during both task-unrelated thought and stimulus-independent thoughts, but their activities were highest during mind wandering. However, unlike other DMN regions, the activities of the lateral temporal area, were associated only with the stimulus-independent thoughts. Based on these results, they concluded that activities of all DMN regions are not necessarily homogeneous.

With regard to the relationship between the DMN and mind wandering, Sonuga-Barke and Castellanos (2007) have proposed a default-mode interference hypothesis. They claim that the activities of DMN show spontaneous periodic fluctuations at a slow rate. DMN activity is typically suppressed during goal-directed cognitive activities, but it might exceed a certain threshold sometime and then mind wandering occurs. There are several potential factors affecting the DMN activation exceeding the threshold, including motivation, worry, and psychological and neurological disorders. Finally mind wandering is not necessarily negative, as research has also shown that it could be related to creative activities (e.g., Baird et al. 2012; Takeuchi et al. 2011).

### *13.3.3 Other DMN Activities*

Recent studies have reported that the DMN shows activities in a variety of tasks besides mind wandering, such as episodic memory, including autobiographical and prospective memory, (e.g., Ino et al. 2011; Schacter et al. 2007; Sestieri et al. 2011; Spreng et al. 2008), problem-solving simulation (Gerlach et al. 2011), evaluation of creative generation activities (Ellamil et al. 2012), task switching (Crittenden et al. 2015), working memory (e.g., Bluhm et al. 2011; Cocchi et al. 2013; Hampson et al. 2006), and social working memory (Meyer et al. 2012). Autobiographical memory is one's memory of the past. Prospective memory is a memory for future appointments and events. For example, we might plan what we need to do on our way home from work, such as buying bread and milk at a supermarket. The DMN is also active in information processing related to self or other people (e.g., D'Argembeau et al. 2005; Gusnard and Raichle 2001). For example, we can think about our roles and functions in our workplace, home, and social relations, or about the characters of famous people, such as politicians and celebrities. Theory of mind and social cognition are also related to DMN activities (e.g., Andrews-Hanna 2012; Iacoboni et al. 2004; Spreng et al. 2008; Spreng and Grady 2010; Young et al. 2010).

Spreng and Grady (2010) investigated roles of the DMN in autobiographical memory, prospective memory, and the theory of mind. In all conditions, participants were given a picture containing at least one person and a word associated with the scene. For example, they might be shown a picture of a family around a dinner table and a word "family." The participants were asked to use the photograph and the word as a cue to remember an event, imagine a future event, or imagine the thoughts and feelings of someone in the photograph. For example, they were given an instruction "Remember a time when you went out with your family" (autobiographical memory), "Imagine a time you will go out with your family" (prospective memory), and "Imagine what the father in the picture was thinking and feeling" (theory of mind). The results showed that the medial DMN regions showed higher activation during the autobiographical memory and prospective memory conditions and the lateral DMN regions showed higher activation during the theory of mind condition. Functional connectivity analyses showed that the medial prefrontal cortex cooperated with the other DMN regions across the three conditions.

The DMN may also be related to passive monitoring of the external environment (Gilbert et al. 2007). For example, when we are tired after concentrating on demanding tasks, we may suddenly realize various sounds that were previously ignored from inside and outside of the room, the sound of the clock, the sound of birds, and so on.

### 13.3.4 Task-Induced Deactivation of the DMN

During various cognitive tasks, task-relevant brain regions show activations, whereas the DMN shows reduction of activation. This reduction of activation of the DMN is a type of task-induced deactivation (TID). TID is basically viewed as an effect of attentional resource allocation. Neural activity of regions that receive attentional resources typically increases, whereas neural activity of regions where attentional resources are not allocated decreases (e.g., Corbetta and Shulman 2002; Corbetta et al. 2008; Hopfinger et al. 2000; Kastner and Ungerleider 2000; McKiernan et al. 2003; Pessoa et al. 2003; Shmuel et al. 2006; Tomasi et al. 2006). For example, Gazzaley et al. (2005) examined the effect of attention using face and landscape pictures as stimuli. The ventral pathway, including the inferior extrastriate, and the inferior temporal regions are closely related to processing these stimuli, especially the fusiform face area (FFA) for face processing and the parahippocampal place area (PPA) for landscape scenes (e.g., Haxby et al. 2000; Kanwisher et al. 1997; O'Craven et al. 1999). In their experiment, two faces and scenes were successively presented one at a time. Participants were told to (1) remember both, (2) remember faces and ignore scenes, (3) remember scenes and ignore faces, and (4) passively view both. Visual stimuli were identical across the four conditions. The left and right PPA activation and right FFA activation were modulated by the instructions that the remember condition produced greater activation than the passive view and the ignore condition produced lower activation (suppression) than the passive view in the corresponding areas. For the left PPA, remembering the scenes resulted in the highest activation, followed by remember both, then passive view, and then ignore scene (remember faces). Because visual stimulation in all these conditions was the same, the differences among the conditions were considered to be due to the effect of top-down control of attention. The brain area that is related to the stimuli received allocation of attention and its activity increased, whereas activity of the region corresponding to the ignored stimuli decreased. This type of effects of attentional allocation on neural activity has been observed in other areas, including the somatosensory cortex (e.g., Drevets et al. 1995; Kastrup et al. 2008; Laurienti et al. 2002; Shmuel et al. 2002), temporoparietal junction (TPJ) (e.g., Shulman et al. 2003; Todd et al. 2005), and the visual cortex (e.g., Kastner et al. 1998; Shmuel et al. 2002; Tootell et al. 1998). In addition, the effect of attentional allocation is not limited to a local area and is seen in global brain regions such as between the left and right hemispheres. Attentional allocation to the right visual field has been found to increase activity in the left hemisphere but to reduce activity in the right hemisphere (Smith et al. 2004).

## 13.4 Relationship Between EN and DMN

### *13.4.1 Competition Between EN and DMN*

The EN and DMN seem to compete with each other for processing resources in many cognitive tasks. The EN typically shows activation and the DMN shows deactivation during cognitive tasks, whereas the EN tends to show deactivation when the DMN is active. In other words, there is an anticorrelation between the EN and DMN. Several studies have led to the conclusion that the EN is a task-positive network and the DMN is a task-negative network (e.g., Fox et al. 2005; Fransson 2005; Greicius et al. 2003; Hampson et al. 2006; Kelly et al. 2008; Levinson et al. 2012; Uddin et al. 2008). The DMN also shows a negative correlation with task resource demands. The DMN activity decreases when the task demands are high (e.g., Mayer et al. 2010; Pyka et al. 2009). For example, Mayer et al. (2010) suggested that reduction of the DMN activity is greater when task difficulty or complexity is high, because task-relevant regions require more resources. Buckner et al. (2008) suggested that the DMN competes with the attention network for external stimuli, and therefore, the DMN activity decreases when attention is focused toward external stimuli, whereas the DMN activity increases when attention is not allocated to anything in particular.

Furthermore, studies have shown that when cognitive load is high, adequate suppression of the DMN is necessary for accurate task execution. (e.g., Daselaar et al. 2004; Greicius and Menon 2004; Kelly et al. 2008; Weissman et al. 2006). For example, Greicius and Menon (2004) reported that greater reduction of activity in task-irrelevant regions was observed for participants who showed higher activity in the task-relevant regions. Weissman et al. (2006) showed that RTs were longer when attentional lapses occurred, and the regions closely related to attentional control (the dorsal ACC and prefrontal regions) showed reduction of activity just before the attentional lapses. Also, during attentional lapses, the DMN, especially the PCC, increased activity. Li et al. (2007) reported that DMN activity increased before error trials in a stop signal task, and the level of deactivation within the DMN could predict performance during goal-directed tasks (e.g., Esposito et al. 2009; Anticevic et al. 2010). It has also been shown that changes in the interregional coupling with the DMN can predict behavioral performance during the processing of external information (Kelly et al. 2008; Sala-Llonch et al. 2012).

The competition between the EN and the DMN occurs because there is only a limited amount of processing resources in the brain. Therefore, when task demands increase, more processing resources are required in task-relevant regions, resources in the task-irrelevant regions such as the DMN may have to be decreased (e.g., Persson et al. 2007; Raichle et al. 2001). In other words, when the task is simple, it can be performed without many resources. However, when the task is more complex, more processing resources are required. When neural activity is increased in task-relevant brain regions, activity in the task-irrelevant regions decreases because the total amount of processing resources in the brain is limited.

### 13.4.2 Cooperation Between the EN and DMN

Do the DMN and other task-relevant networks such as the EN always compete? In other words, is it possible for the DMN and EN to cooperate? As we discussed before, the DMN is active during many tasks, especially those associated with social and self-related information processing. According to recent studies, the DMN and EN are not necessarily always anticorrelated, but, instead, sometimes work together. For example, recent studies have reported that the DMN and the task-relevant networks, such as the EN, seem to cooperate during cognitive and social tasks, including problem-solving simulation (Gerlach et al. 2011), evaluation of creative generation activities (Ellamil et al. 2012), episodic memory (e.g., Chadick and Gazzaley 2011; Fornito et al. 2012; Sestieri et al. 2011; Spreng et al. 2014), working memory (e.g., Bluhm et al. 2011; Cocchi et al. 2013; Hampson et al. 2006), and social working memory (Meyer et al. 2012).

For example, in the experiment by Gerlach et al. (2011), participants were given a scenario and required to solve a problem associated with it. For instance, they were asked to imagine being left alone in a friend's dorm room, trying on the friend's class ring, and being unable to remove it. In this case, soap was suggested as a possible object that could help them solve the problem, and they simulated removing the class ring using soap. Other scenarios included volunteering in a local retirement community, navigating a new neighborhood, organizing a camping trip, house-sitting for family friends, and planning a class research project. Results showed that both the DMN and the DLPFC were involved together in the simulation of problem solving. In addition, functional connectivity was calculated with the posterior cingulate and DLPFC seeds, and the results revealed that activity in these regions was synchronized with other DMN and EN regions, including the MPFC, medial temporal, and parietal regions, during the simulation period.

Ellamil et al. (2012) examined the brain network involved in the evaluation and generation elements of creativity. Participants were required to design book cover illustrations while alternating between the generation and evaluation of ideas. They were given a book description and, then asked to generate ideas and to evaluate them. Creative generation was associated with activity in the medial temporal regions, whereas evaluation was associated with activation in the EN and DMN regions, as well as activation in the rostrolateral PFC, insula, and temporopolar cortex. The EN and DMN regions also showed positive functional connectivity during the task.

Furthermore Sestieri et al. (2011) examined how the DMN interacts with other networks during an episodic memory task. Episodic memory retrieval activated posterior regions of the DMN, including the angular gyrus and the posterior cingulate/precuneus, whereas the anterior medial prefrontal cortex was deactivated. The results suggest functional heterogeneity within the DMN during episodic memory retrieval. The DMN regions are not necessarily always activated together, but rather some regions may be connected with non-DMN regions depending on task demands.

Chadick and Gazzaley (2011) used a working memory task with faces and scenes to investigate how the fusiform face area (FFA) and the parahippocampal place area (PPA) are connected with other networks depending on task goals. The results showed that the areas that process relevant information are functionally synchronized with the frontoparietal network, whereas those that process irrelevant information are connected with the DMN. When the task goal required face processing, the FFA was activated and connected with the EN, including the right middle frontal gyrus (MFG) and bilateral inferior frontal junction (IFJ), located in the vicinity of the junction of the inferior frontal sulcus and the inferior precentral sulcus. In this condition, PPA was deactivated and connected with the DMN regions, especially the medial prefrontal cortex (mPFC) and posterior cingulate cortex (PCC). The reverse pattern was observed when scene memory was required. They suggested that visual perception areas are dynamically synchronized with the EN and DMN and sometimes compete and other times cooperate, depending on the task goals.

Spreng et al. (2014) investigated whether DMN activity enhances performance on an executive function task when the cognitive control processes engage long-term memory representations that are supported by the DMN. They used 2-back task, which requires involvement of the EN (e.g., Owen et al. 2005), with pictures of famous faces, which are associated with the DMN (Leveroni et al. 2000; Eger et al. 2005). The results showed greater activation of the DMN during processing of famous faces compared to anonymous faces, suggesting that the DMN activity could facilitate task performance during an executive function task.

Meyer et al. (2012) examined effects of social working memory load on the medial prefrontoparietal regions. Participants evaluated characteristics of their ten close friends before the experiment, and the results were used as stimuli in an fMRI experiment. In the social WM task, participants were shown names of their friends (two to four people), and after a delay period, a word that represents the characteristics of a friend was presented. During the subsequent delay period, the participants were asked to consider how well the word described characteristics of the friends, then to rank the friends accordingly. For example, if the word is "funny," the participants were instructed to consider how funny each of the friend was and then rank them from most funny to least funny. Finally, participants were given a true/false probe question regarding the ranked position of a friend from the list. For example, if a trial contained three names (Claire, Kristin, Rebecca), participants might receive a question such as "second funniest? Rebecca," and they needed to indicate whether Rebecca was the second funniest among the friends in the list. The result showed that social WM load increased the activity of the medial as well as lateral frontoparietal regions. Therefore, they suggested that medial and lateral frontoparietal regions do not necessarily show anticorrelation, but instead they might cooperate with each other depending on task demands.

### *13.4.3 Dynamic Changes of Competition and Coactivation Between EN and DMN*

So far the reviewed studies have been mainly based on the comparison between different task conditions including rest. However, cognitive activities in our daily lives are constantly changing from time to time. Therefore, the DMN also might show dynamic changes of activation and deactivation depending on task demands, and the relationship between the EN and the DMN might also change dynamically. For example, Koshino et al. (2011) investigated dynamic changes in the activity of the anterior medial prefrontal cortex (MPFC, BA10) using a face WM task. The anterior MPFC is one of the core hubs of the DMN, but it is also known to be involved in task preparation and task set formation (e.g., Sakai 2008). Also, as discussed before, the DMN and EN show anticorrelations in various tasks. Therefore, it was hypothesized that the anterior MPFC would show activation during a task preparation period but deactivation during a task execution period. In this experiment, the preparation and execution phases were separated. During the preparation period, participants were asked to form a task set to prepare for incoming face stimuli. During the execution period, three pictures of faces were presented, and the participants were asked to remember them. Three digits were also presented at the center of the display one at a time, and they were required to add the digits (Fig. 13.1). In a dual-task condition, participants were asked to perform the face WM and addition simultaneously. In a single-task condition, participants were asked to ignore the face stimuli and just to perform the addition.

The results showed that the anterior MPFC and the inferior extrastriate were activated during the preparation period, consistent with previous research that the anterior MPFC is involved in task preparation (Fig. 13.2). Also the inferior extrastriate area includes the fusiform face area, and therefore, it was part of the task set that participants formed. During the execution period, the anterior MPFC showed deactivation, but the regions of the EN that are closely related to face working memory showed activation, including the lateral PFC, bilateral intraparietal sulcus, and bilateral inferior extrastriate regions. The anterior MPFC

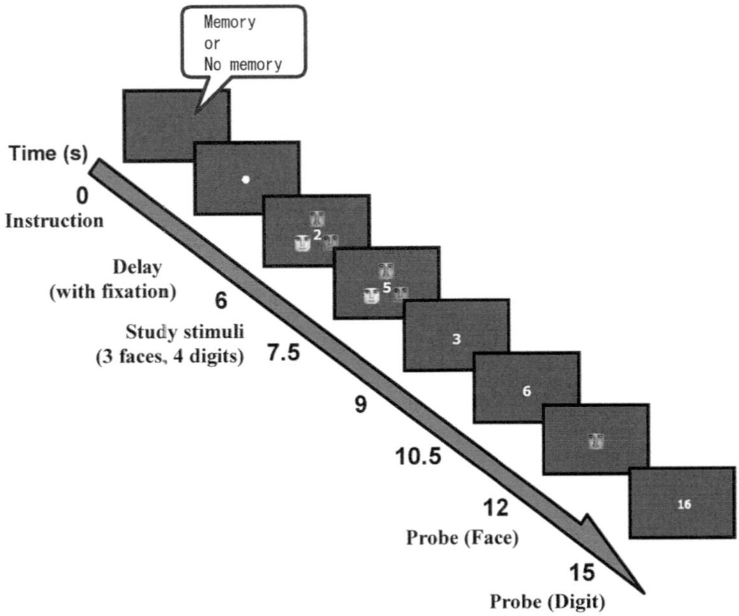


**Fig. 13.1** An example trial sequence in Koshino et al. (2011)

showed deactivation during the task execution period because resource demands in the task-relevant regions increased, and as a result, attentional resources were reallocated to those regions. The results suggested that activity in the anterior MPFC is determined dynamically by allocation of processing resources depending on task demands during different phases within a single task.

Similar results were also observed in a verbal WM task (Koshino et al. 2014), in which participants were required to remember words instead of face stimuli. In this experiment, DMN regions other than the anterior MPFC (especially posterior cingulate times/precuneus) also showed increased activity during the preparation period and decreased activity during the execution period (Fig. 13.3), suggesting that the DMN regions tend to cooperate together. In our task, participants were not required to do anything but form the task set and get ready for the working memory task during the preparation period, and therefore, the DMN regions other than the anterior MPFC were relatively free from any task demands. Therefore, one possibility is that when the anterior MPFC became active, the other DMN regions, especially the posterior ACC/precuneus, were also activated because of strong connectivity with the anterior MPFC. However, the hippocampal formation (HF) showed coactivation with the DMN regions during the preparation period, whereas it was coactivated with the EN during the execution period, suggesting functional heterogeneity among the DMN regions.

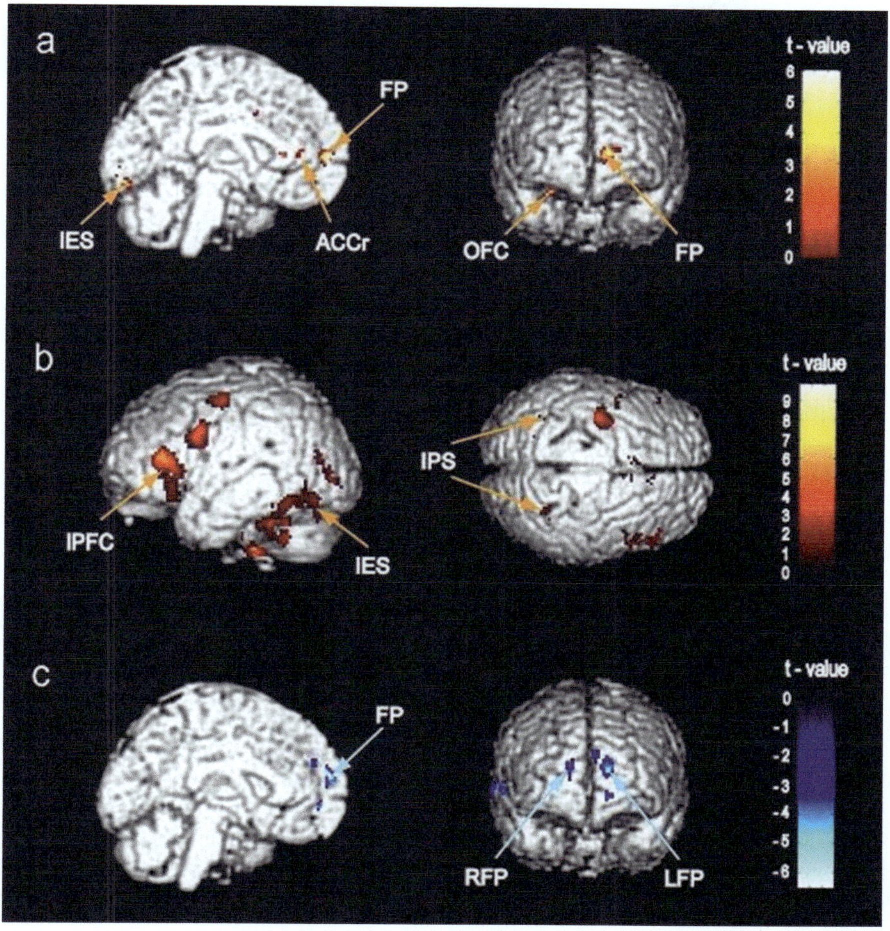


**Fig. 13.2** Brain activation for the task preparation and task execution periods (Koshino et al. 2011). (**a**) Regions that showed activation during the preparation period in the contrast of face memory condition > no face memory condition, including the left frontal pole (FP), left rostral anterior cingulate cortex (ACCr), right orbital frontal cortex (OFC), and left inferior extrastriate (IES). No brain regions showed activation in the contrast of no face memory condition > face memory condition for the preparation period. (**b**) Regions that showed activation during the execution period, in the contrast of face memory condition > no face memory condition, including the lateral prefrontal cortex (LPFC), intraparietal sulcus (IPS), and IES. (**c**) Regions that showed activation during the execution period in the contrast of no face memory condition > face memory condition, including bilateral FP. An uncorrected height threshold ($p = 0.001$) and an extent threshold (ten voxels) were used

Piccoli et al. (2015) also examined whether the DMN and working memory network (WMN) are anticorrelated during three phases (encoding, maintenance, and retrieval) of a working memory task. They reported that the DMN and WMN were anticorrelated during the maintenance phase, but the two networks were positively correlated during the encoding and retrieval phases.

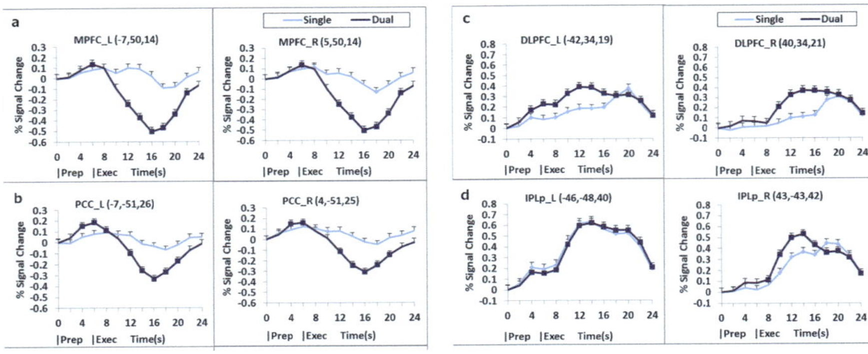


**Fig. 13.3** Signal change (%) of the DMN and WMN regions across the time course with the stimulus onset as a baseline (Koshino et al. 2014). (**a**) Anterior medial prefrontal cortex (MPFC), (**b**) posterior cingulate cortex (PCC), (**c**) dorsolateral prefrontal cortex (DLPFC), and (**d**) posterior IPL (IPLp). All WMN regions showed activation for the dual-task condition during the execution period. The central coordinates for each ROI are shown inside the parentheses. *Error bars* denote the standard error. The *large data points* indicate the points that the 99% confidence interval did not include zero. The *small data points* indicate those points that the 99% confidence interval included zero

### *13.4.4 Functional Heterogeneity of Networks*

Functional heterogeneity refers to the idea that a brain region that is anatomically classified as a single area can be divided into a number of functionally different subregions. Functional heterogeneity has been discussed with respect to large brain areas as well as networks such as the EN and DMN. With regard to the relationships between the structure and functions in the brain, there are many-to-many mappings, especially in higher-level cognition. In other words, there are many brain regions working together as a network to perform a function, and at the same time, a single region, especially the association cortex, is related to many functions. Therefore, a region might function as part of one network in a task, while the same region might function as part of a different network under a different task. As discussed before, the regions of DMN may not always work together, and instead task demands might determine which brain regions are synchronized with which network (e.g., Laird et al. 2009; Seghier and Price 2012). In general, brain networks may not be necessarily fixed, and their components may change dynamically, depending on task requirements and available processing resources. Individual regions, especially regions such as the association cortex, may have their own functions, but they are also connected with many other brain regions. In other words, each brain region does not belong to a single network, but rather belongs to multiple potential networks, and its membership is determined by factors such as task demands and resource availability. During rest, the regions that have strong intrinsic connections tend to be synchronized as resting state networks, such the DAN, EN, and DMN.

However, during task periods, some regions could be synchronized with other networks, depending on task demands and resources at that time.

Seghier and Price (2012) used semantic decision tasks for familiar objects and written names of the objects, perceptual decision tasks on unfamiliar pictures of meaningless non-objects and unfamiliar Greek letter strings, speech production tasks that involved naming pictures of familiar objects, and a digit-naming task which involved saying "1, 2, 3" to unfamiliar pictures of meaningless objects and unfamiliar Greek letters. They showed different patterns of deactivation in core DMN regions during perceptual, semantic, and speech production tasks. These results showed that the DMN overlaps with task-relevant networks. Instead of describing the DMN as a homogeneous network, other studies have segregated the DMN into different functional components (e.g., Andrews-Hanna 2012; Andrews-Hanna et al. 2010; Laird et al. 2009; Mayer et al. 2010; Stawarczyk et al. 2011). Also, other studies have characterized the differences in connectivity within the DMN or between the DMN and other networks (e.g., Jiao et al. 2011; Mennes et al. 2010; Uddin et al. 2009).

Cole et al. (2013) found that the brain-wide functional connectivity pattern of the frontoparietal network (FPN) shifted across a variety of task states. They also showed that these connectivity patterns could be used to identify the current task. Furthermore, these patterns were consistent across practiced and novel tasks, suggesting that reuse of flexible hub connectivity patterns facilitates novel task performance. Together, these findings support a central role for frontoparietal flexible hubs in cognitive control and adaptive implementation of task demands.

### 13.4.5 *Switching Between the EN and DMN*

In recent years, it is suggested that there is a network that controls the switching between the EN and DMN (e.g., Bressler and Menon 2010; Seeley et al. 2007; Sridharan et al. 2008), called a saliency network. This network is constructed around the ACC and anterior insular (AI) cortex, and therefore, it is part of the salience network. The DMN is active when our mind is wandering, but when something happened suddenly in the environment, the saliency network may detect significance of the event. If it is judged important, then the EN is activated to deal with it.

How is the switch made from the EN to the DMN? We can typically concentrate on demanding cognitive tasks for a limited amount of time. After a while, we may start feeling "our brain is tired," and we may find it difficult to focus on the task any more and may start mind wandering. In this case, processing is switched from the EN to the DMN. In other words, a switch between two antagonistic processes might occur when one process is fatigued. This mechanism seems to be common in human perception and cognition; for instance, such as color aftereffects and motion aftereffects. In the color aftereffect, according to the opponent process theory, the red and green channels are opponent processes. When we stare at one color (e.g., red) for a while, the red channel is fatigued. Then when we look at a white background, the red channel cannot be activated as much as it used to, whereas the green channel is activated as usual. Therefore, the green channel wins the competition, resulting in a color after image of green.

In the brain, the EN and DMN might compete with each other in cognitive tasks. When the EN is fatigued and the DMN is not, the DMN might win the competition, and the DMN activity becomes more dominant. This also seems to be related to the default-mode interference hypothesis (Sonuga-Barke and Castellanos 2007) previously described.

## 13.5 Conclusion

The executive network (EN) plays a critical role in higher-level cognition, whereas the default mode network (DMN) was discovered as a network that showed deactivation during cognitive tasks. Therefore, the EN was viewed as task-positive network and the DMN was once considered as task-negative network, and their relationship was described as anticorrelational. However, recent research has shown that the two networks are not necessarily anticorrelated, and instead they actually show cooperation during some cognitive tasks. The relationship between the two networks also changes dynamically depending on task demands (e.g., Vatansever et al. 2015).

The EN and DMN are also are associated with psychological and neurological disorders (e.g., Anticevic et al. 2012; Bassett and Bullmore 2009; Broyd et al. 2009; Stam 2014). For example, in schizophrenia, the DMN shows (1) hyperconnectivity at rest (e.g., Jafri et al. 2008; Liu et al. 2010; Whitfield-Gabrieli et al. 2009), (2) hypoactivation of the EN and hyperactivation of the DMN during cognitive tasks (e.g., Meyer-Lindenberg et al. 2005; Pomarol-Clotet et al. 2008; Whitfield-Gabrieli et al. 2009), and (3) reduction of anticorrelation between the EN and the DMN (e.g., Liu et al. 2010; Whitfield-Gabrieli et al. 2009; Williamson 2007). In depression, it has been reported that the DMN shows hyperactivation (e.g., Hamilton et al. 2011) and hyperconnectivity within the DMN as well as with other regions associated with depression (e.g., Greicius et al. 2007). The higher the connectivity, the higher the degree of depression (e.g., Berman et al. 2011). People with depression also tend to be unable to suppress the DMN activity (e.g., Grimm et al. 2009; Sheline et al. 2009). These results suggest that depressed people have difficulties in focusing on current tasks because they are not able to suppress negative thoughts and feelings (e.g., Buckner et al. 2008; Whitfield-Gabrieli and Ford 2012). In Alzheimer's disease (AD), the DMN shows lower functional connectivity at rest (e.g., Buckner et al. 2008; Delbeuck et al. 2003; Sorg et al. 2007; Wang et al. 2006). Also, the patients with Alzheimer's disease have difficulties in suppressing DMN activity during working memory tasks (Rombouts et al. 2005). In Autism, functional connectivity between the anterior and posterior regions of the DMN is lower, even though the connectivity of the task-positive network is not different from that in a normal control group (Cherkassky et al. 2006; Kennedy et al. 2006; Kennedy and Courchesne 2008). People with autism also showed lack of suppression of the DMN during cognitive tasks, and they did not exhibit anticorrelation between the DMN and EN (e.g., Kennedy et al. 2006; Kennedy and Courchesne 2008). Individuals with ADHD showed anomalous connectivity between the anterior and posterior hubs of the DMN (e.g., Castellanos et al. 2008; Uddin et al. 2008). The EN and DMN show competition and cooperation depending on task demands, and the relationship between the two networks also changes dynamically depending on different phases of a task. The abnormality in the relationship between the two networks seems to be closely associated with various psychological disruptions. Further converging evidence from brain imaging and neurological cases would help enhance our understanding of the EN and DMN in the future.

## References

Adcock RA, Constable RT, Gore JC, Goldman-Rakic PS (2000) Functional neuroanatomy of executive processes involved in dual-task performance. Neurobiology 97:3567–3572

Andrews-Hanna JR (2012) The brain's default network and its adaptive role in internal mentation. Neuroscientist 18:251–270

Andrews-Hanna JR, Reidler JS, Sepulcre J, Poulin R, Buckner RL (2010) Functional-anatomic fractionation of the brain's default network. Neuron 65:550–562

Anticevic A, Repovs G, Shulman GL, Barch DM (2010) When less is more: TPJ and default network deactivation during encoding predicts working memory performance. NeuroImage 49:2638–2648

Anticevic A, Cole MW, Murray JD, Corlett PR, Wang X-J, Krystal JH (2012) The role of default network deactivation in cognition and disease. Trends Cogn Sci 16:584–592

Ashcraft MH, Kirk EP (2001) The relationships among working memory, math anxiety, and performance. J Exp Psychol Gen 130:224–237

Baddeley A (2012) Working memory: theories, models, and controversies. Annu Rev Psychol 63:1–29

Baddeley AD, Hitch GJ (1974) Working memory. In: Bower GH (ed) The psychology of learning and motivation: advances in research and theory, vol 8. Academic, New York, pp 47–89

Badre D (2008) Cognitive control, hierarchy, and the rostro–caudal organization of the frontal lobes. Trends Cogn Sci 12(5):193–200

Baird B, Smallwood J, Mrazek MD, Kam JWY, Franklin MS, Schooler JW (2012) Inspired by distraction: mind wandering facilitates creative incubation. Psychol Sci 23:1117–1122

Barch DM, Braver TS, Akbudak E, Conturo T, Ollinger J, Snyder A (2001) Anterior cingulate cortex and response conflict: effects of response modality and processing domain. Cereb Cortex 11(9):837–848

Bassett DS, Bullmore ET (2009) Human brain networks in health and disease. Curr Opin Neurol 22:340–347

Beilock SL (2008) Math performance in stressful situations. Curr Dir Psychol Sci 17:339–343

Berman MG, Peltier S, Nee DE, Kross E, Deldin PJ, Jonides J (2011) Depression, rumination and the default network. Soc Cogn Affect Neurosci 6:548–555

Binder JR, Frost JA, Hammeke TA, Bellgowan PSF, Rao SM, Cox RW (1999) Conceptual processing during the conscious resting state: a functional MRI study. J Cogn Neurosci 11:80–95

Biswal B, Yetkin FZ, Haughton VM, Hyde JS (1995) Functional connectivity in the motor cortex of resting human brain using echo-planar MRI. Magn Reson Med 34:537–541
Bluhm RL, Clark CR, McFarlane AC, Moores KA, Shaw ME, Lanius RA (2011) Default network connectivity during a working memory task. Hum Brain Mapp 32:1029–1035
Botvinick MM, Nystrom L, Fissell K, Carter CS, Cohen JD (1999) Conflict monitoring versus selection for action in anterior cingulate cortex. Nature 402:179–181
Braver TS, Barch DM, Gray JR, Molfese DL, Snyder A (2001) Anterior cingulate cortex and response conflict: effects of frequency, inhibition and errors. Cereb Cortex 11(9):825–836
Braver TS, West R (2008) Working memory, executive control, and aging. The handbook of aging and cognition, 3rd edn. Psychology Press, New York, pp 311–372
Braver TS, Raynolds JR, Donaldson DI (2003) Neural mechanisms of transient and sustained cognitive control during task switching. Neuron 39:713–726
Bressler SL, Menon V (2010) Large-scale brain networks in cognition: emerging methods and principles. Trends Cogn Sci 14:277–290
Broyd SJ, Demanuele C, Debener S, Helps SK, James CJ, Sonuga-Barke EJ (2009) Default-mode brain dysfunction in mental disorders: a systematic review. Neurosci Biobehav Rev 33:279–296
Buckner RL, Andrews-Hanna JR, Schacter DL (2008) The brain's default network: anatomy, function, and relevance to disease. Ann N Y Acad Sci 1124:1–38
Bullmore E, Sporns O (2012) The economy of brain network organization. Nat Rev Neurosci 13:336–349
Bunge SA, Klingberg T, Jacobson RB, Gabrieli JDE (2000) A resource model of the neural basis of executive working memory. Proc Natl Acad Sci U S A 97:3573–3578
Castellanos FX, Margulies DS, Kelly AMC, Uddin LQ, Ghaffari M, Kirsch A, Shaw D, Shehzad Z, Di Martino A, Biswal BB, Sonuga-Barke EJS, Rotrosen J, Adler LA, Milham MP (2008) Cingulate-precuneus interactions: a new locus of dysfunction in adult attention-deficit/hyperactivity disorder. Biol Psychiatry 63:332–337
Chadick JZ, Gazzaley A (2011) Differential coupling of visual cortex with default or frontal-parietal network based on goals. Nat Neurosci 14:830–832
Cherkassky VL, Kana RK, Keller TA, Just MA (2006) Functional connectivity in baseline resting-state network in autism. NeuroReport 17:1687–1690
Christoff K (2012) Undirected thought: neural determinants and correlates. Brain Res 1428:51–59
Christoff K, Gabrieli JDE (2000) The frontopolar cortex and human cognition: evidence for a rostrocaudal hierarchical organization within the human prefrontal cortex. Psychobiology 28:168–186
Christoff K, Gordon AM, Smallwood J, Smith R, Schooler JW (2009) Experience sampling during fMRI reveals default network and executive system contributions to mind wandering. Proc Natl Acad Sci U S A 106:8719–8724
Cocchi L, Zalesky A, Fornito A, Mattingley JB (2013) Dynamic cooperation and competition between brain systems during cognitive control. Trends Cogn Sci 17:493–501
Cole MW, Reynolds JR, Power JD, Repovs G, Anticevic A, Braver TS (2013) Multi-task connectivity reveals flexible hubs for adaptive task control. Nat Neurosci 16:1348–1355
Corbetta M, Shulman GL (2002) Control of goal-directed and stimulus driven attention in the brain. Nat Rev Neurosci 3:201–215
Corbetta M, Patel G, Shulman GL (2008) The reorienting system of the human brain: from environment to theory of mind. Neuron 58:306–324
Courtney SM (2004) Attention and cognitive control as emergent properties of information representation in working memory. Cogn Affect Behav Neurosci 4:501–516
Critchley HD, Wiens S, Rotshtein P, Öhman A, Dolan RJ (2004) Neural systems supporting interoceptive awareness. Nat Neurosci 7:189–195
Crittenden BM, Mitchell DJ, Duncan J (2015) Recruitment of the default mode network during a demanding act of executive control. eLife 4:e06481

D'Argembeau A, Collette F, Van der Linden M, Laureys S, Del Fiore G, Degueldre C (2005) Self-referential reflective activity and its relationship with rest: a PET study. NeuroImage 25:616–624
D'Esposito M, Detre JA, Alsop DC, Shin RK, Atlas S, Grossman M (1995) The neural basis of the central executive system of working memory. Nature 378:279–281
Dagher A, Owen AM, Boecker H, Brooks DJ (1999) Mapping the network for planning: a correlational PET activation study with the Tower of London task. Brain 122:1973–1987
Damoiseaux JS, Rombouts SARB, Barkhof F, Scheltens P, Stam CJ, Smith SM, Beckmann CF (2006) Consistent resting-state networks across healthy subjects. Proc Natl Acad Sci U S A 103:13848–13853
Daneman M, Carpenter PA (1980) Individual differences in working memory and reading. J Verbal Learn Verbal Behav 19:450–466
Daselaar SM, Prince SE, Cabeza R (2004) When less means more: deactivations during encoding that predict subsequent memory. NeuroImage 23:921–927
De Luca M, Beckmann CF, De Stefano N, Matthews PM, Smith SM (2006) fMRI resting state networks define distinct modes of long-distance interactions in the human brain. NeuroImage 29:1359–1367
Deco G, Corbetta M (2011) The dynamical balance of the brain at rest. Neuroscientist 17:107–123
Delbeuck X, Van der Linden M, Collette F (2003) Alzheimer's disease as a disconnection syndrome? Neuropsychol Rev 13:79–92
Dosenbach NU, Visscher KM, Palmer ED, Miezin FM, Wenger KK, Kang HC, Burfund ED, Grimes AL, Schlaggar BL, Petersen SE (2006) A core system for the implementation of task sets. Neuron 50:799–812
Dosenbach NUF, Fair DA, Miezin FM, Cohen AL, Wenger KK, Dosenbach RAT, Fox MD, Snyder AZ, Vincent JL, Raichle ME et al (2007) Distinct brain networks for adaptive and stable task control in humans. Proc Natl Acad Sci U S A 104:11073–11078
Dove A, Pollmann S, Schubert T, Wiggins CJ, von Cramon DY (2000) Prefrontal cortex activation in task switching: an event-related fMRI study. Cogn Brain Res 9:103–109
Dreher J-C, Grafman J (2003) Dissociating the roles of the rostral anterior cingulated and the lateral prefrontal cortices in performing two tasks simultaneously or successively. Cereb Cortex 13:329–339
Drevets WC, Burton H, Videen TO, Snyder AZ, Simpson JR, Raichle M (1995) Blood flow changes in human somatosensory cortex during anticipated stimulation. Nature 373:249–252
Durston S, Thomas KM, Worden MS, Yang Y, Casey BJ (2002) The effect of preceding context on inhibition: an event-related fMRI study. NeuroImage 16(2):449–453
Eger E, Schweinberger SR, Dolan RJ, Henson RN (2005) Familiarity enhances invariance of face representations in human ventral visual cortex: fMRI evidence. NeuroImage 26:1128–1139
Ellamil M, Dobson C, Beeman M, Christoff K (2012) Evaluative and generative modes of thought during the creative process. NeuroImage 59:1783–1794
Engle RW (2002) Working memory capacity as executive attention. Curr Dir Psychol Sci 11:19–23
Eriksen BA, Eriksen CW (1974) Effects of noise letters upon identification of a target letter in a non- search task. Percept Psychophys 16:143–149
Esposito F, Aragri A, Latorre V, Popolizio T, Scarabino T, Cirillo S, … Di Salle F (2009) Does the default-mode functional connectivity of the brain correlate with working-memory performances? Arch Ital Biol 147(1/2):11–20
Fincham JM, Carter CS, van Veen V, Stenger VA, Anderson JR (2002) Neural mechanisms of planning: a computational analysis using event-related fMRI. Proc Natl Acad Sci U S A 99:53346–53351
Fornito A, Harrison BJ, Zalesky A, Simons JS (2012) Competitive and cooperative dynamics of large-scale brain functional networks supporting recollection. Proc Natl Acad Sci 109:12788–12793

Fox MD, Snyder AZ, Vincent JL, Corbetta M, Van Essen DC, Raichle ME (2005) The human brain is intrinsically organized into dynamic, anticorrelated functional networks. Proc Natl Acad Sci U S A 102:9673–9678
Fox MD, Corbetta M, Snyder AZ, Vincent JL, Raichle ME (2006) Spontaneous neuronal activity distinguishes human dorsal and ventral attention systems. Proc Natl Acad Sci U S A 103:10046–10051
Fox KCR, Spreng RN, Ellamil M, Andrews-Henna JR, Christoff K (2015) The wandering brain: meta-analysis of functional neuroimaging studies of mind-wandering and related spontaneous thought processes. NeuroImage 111:611–621
Fransson P (2005) Spontaneous low-frequency BOLD signal fluctuations: an fMRI investigation of the resting-state default mode of brain function hypothesis. Hum Brain Mapp 26:15–29
Friston KJ (1994) Functional and effective connectivity in neuroimaging: a synthesis. Hum Brain Mapp 2:56–78
Fuster JM (1997) Network memory. Trends Neurosci 20:451–459
Gazzaley A, Cooney JW, McEvoy K, Knight RT, D'Esposito M (2005) Top-down enhancement and suppression of the magnitude and speed of neural activity. J Cogn Neurosci 17:507–517
Gerlach KD, Spreng RN, Gilmore AW, Schacter DL (2011) Solving future problems: default network and executive activity associated with goal-directed mental simulations. NeuroImage 55:1816–1824
Gilbert SJ, Dumontheil I, Simons JS, Frith CD, Burgess PW (2007) Comment on "Wandering Minds: The Default Network and Stimulus-Independent Thought". Science 317:43b
Greicius MD, Menon V (2004) Default-mode activity during a passive sensory task: uncoupled from deactivation but impacting activation. J Cogn Neurosci 16:1484–1492
Greicius MD, Krasnow B, Reiss AL, Menon V (2003) Functional connectivity in the resting brain: a network analysis of the default mode hypothesis. Proc Natl Acad Sci U S A 100:253–258
Greicius MD, Flores BH, Menon V, Glover GH, Solvason HB, Kenna H, Reiss AL, Schatzberg AF (2007) Resting-state functional connectivity in major depression: abnormally increased contributions from subgenual cingulate cortex and thalamus. Biol Psychiatry 62:429–437
Grenard JL, Ames SL, Wiers RW, Thush C, Sussman S, Stacy AW (2008) Working memory moderates the predictive effects of drug-related associations on substance use. Psychol Addict Behav 22:426–432
Grimm S, Boesiger P, Beck J, Schuepbach D, Bermpohl F, Walter M, Ernst J, Hell D, Boeker H, Northoff G (2009) Altered negative BOLD responses in the default-mode network during emotion processing in depressed subjects. Neuropsychopharmacology 34:932–943
Gruber O, Goschke T (2004) Executive control emerging from dynamic interactions between brain systems mediating language, working memory and attentional processes. Acta Psychol 115:105–121
Gusnard DA, Raichle ME (2001) Searching for a baseline: functional imaging and the resting human brain. Nat Rev Neurosci 2:685–694
Hamilton JP, Furman DJ, Chang C, Thomason ME, Dennis E, Gotlib IH (2011) Default-mode and task positive network activity in major depressive disorder: implications for adaptive and maladaptive rumination. Biol Psychiatry 70:327–333
Hampson M, Driesen NR, Skudlarski P, Gore JC, Constable RT (2006) Brain connectivity related to working memory performance. J Neurosci 26:13338–13343
Hasher L, Zacks RT (1988) Working memory, comprehension, and aging: a review and a new view. In: Bower GH (ed) The psychology of learning and motivation, vol 22. Academic Press, New York, pp 193–225
Haxby JV, Hoffman EA, Bobbini MI (2000) The distributed human neural system for face perception. Trends Cogn Sci 4:223–233
Herath P, Klingberg T, Young J, Amunts K, Roland P (2001) Neural correlates of dual task interference can be dissociated from those of divided attention: an fMRI study. Cereb Cortex 11:796–805
Hill EL (2004) Executive dysfunction in autism. Trends Cogn Sci 8:26–32

Hommel B (2011) The Simon effect as tool and heuristic. Acta Psychol 136:189–202
Hopfinger JB, Buonocore MH, Mangun GR (2000) The neural mechanisms of top-down attentional control. Nat Neurosci 3:284–291
Horwitz B, Rumsey JM, Donohue BC (1998) Functional connectivity of the angular gyrus in normal reading and dyslexia. Proc Natl Acad Sci U S A 95:8939–8944
Iacoboni M, Lieberman MD, Knowlton BJ, Molnar-Szakacs I, Moritz M, Throop CJ, Fiske AP (2004) Watching social interactions produces dorsomedial prefrontal and medial parietal BOLD fMRI signal increases compared to a resting baseline. NeuroImage 21:1167–1173
Ino T, Nakai R, Azuma T, Kimura T, Fukuyama H (2011) Brain activation during autobiographical memory retrieval with special reference to default mode network. Open Neuroimaging J 5:14–23
Jafri MJ, Pearlson GD, Stevens M, Calhoun VD (2008) A method for functional network connectivity among spatially independent resting-state components in schizophrenia. NeuroImage 39:1666–1681
Jiang Y (2004) Resolving dual-task interference: an FMRI study. NeuroImage 22:748–754
Jiao Q, Lu G, Zhang Z, Zhong Y, Wang Z, Guo Y, Li K, Ding M, Liu Y (2011) Granger causal influence predicts BOLD activity levels in the default mode network. Hum Brain Mapp 32 (1):154–161
Jonides J, Lacey SC, Nee DE (2005) Processes of working memory in mind and brain. Curr Dir Psychol Sci 14:2–5
Just MA, Carpenter PA, Keller TA, Emery L, Zajac H, Thulborn KR (2001) Interdependence of nonoverlapping cortical systems in dual cognitive tasks. NeuroImage 14:417–426
Just MA, Cherkassky V, Keller TA, Minshew NJ (2004) Cortical activation and synchronization during sentence comprehension in high-functioning autism: evidence of underconnectivity. Brain 127:1811–1821
Kanwisher N, McDermott J, Chun M (1997) The Fusiform Face Area: a module in human extrastriate cortex specialized for the perception of faces. J Neurosci 17:4302–4311
Kastner S, Ungerleider LG (2000) Mechanisms of visual attention in the human cortex. Annu Rev Neurosci 23:315–341
Kastner S, De Weerd P, Desimone R, Ungerleider LG (1998) Mechanisms of directed attention in the human extrastriate cortex as revealed by functional MRI. Science 282:108–111
Kastrup A, Baudewig J, Schnaudigel S, Huonker R, Becker L, Sohns JM, Dechent P, Klingner C, Witte OW (2008) Behavioral correlates of negative BOLD signal changes in the primary somatosensory cortex. NeuroImage 41:1364–1371
Kaufman S (1995) At home in the universe: the search for laws of self-organization and complexity. Oxford University Press, New York
Kelly AMC, Uddin LQ, Biswal BB, Castellanos FX, Milham MP (2008) Competition between functional brain networks mediates behavioral variability. NeuroImage 39:527–537
Kennedy DP, Courchesne E (2008) The intrinsic functional organization of the brain is altered in autism. NeuroImage 39:1877–1885
Kennedy DP, Redcay E, Courchesne E (2006) Failing to deactivate: resting functional abnormalities in autism. Proc Natl Acad Sci U S A 103:8275–8280
Kim C, Cilles SE, Johnson NF, Gold BT (2012) Domain general and domain preferential brain regions associated with different types of task switching: a meta-analysis. Hum Brain Mapp 33:130–142
Klingberg T (1998) Concurrent performance e of two working memory tasks: potential mechanisms of interference. Cereb Cortex 8:593–601
Klingberg T, Forssberg H, Westerberg H (2002) Training of working memory in children with ADHD. J Clin Exp Neuropsychol 24:781–791
Klingberg T, Fernell E, Olesen PJ, Johnson M, Gustafsson P, Dahlström K, Westerberg H (2005) Computerized training of working memory in children with ADHD – a randomized, controlled trial. J Am Acad Child Adolesc Psychiatry 44:177–186

Klinger E (2009). Daydreaming and fantasizing: thought flow and motivation. In: Markman KD, Klein WMP, Suhr JA (eds) Handbook of imagination and mental stimulation, xix. Psychology Press, New York, pp 225–239
Koechlin E, Basso G, Pietrini P, Panzer S, Grafman J (1999) Exploring the role of the anterior prefrontal cortex in human cognition. Nature 399:148–151
Kolb B, Whishaw I (2003) Fundamentals of human neuropsychology, 5th edn. Worth Publishers
Kondo H, Osaka N, Osaka M (2004) Cooperation of the anterior cingulate cortex and dorsolateral prefrontal cortex for attention shifting. NeuroImage 23:670–679
Koshino H, Carpenter PA, Minshew NJ, Cherkassky VL, Keller TA, Just MA (2005) Functional connectivity in an fMRI working memory task in high-functioning autism. NeuroImage 24:810–821
Koshino H, Kana RK, Keller TA, Cherkassky VL, Minshew NJ, Just MA (2008) fMRI investigation of working memory for faces in autism: visual coding and underconnectivity with frontal areas. Cereb Cortex 18:289–300
Koshino H, Minamoto T, Ikeda T, Osaka M, Otsuka Y, Osaka N (2011) Anterior medial prefrontal cortex exhibits activation during task preparation but deactivation during task execution. PLoS ONE 6(8):e22909
Koshino H, Minamoto T, Yaoi K, Osaka M, Osaka N (2014) Coactivation of the default mode network and working memory network regions during task preparation: an event-related fMRI study. Sci Rep 4:5954
Laird AR, Eickhoff SB, Li K, Robin DA, Glahn DC, Fox PT (2009) Investigating the functional heterogeneity of the default mode network using coordinate-based meta-analytic modeling. J Neurosci 29:14496–14505
Laurienti PJ, Burdette JH, Wallace MT, Yen Y-F, Field AS, Stein BE (2002) Deactivation of sensory-specific cortex by cross-modal stimuli. J Cogn Neurosci 14:420–429
Leveroni CL, Seidenberg M, Mayer AR, Mead LA, Binder JR, Rao SM (2000) Neural systems underlying the recognition of familiar and newly learned faces. J Neurosci 20:878–886
Levinson DB, Smallwood J, Davidson RJ (2012) The persistence of thought: evidence for a role of working memory in the maintenance of task-unrelated thinking. Psychol Sci 23:375–380
Li C-SR, Yan P, Bergquist KL, Sinha R (2007) Greater activation of the "default" brain regions predicts stop signal errors. NeuroImage 38:640–648
Lieberman MD (2013) Social: why our brains are wired to connect. Crown Publishers, New York
Liu H, Kaneko Y, Ouyang X, Li L, Hao Y, Chen EYH, Jiang T, Zhou Y, Liu Z (2010) Schizophrenic patients and their unaffected siblings share increased resting-state connectivity in the task-negative network but not its anticorrelated task-positive network. Sch Bull, sbq074
Lux S, Marshall JC, Ritzl A, Weiss PH, Pietrzyk U, Shah NJ, … Fink GR (2004) A functional magnetic resonance imaging study of local/global processing with stimulus presentation in the peripheral visual hemifields. Neuroscience 124(1):113–120
MacLeod CM (1991) Half a century of research on the Stroop effect: an integrative review. Psychol Bull 109:163–203
MacLeod CM, MacDonald AW (2000) Interdimensional interference in the Stroop effect: uncovering the cognitive and neural anatomy of attention. Trends Cogn Sci 4:383–391
Mars RB, Neubert F-X, Noonan MAP, Sallet J, Toni I, Rushworth MFS (2012) On the relationship between the "default mode network" and the "social brain". Front Hum Neurosci 6:189
Mason MF, Norton MI, Van Horn JD, Wegner DM, Grafton ST, Macrae CN (2007) Wandering minds: the default network and stimulus-independent thought. Science 315:393–395
Mayer JS, Roebroeck A, Maurer K, Linden DEJ (2010) Specialization in the default mode: task-induced brain deactivations dissociate between visual working memory and attention. Hum Brain Mapp 31:126–139
Mazoyer B, Zago L, Mellet E, Bricogne S, Etard O, Houde O, Crivello F, Joliot M, Petit L, Tzourio-Mazoyer N (2001) Cortical networks for working memory and executive functions sustain the conscious resting state in man. Brain Res Bull 54:287–298

McIntosh AR, Gonzalez-Lima F (1994) Structural equation modeling and its application to network analysis in functional brain imaging. Hum Brain Mapp 2:2–22
McKiernan KA, Kaufman JN, Kucera-Thompson J, Binder JR (2003) A parametric manipulation of factors affecting task-induced deactivation in functional neuroimaging. J Cogn Neurosci 15:394–408
Meiran N (1996) Reconfiguration of processing mode prior to task performance. J Exp Psychol Learn Mem Cogn 22:1423–1442
Mennes M, Kelly C, Zuo XN, Di Martino A, Biswal BB, Castellanos FX et al (2010) Inter-individual differences in resting-state functional connectivity predict task-induced BOLD activity. NeuroImage 50:1690–1701
Menon V (2011) Large-scale brain networks and psychopathology: a unifying triple network model. Trends Cogn Sci 15:483–506
Mesulam M-M (1990) Large-scale neurocognitive networks and distributed processing for attention, language and memory. Ann Neurol 28:597–613
Mesulam M-M (1998) From sensation to cognition. Brain 121:1013–1052
Meunier D, Lambiotte R, Fornito A, Ershe KD, Bullmore ET (2009) Hierarchical modularity in human brain functional networks. Front Neuroinformatics 3:1–12
Meyer ML, Spunt RP, Berkman ET, Taylor SE, Lieberman MD (2012) Evidence for social working memory from a parametric functional MRI study. Proc Natl Acad Sci U S A 109:1883–1888
Meyer-Lindenberg AS, Olsen RK, Kohn PD, Brown T, Egan MF, Weinberger DR, Berman KF (2005) Regionally specific disturbance of dorsolateral prefrontal-hippocampal functional connectivity in schizophrenia. Arch Gen Psychiatry 62:379–386
Miller EK, Cohen JD (2001) An integrative theory of prefrontal cortex function. Annu Rev Neurosci 24:167–202
Miyake A, Friedman NP, Emerson MJ, Witzki AH, Howerter A, Wagner TD (2000) The unity and diversity of executive functions and their contributions to complex 'frontal lobe' tasks: a latent variable analysis. Cogn Psychol 41:49–100
Monsell S (2003) Task switching. Trends Cogn Sci 7:134–140
Nebel K, Wiese H, Stude P, de Greiff A, Diener H-C, Keidel M (2005) On the neural basis of focused and divided attention. Cogn Brain Res 25:760–776
Niendam TA, Laird AR, Ray KL, Dean YM, Glahn DC, Carter CS (2012) Meta-analytic evidence for a superordinate cognitive control network subserving diverse executive functions. Cogn Affect Behav Neurosci 12:241–268
Norman DA, Bobrow DG (1975) On data-limited and resource-limited processes. Cogn Psychol 7:44–64
O'Craven KM, Downing PE, Kanwisher N (1999) fMRI evidence for objects as the units of attentional selection. Nature 401:584–587
Owen AM, Stern CE, Look RB, Tracey I, Rosen BR, Petrides M (1998) Functional organization of spatial and nonspatial working memory processing within the human lateral frontal cortex. Proc Natl Acad Sci 95:7721
Owen AM, McMillan KM, Laird AR, Bullmore E (2005) N-back working memory paradigm: a meta-analysis of normative functional neuroimaging studies. Hum Brain Mapp 25:46–59
Pashler H (1998) The psychology of attention. The MIT Press, Cambridge
Pashler H, Johnston JC, Ruthruff E (2001) Attention and performance. Annu Rev Psychol 52:629–651
Persson J, Lustig C, Nelson JK, Reuter-Lorenz PA (2007) Age differences in deactivation: a link to cognitive control? J Cogn Neurosci 19:1021–1032
Pessoa L, Kastner S, Ungerleider LG (2003) Neuroimaging studies of attention: from modulation of sensory processing to top-down control. J Neurosci 23:3990–3998
Peterson BS, Kane MJ, Alexander GM, Lacadie C, Skudlarski P, Leung HC, … Gore JC (2002) An event-related functional MRI study comparing interference effects in the Simon and Stroop tasks. Cogn Brain Res 13(3):427–440

Piccoli T, Valente G, Linden DE, Re M, Esposito F, Sack AT, Di Salle F (2015) The default mode network and the working memory network are not anti-correlated during all phases of a working memory task. PLoS One 10(4):e0123354
Pomarol-Clotet E, Salvador R, Sarro S, Gomar J, Vila F, Martinez A, Guerrero A, Ortiz-Gil J, Sans-Sansa B, Capdevila A, Cebamanos JM, McKenna PJ (2008) Failure to deactivate in the prefrontal cortex in schizophrenia: dysfunction of the default mode network? Psychol Med 38:1185–1193
Postle BR (2006) Working memory as an emergent property of the mind and brain. Neuroscience 139:23–38
Power JD, Cohen AL, Nelson SM, Wig GS, Barnes KA, Church JA et al (2011) Functional network organization of the human brain. Neuron 72:665–678
Pyka M, Beckmann CF, Schöning S, Hauke S, Heider D, Kugel H, Arolt V, Konrad C (2009) Impact of working memory load on fMRI resting state pattern in subsequent resting phases. PLoS One 4(9):e7198
Raichle ME (2015) The brain's default mode network. Annu Rev Neurosci 38:433–447
Raichle ME, MacLeod AM, Snyder AZ, Powers WJ, Gusnard DA, Shulman GL (2001) A default mode of brain function. Proc Natl Acad Sci U S A 98:676–682
Ramsey NF, Jansma JM, Jager G, Van Raalten T, Kahn RS (2004) Neurophysiological factors in human information processing capacity. Brain 127:517–525
Reuter-Lorenz PA, Sylvester C-YC (2005) The cognitive neuroscience of working memory and aging. In: Cabeza R, Nyberg L, Park D (eds) Cognitive neuroscience of aging: linking cognitive and cerebral aging. Oxford University Press, New York, pp 186–217
Rombouts SA, Barkhof F, Goekoop R, Stam CJ, Scheltens P (2005) Altered resting state networks in mild cognitive impairment and mild Alzheimer's disease: an fMRI study. Hum Brain Mapp 26:231–239
Rushworth MFS, Hadland KA, Paus T, Sipila PK (2002) Role of the human medial frontal cortex in task switching: a combined fMRI and TMS study. J Neurophysiol 87:2577–2592
Sakai K (2008) Task set and prefrontal cortex. Annu Rev Neurosci 31:219–245
Sala-Llonch R, Pena-Gomez C, Arenaza-Urquijo EM, Vidal-Piñeiro D, Bargallo N, Junque C, Bartres-Faz D (2012) Brain connectivity during resting state and subsequent working memory task predicts behavioural performance. Cortex 48(9):1187–1196
Schacter DL, Addis DR, Buckner RL (2007) Remembering the past to imagine the future: the prospective brain. Nat Rev Neurosci 8:657–661
Schubert T, Szameitat AJ (2003) Functional neuroanatomy of interference in overlapping dual tasks: an fMRI study. Cogn Brain Res 17:733–746
Seeley WW, Menon V, Schatzberg AF, Keller J, Glover GH, Kenna H, Reiss AL, Greicius MD (2007) Dissociable intrinsic connectivity networks for salience processing and executive control. J Neurosci 27:2349–2356
Seghier ML, Price CJ (2012) Functional heterogeneity within the default network during semantic processing and speech production. Front Psychol 3:281
Sestieri C, Corbetta M, Romani GL, Shulman GL (2011) Episodic memory retrieval, parietal cortex, and the default mode network: functional and topographic analyses. J Neurosci 31:4407–4420
Seung S (2013) Connectome: how the brain's wiring makes us who we are. Mariner Books, New York
Sheline YI, Barch DM, Price JL, Rundle MM, Vaishnavi SN, Snyder AZ, Mintun MA, Wang S, Coalson RS, Raichle ME (2009) The default mode network and self-referential processes in depression. Proc Natl Acad Sci U S A 106:1942–1947
Shmuel A, Yacoub E, Pfeuffer J, Van de Moortele P-F, Adriany G, Hu X, Ugurbil K (2002) Sustained negative BOLD, blood flow and oxygen consumption response and its coupling to the positive response in the human brain. Neuron 36:1195–1210
Shmuel A, Augath M, Oeltermann A, Logothetis NK (2006) Negative functional MRI response correlates with decreases in neuronal activity in monkey visual area V1. Nat Neurosci 9:569–577

Shulman GL, Fiez JA, Corbetta M, Buckner RL, Miezin FM, Raichle ME, Petersen SE (1997) Common blood flow changes across visual tasks: II. Decreases in cerebral cortex. J Cogn Neurosci 9:648–663
Shulman GL, McAvoy MP, Cowan MC, Astafiev SV, Tansy AP, d'Avossa G, Corbetta M (2003) Quantitative analysis of attention and detection signals during visual search. J Neurophysiol 90:3384–3397
Simon JR (1969) Reactions towards the source of stimulation. J Exp Psychol 81:174–176
Smallwood J (2013) Distinguishing how from why the mind wanders: a process-occurrence framework for self-generated mental activity. Psychol Bull 139:519–535
Smallwood J, Schooler JW (2006) The restless mind. Psychol Bull 132:946–958
Smith EE, Jonides J (1999) Storage and executive processes in the frontal lobes. Science 283:1657–1661
Smith EE, Geva J, Jonides J, Miller A, Reuter-Lorenz P, Koeppe R (2001) The neural basis of task-switching in working memory: effects of performance and aging. Proc Natl Acad Sci U S A 98:2095–2100
Smith AT, Williams AL, Singh KD (2004) Negative BOLD in the visual cortex: evidence against blood stealing. Hum Brain Mapp 21:213–220
Smith SM, Fox PT, Miller KL, Glahn DC, Fox PM, Mackay CE et al (2009) Correspondence of the brain's functional architecture during activation and rest. Proc Natl Acad Sci 106:13040–13045
Sonuga-Barke EJS, Castellanos FX (2007) Spontaneous attentional fluctuations in impaired states and pathological conditions: a neurobiological hypothesis. Neurosci Biobehav Rev 31:977–986
Sorg C, Riedl V, Mühlau M, Calhoun VD, Eichele T, Läer L, Drzezga A, Förstl H, Kurz A, Zimmer C, Wohlschläger AM (2007) Selective changes of resting-state networks in individuals at risk for Alzheimer's disease. Proc Natl Acad Sci 104:18760–18765
Sporns O (2012) Discovering the human connectome. MIT press, Cambridge, MA
Spreng RN (2012) The fallacy of a "task-negative" network. Front Psychol 3:1–5
Spreng RN, DuPre E, Selarka D, Garcia J, Gojkovic S, Mildner J, Luh WM, Turner GR (2014) Goal-congruent default network activity facilitates cognitive control. J Neurosci 34(42):14108–14114
Spreng RN, Grady C (2010) Patterns of brain activity supporting autobiographical memory, prospection and theory-of-mind and their relationship to the default mode network. J Cogn Neurosci 22:1112–1123
Spreng RN, Mar RA, Kim ASN (2008) The common neural basis of autobiographical memory, prospection, navigation, theory of mind, and the default mode: a quantitative meta-analysis. J Cogn Neurosci 21:489–510
Sridharan D, Levitin DJ, Menon V (2008) A critical role for the right fronto-insular cortex in switching between central-executive and default-mode networks. Proc Natl Acad Sci 105:12569–12574
Stam CJ (2014) Modern network science of neurological disorders. Nat Rev Neurosci 15:683–695
Stawarczyk D, Majerus S, Maquet P, D'Argembeau A (2011) Neural correlates of ongoing conscious experience: both task-unrelatedness and stimulus-independence are related to default network activity. PLoS ONE 6(2):e16997
Stroop JR (1935) Studies of interference in serial verbal reactions. J Exp Psychol 18:643–662
Suchy Y (2009) Executive functioning: overview, assessment, and research issues for non-neuropsychologists. Ann Behav Med 37:106–116
Sudevan P, Taylor DA (1987) The cuing and priming of cognitive operations. J Exp Psychol Hum Percept Perform 13:89–103
Szameitat AJ, Schubert T, Muller K, von Cramon DY (2002) Localization of executive functions in dual-task performance with fMRI. J Cogn Neurosci 14:1184–1199
Takeuchi H, Taki Y, Hashizume H, Sassa Y, Nagase T, Nouchi R, Kawashima R (2011) Cerebral blood flow during rest associates with general intelligence and creativity. PLoS ONE 6:e25532

Todd JJ, Fougnie D, Marois R (2005) Visual-short term memory load suppresses temporo-parietal junction activity and induces inattentional blindness. Psychol Sci 16:965–972
Tomasi D, Ernst T, Caparelli EC, Chang L (2006) Common deactivation patterns during working memory and visual attention tasks: an intrasubject fMRI study at 4 Tesla. Hum Brain Mapp 27:694–705
Tootell RBH, Hadjikhani N, Hall EK, Marrett S, Vanduffel W, Vaughan JT, Dale AM (1998) The retinotopy of visual spatial attention. Neuron 21:1409–1422
Toro R, Fox PT, Paus T (2008) Functional coactivation map of the human brain. Cereb Cortex 18:2553–2559
Turner ML, Engle RW (1989) Is working memory capacity task dependent? J Mem Lang 28:127–154
Uddin LQ, Kelly AMC, Biswal BB, Castellanos FX, Milham MP (2009) Functional connectivity of default mode network components: correlation, anticorrelation, and causality. Hum Brain Mapp 30(2):625–637
Uddin LQ, Kelly AMC, Biswal BB, Margulies DS, Shehzad Z, Shaw D, Ghaffari M, Rotrosen J, Adler LA, Castellanos FX, Milham MP (2008) Network homogeneity reveals decreased integrity of default-mode network in ADHD. J Neurosci Methods 169:249–254
van den Heuvel MP, Hulshoff Pol HE (2010) Exploring the brain network: a review on resting-state fMRI functional connectivity. Eur Neuropsychopharmacol 20:519–534
Vatansever D, Menon DK, Manktelow AE, Sahakian BJ, Stamatakis EA (2015) Default mode network connectivity during task execution. NeuroImage 122:96–104
Wang L, Zang Y, He Y, Liang M, Zhang X, Tian L, Wu T, Jiang T, Li K (2006) Changes in hippocampal connectivity in the early stages of Alzheimer's disease: evidence from resting state fMRI. NeuroImage 31:496–504
Weissman DH, Giesbrecht B, Song AW, Mangun GR, Woldorff MG (2003) Conflict monitoring in the human anterior cingulate cortex during selective attention to global and local object features. NeuroImage 19(4):1361–1368
Weissman DH, Roberts KC, Visscher KM, Woldorff MG (2006) The neural basis of momentary lapses in attention. Nat Neurosci 9:971–978
Whitfield-Gabrieli S, Ford JM (2012) Default mode network activity and connectivity in psychopathology. Annu Rev Clin Psychol 8:49–76
Whitfield-Gabrieli S, Thermenos HW, Milanovic S, Tsuang MT, Faraone SV, McCarley RW, Shenton ME, Green AI, Nieto-Castanon A, LaViolette P, Wojcik J, Gabrieli JDE, Seidman LJ (2009) Hyperactivity and hyperconnectivity of the default network in schizophrenia and in first-degree relatives of persons with schizophrenia. Proc Natl Acad Sci U S A 106:1279–1284
Wickens CD (2002) Multiple resources and performance prediction. Theor Issues Ergon Sci 3:159–177
Williamson P (2007) Are anticorrelated networks in the brain relevant to schizophrenia? Schizophr Bull 33:994–1003
Young L, Dodell-Feder D, Saxe R (2010) What gets the attention of the temporo-parietal junction? An fMRI investigation of attention and theory of mind. Neuropsychologia 48:2658–2664

MIX
Papier aus verantwortungsvollen Quellen
Paper from responsible sources
FSC® C105338

If you have any concerns about our products,
you can contact us on
**ProductSafety@springernature.com**

In case Publisher is established outside the EU,
the EU authorized representative is:
**Springer Nature Customer Service Center GmbH**
**Europaplatz 3, 69115 Heidelberg, Germany**

Printed by Libri Plureos GmbH
in Hamburg, Germany